# RECENT TRENDS IN ALGAL TAXONOMY

## *Vol. I. Taxonomical Issues*

*Editors*
**Prof. Vidyavati**
Hydrobiology Laboratory
Department of Botany, Kakatiya Univerisity
Warangal 506 009

*and*

**Dr. A.K. Mahato**
Department of Botany
T.M. Bhagalpur University
Bhagalpur 812 007

**2017**
**Associated Publishing Company**
*A Division of*
**Astral International (P) Ltd**
**New Delhi 110 002**

*First Published, 1999*
*Reprinted,* 2017

ISBN: 978-81-86580-33-2 (Hardbound)

*Publisher's note:*

*Every possible effort has been made to ensure that the information contained in this book is accurate at the time of going to press, and the publisher and author cannot accept responsibility for any errors or omissions, however caused. No responsibility for loss or damage occasioned to any person acting, or refraining from action, as a result of the material in this publication can be accepted by the editor, the publisher or the author. The Publisher is not associated with any product or vendor mentioned in the book. The contents of this work are intended to further general scientific research, understanding and discussion only. Readers should consult with a specialist where appropriate.*

*Every effort has been made to trace the owners of copyright material used in this book, if any. The author and the publisher will be grateful for any omission brought to their notice for acknowledgement in the future editions of the book.*



*Published by* : **Associated Publishing Company**
*A Division of*
**Astral International Pvt. Ltd.**
– ISO 9001:2008 Certified Company –
4760-61/23, Ansari Road, Darya Ganj
New Delhi-110 002
Ph. 011-43549197, 23278134
E-mail: info@astralint.com
Website: www.astralint.com

*Digitally Printed at* : Replika Press Pvt. Ltd

# Contents

**Section - 5**



**Section - 6**

# Foreword

There are various views expressed about the recent trends in algal taxonomy. The present volume contains papers from leading phycologists regarding the taxonomical issues.

Dr. Ajith Kumar, Dr. Ravishankar and Dr. L.V. Venkataraman have discussed the fatty acid profiles of algae belonging to various classes such as Cyanophyceae, Bacillariophyceae, Phaeophyceae and Rhodophyceae in the light of modern biotechnological techniques. In their article on *Nuclear cytology of algae—A promising adjunct to taxonomy*, Dr. S.B. Agrawal and Dr. B.R. Chaudhary considered cytological features being more stable and consistent in providing the reliable information of taxonomy. Dr. V. Sankaran has dealt with the Phycotaxonomy in Tamilnadu. Dr. F.R. Trainor and Dr. E.A. Morales suggested three possible avenues for identification of members of either *Neochloris* and *Scenedesmus*. They further proposed to use a multidisciplinary approach before attempting to identify a member of either of these genera to species, Dr. B.N. Verma and Dr. Vidyavati have given an account of morphotaxonomy vs. cytotaxonomy of Cladoporales. Dr. Vidyavati in her article on ultrastructure versus taxonomy of desmids has tried to show that ultrastructural studies might provide more vital information in understanding specially, the taxonomic features of this group of algae. Dr. A.N. Kargupta has dealt with the taxonomal problems and prospects of Zygnemataceae. One more article on Zygnemacean algae is contributed by Dr. A.K. Mahato and Pushpa Mahato elucidating taxonomic proposals. They emphasized that each character plays a significant role to interpret a taxon.

Among the algal groups, Charophytes, commonly known as stoneworts are marked out for their isolation, having no clear affinity to any other group. Dr. Sandhya Sinha and Dr. Ichhapurak have given an account of taxonomy of Charophyta taking into consideration their trends and perspectives. Dr. S. Bharathan mentioned a reappraisal on the taxonomy of *Lamprothamnium*. Dr. H.S. Pundhir and Dr. M. Khan have given an account of bibliography of Indian Charophyta. Dr. Daniel E. Wujek has enlightened the recent and emerging trends in Chrysophycean taxonomy. Dr. N. Jeeji Bai enumerated the taxonomic appraisal of the genera *Spirulina* and *Arthrospira*.

Dr. N. Kaliaperumal has contributed an article on marine algal taxonomical account in India. A cytotaxonomical approach to Red Algal members is enumerated by Dr. K. Rama Rao. The modern trends in the taxonomy of Gelidiales is described by Dr. M.V.N. Panikkar and Dr. P. Ampili.

The editors have done a commendable job in compiling these articles in a book form which has dealt with taxonomic aspects of various algal forms. The references cited at the end of each article is a good source of information. I hope that this book will be an asset to the students, Researchers and teachers of Phycology.

**Prof. Jafar Nizam**
*Former Vice-Chancellor*
*Kakatiya University*
*Warangal : 506 009*

# List of Contributors

**S.B. Agrawal**
Centre of Advanced Study in Botany
Banaras Hindu University
Varanasi 221 005, India

**T. Ajith Kumar**
Oska National Research Institute
1-8-31, Midorigaoka Ikeda
Oska 563, Japan

**P. Ampili**
Department of Botany, S.N. College
Kollam 691 001, India

**S. Bharathan**
Department of Botany
Pachaiyappa's College
Chennai 600 030, India

**B.R. Chaudhary**
Centre of Advanced Study in Botany
Banaras Hindu University
Varanasi 221 005, India

**Daniel E. Wujek**
Department of Biologia
Central Michigan University
Mt. Pleasant, Michigan 48859, USA

**Eduardo A. Morales**
Department of Ecology & Evolutionary
Biology, University of Connecticut
Storrs, Connecticut 06268-3042, USA

**Francis R. Trainor**
Department of Ecology & Evolutionary
Biology, University of Connecticut
Storrs, Connecticut 06268-3042, USA

**Ichha Purak**
Department of Botany
Ranchi Women's College
Ranchi 834 001, India

**N. Jeeji Bai**
Shri A.M.M. Murugappa Chattiar
Research Centre, Taramani
Chennai 600 113, India

**M. Khan**
Department of Botany
Faculty of Life Sciences
Aligarh Muslim University
Aligarh 202 002, India

**N. Kaliaperumal**
Regional Centre, Central Marine
Fisheries Research Institute
Tamil Nadu 623 520, India

**A.N. Kargupta**
University Department of Botany
L.N. Mithila University, Kameshwarnagar
Darbhanga 846 004, India

**A.K. Mahato**
Department of Botany, J.R.S. College
Jamalpur, Munger 811 214, India

**M.V.N. Panikkar**
Department of Botany, S.N. College
Kollam 691 001, India

**H.S. Pundhir**
Department of Botany
Faculty of Life Sciences
Aligarh Muslim University
Aligarh 202 002, India

**Pushpa Mahato**
Department of Botany
P.T.P.S. College
Patratu 829 119, India

**K. Rama Rao**
CSMCRI Marine Algal Research Station
Mandapam Camp 623 519, India

**G.A. Ravishankar**
Central Food Technological Research
Institute, Mysore 570 013, India

**Sandhya Sinha**
Department of Botany
Ranchi Women's College
Ranchi 834 001, India

**V. Sankaran**
Department of Botany
N.G.M. College
Pollachi 642 001, India

**L.V. Venkataraman**
'Sudarshana' 236, 8th Cross Gokulam
III Stage, Mysore 570 002, India

**Vidyavati**
Vice-Chancellor
Kakatiya University
Warangal 506 009, India

**B.N. Verma**
University Department of Botany
BRA Bihar University
Muzaffarpur 842 001, India

# Section - 1

Recent Trends in Algal Taxonomy
Vol. I. Taxonomical Issues
Editors : Vidyavati and A.K. Mahato
APC Publications Pvt. Ltd., New Delhi, 1999, pp. 3-20

# *Profile of Lipids in Algae*

T. Ajith Kumar[1], G.A. Ravishankar and L.V. Venkataraman[2]
Central Food Technological Research Institute
Mysore-570 013, India

**ABSTRACT**

Polyunsaturated fatty acids with 18 to 22 carbon atoms have been known to have potential health benefits. These include arachidonic acid (ARA), eicosapentaenoic acid (EPA), docosahexaenoic acid (DHA) and linolenic acid (GLA). Several important physiological activities in humans are attributed to the role played by these essential fatty acids. Various classes of micro and macro algae contain relatively high proportions of these fatty acids. The fatty acid profiles of algae belonging to various classes such as Cyanophyceae, Bacillariophyceae, Rhodophyceae and Phaeophyceae are discussed. Fatty acid composition in algae, both qualitatively and quantitatively, can be manipulated by modern biotechnological techniques. Identification of a suitable strain with high levels of essential fatty acids will lead to large scale production of essential PUFA in the most economical and unconventional way.

## INTRODUCTION

Algae as a group display a tremendous diversity in morphologic form, ranging from microscopic prokaryotic cells (the cyanobacteria) to huge seaweeds such as the 70m long giant kelp *Macrocystis*, which are among the largest plants on earth. Within these extremes are unicellular forms from microns to millimeters in diameter, colonies of cells, filaments, sheets and complex thalli with differentiation into leaf-like blades and stem-like spikes (Schopf, 1976). While most of the algae are aquatic in either fresh or marine waters, some species may be found growing in such diverse habitats as treetrunks (Graham *et al.* 1981), snow banks (Marchant, 1982; Friedman, 1982), hotsprings (Bauld and Brock 1974) or even within minute cavities in desert rocks (Friedman, 1974). A number of algae live symbiotically with (or within) animals,fungi or other plants (Ahmadjian 1967; Albers *et al.* 1982; Chapman and Good 1982; Cooke, 1975; Reisser, 1981).

1. Osaka National Research Institute, 1-8-31, Midorigaoka, Ikeda, Osaka 563, Japan.
2. 'Sudarshana', 236, 8th Cross, Gokulam III Stage, Mysore-570002.

Microalgae constitute a reservoir of natural substances of commercial importance. The production of valuable biochemicals from microalgae is based on the exploitation of their relatively efficient photosynthetic system. Since the production of biomass is dependent on the availability of salt, water, high light irradiation, high temperature and land that may not be suitable for conventional agriculture. Microalgal biotechnology constitutes an agro-industry that is suitable for areas which cannot support conventional crops (Venkataraman and Becker, 1986).

Many industries throughout the world are engaged in developing new products of algal origin and testing a new generation of natural products derived form microalgae, including natural colourants, health foods, polyunsaturated fatty acids etc.

## POLYUNSATURATED FATTY ACIDS

Polyunsaturated fatty acids, especially those with 20 carbon atoms, are currently receiving much attention in view of their physiological, industrial and pharmaceutical potential. It has been reported that marine microalgae offer an important source of dietary lipids to marine life and inturn, directly or indirectly to humans as well (Kyle, 1992). The active ingredients of fish oil are long chain polyunsaturated fatty acids, which ultimately originate from the microalga in the food chain.

Polyunsaturated fatty acids (PUFA) are classified according to the position of double bond either from the methyl end or carboxyl end. In the 'omega' or 'n' convention, the double bonds are counted from the methyl end of the carbon chain. A majority of PUFAs of biological importance belong to n-6 (e.g., arachidonic acid) and n-3 (e.g., eicosapentaeonic acid) groups (Fig. 1). The first double bond in n-3 fatty acids starts at the third carbon atom from the methyl end. The fatty acids belonging to this family are—linolenic acid (ALA, 18:3), eicosapentaenoic acid (EPA, 20:5), docosahexaenoic acid (DHA, 22:6) and docosapentaenoic acid (DPA, 22:5).—linolenic acid (18:3) and the n-6 fatty acid, linoleic acid (18:2) are essential dietary fatty acids because they cannot be synthesised by humans (Radwan, 1991). Several important physiological activities are being attributed to role played by the essential fatty acids and to the C20-polyunsatu-rated fatty acids derived from them. These include lowering of plasma cholesterol (Sugano *et al.*, 1986; Horrobin an Huang 1987) and tryacylglycerols (Adam, 1990), prevention of atherosclerosis (Dyerberg 1986) and cardiovascular diseases (Horrobin and Huang 1987; Singer 1990) as well as reduction of collagen-induced thrombocyte-agregation (Stroh *et al.*, 1990). The effects of these fatty acids on free radical generation in tumor cells (Das *et al.*, 1987), tocopherol during marginal protein nutrition (Baus *et al.*, 1990) and liver metabolism during cardiovascular diseases (Metzner *et al.*, 1990) have also been reported.

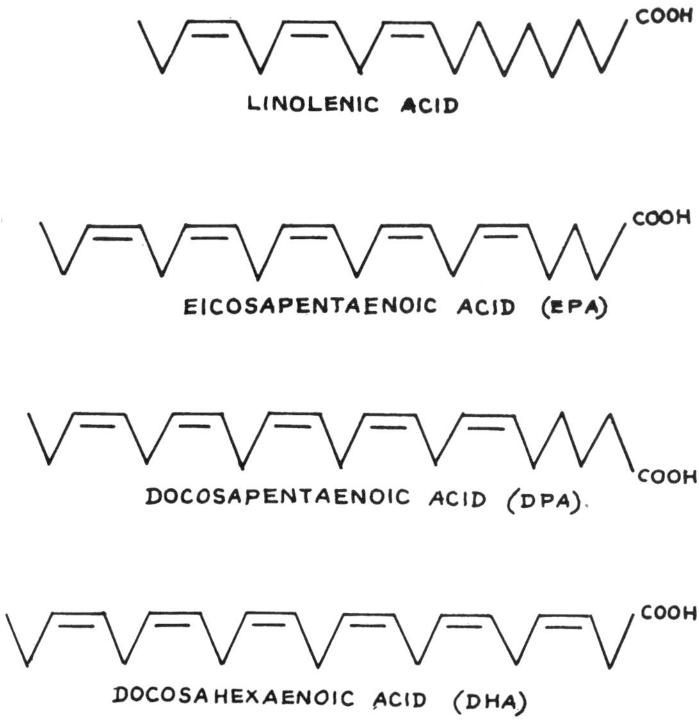


Fig. 1. Members of *n*-3 fatty acid family

## BIOSYNTHETIC PATHWAY OF POLYUNSATURATED FATTY ACIDS IN ALGAE

The same enzyme system found in higher organisms is responsible for the polyunsaturated fatty acid (PUFA) biosyn-thesis in algae also, the starting material being the monounsaturated fatty acid, oleic acid (Harwood and Russel 1984). Mainly, two reactions are involved in the biosynthetic pathway of polyunsaturated fatty acids in algae i.e., (a) chain elongation and (b) desaturation (Fig. 2). In chain elongation two carbon atoms from the donor (either acetyl CoA or malonyl CoA) are introduced to increase the chain length. The desaturation of fatty acids may proceed in two different routes. (1) The omega-3 route where desaturation usually produces linoleic acid (18:3, n-6) and linolenic acid (18:3, n-3) which will yield all-cis eicosapentaenoic acid by further chain elongation and two more desaturation reactions. Many micro algae, such as *Porphyridium yezoensis narawaensis* (Kayama *et al.*, 1986), *Chetoceros gracilis*, *Nan-*

*nochloropsis salina*, *Chlorella stigmatophora* (Ben-Amtoz *et al.*, 1987) and *Crypthecodinium* (Henderson *et al.*, 1988) follow this route. (2) In the omega-6 route the additional double bonds are usually introduced towards the carboxyl end to produce arachidonic acid (20:4, n-6) through the desaturation of linoleic acid to linolenic acid (18:3, n-3) followed by chain elongation as reported in *Porphyridium cruentum* (Nichols and Appleby 1969).

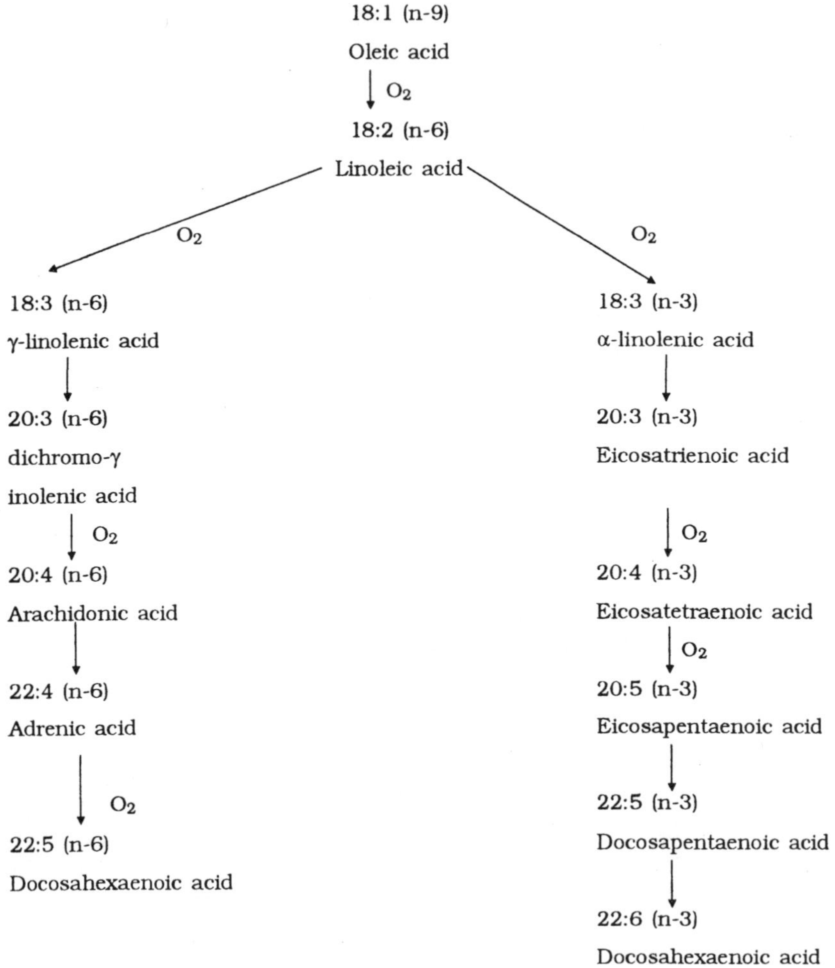


Fig. 2. Polyunsaturated fatty acid biosynthetic pathways in algae (Ref. Korn *et al.*, 1965).

## LOCATION AND FUNCTION

Fatty acids are important constituents of algae, where they are located principally in the chloroplast and membrane constituents such as galactolipids and phospholipids. They have many other functions and it has been reported, for example, that the unsaturated fatty acids from a red alga, *Porphyra* sp. have pronounced antifungal activity (Noda *et al.*, 1983). It is evident that different species have very different fatty acid compositions, which value for making taxonomic distinction. The fatty acid distributions of a large number of algal species have been reported (Aknin *et al.*, 1992; Cohen *et al.*, 1987; Dembitzky *et al.*, 1991; Dunstan *et al.*, 1994; Jones and Harwood 1992; Khotimchenko 1993; Viso and Marty 1993). The objectives of these studies include: (i) distinction of taxonomic features (Gillan *et al.*, 1981; Volkman *et al.*, 1993), (ii) production of essential fatty acids such as 20:4 (n-6), 20:5 (n-3) and 22:6 (n-3) and search of algae suitable for large scale cultivation (Ahren *et al.*, 1983; Ben-Amtoz *et al.*, 1987; Fraser *et al.*, 1989; Volkman *et al.*, 1993). This review gives a brief account of the occurrence and distribution of fatty acids in selected taxonomic groups of algae vis-a-vis its future biotechnological potential.

## CYANOPHYCEAE

The blue-green algae (Cyanobacteria) morphologically resemble the bacteria, but their physiological activities are similar to eukaryotic algae. Their fatty acid contents and profile are intermediate between bacteria and eukaryotic algae (Pohl *et al.*, 1981; Kenyon 1972; Oren *et al.*, 1985). These algae produce C12 to C18 fatty acids with the maximum of four double bonds. Therefore, they are considered as potential sources of EPA and DHA.

Cohen *et al.*, (1987) studied the fatty acid composition of various strains of *Spirulina*. Of the 19 strains tested, all but one contained significantly high amounts of GLA. The predominant fatty acids of *Spirulina* were palmitic acid (16:0), GLA, linoleic acid (18:2) and oleic acid (18:1) (Table 1). They also found out that the maximum production of GLA was identical with the conditions for the alga's optimal growth.

Since the GLA is quite rare and is found only in human milk among foods, the production of this fatty acid through other means is a valuable proposition. The oil of evening primrose was found to contain 8-12% of GLA and is presently its source commercially. *Spirulina* with its high levels of GLA (about 25% of the total fatty acids) could be exploited commercially since the technology for its large scale cultivation is already available.

The cyanobacterium *Spirulina* is the classic example of the

**Table 1. Fatty acid composition of various species of algae (% of total fatty acids).**

| Species | 16:0 | 18:0 | 18:1 | 18:2 | 18:3 | 18:4 | 20:4 | 20:5 | 22:6 | Others | Reference |
|---|---|---|---|---|---|---|---|---|---|---|---|
| **Cyanophyceae** | | | | | | | | | | | |
| *Aqmenellum quadruplicatum* | 23.9 | 0.7 | 6.7 | 3.3 | 45.6 | - | - | - | - | 19.8 | Viso & Marty, 1993. |
| *Anacystis marina* | 24.7 | 2.2 | 3.7 | 0.3 | 2.0 | - | - | - | - | 67.1 | Viso & Marty, 1993. |
| *Synechocystis* sp. | 25.2 | 3.4 | 7.6 | 0.2 | 4.3 | - | - | - | - | 59.3 | Viso & Marty, 1993. |
| *Spirulina platensis* | 44.2 | 1.0 | 2.6 | 25.6 | 19.6 | - | - | - | - | 7.0 | Cohen *et al.*, 1993. |
| **Bacillariophyceae** | | | | | | | | | | | |
| *Coscinodiscus* sp. | 16.5 | 0.4 | 4.4 | 1.6 | 1.4 | 3.5 | 1.6 | 26.0 | 4.6 | 40.0 | Dunstan *et al.*, 1994. |
| *Skelotonema costatum* | 5.1 | 0.1 | 0.5 | 1.4 | 0.8 | 2.8 | 0.2 | 26.1 | 4.7 | 58.3 | Dunstan *et al.*, 1994. |
| *Amphiprora hyalina* | 10.7 | 0.2 | 3.4 | 1.4 | 0.8 | 5.5 | 1.0 | 30.0 | 1.9 | 45.1 | Dunstan *et al.*, 1994. |
| *Amphora* sp. | 14.6 | 0.3 | 0.8 | 0.8 | 0.4 | 0.9 | 1.3 | 30.2 | 0.7 | 50.0 | Dunstan *et al.*, 1994. |
| *Nitzschia closterium* | 7.2 | 0.1 | 0.3 | 0.5 | 0.5 | 3.5 | 0.1 | 24.2 | 2.4 | 61.2 | Dunstan *et al.*, 1994. |
| *Pheodactylum tricornutum* | 11.6 | 0.0 | 1.1 | 3.2 | 0.9 | 0.0 | 1.3 | 30.5 | 2.5 | 48.9 | Yongmanitchai & Ward, 1991. |
| **Rhodophyceae** | | | | | | | | | | | |
| *Porphyridium cruentum* | 35.2 | 0.0 | 0.0 | 7.7 | 0.0 | 0.0 | 21.5 | 22.7 | 0.0 | 12.9 | Yongmanitchai & Ward, 1991. |
| *Corallina granifera* | 18.4 | 0.9 | 6.6 | 3.4 | 3.7 | 0.8 | 5.1 | 56.1 | 0.0 | 5.0 | Dembitsky *et al.*, 1991. |
| *Gelidium latifolium* | 24.3 | 3.2 | 10.1 | 2.3 | 2.0 | 0.6 | 21.5 | 29.5 | 0.0 | 6.5 | Dembitsky *et al.*, 1991. |
| *Laurencia coronopus* | 31.0 | 2.2 | 9.3 | 1.7 | 0.9 | 0.7 | 11.6 | 31.2 | 0.0 | 11.4 | Dembitsky *et al.*, 1991. |
| *Polysiphonia elongata* | 29.4 | 1.6 | 11.3 | 3.0 | 1.4 | 0.0 | 3.4 | 34.5 | 0.0 | 15.4 | Dembitsky *et al.*, 1991. |
| **Chlorophyceae** | | | | | | | | | | | |
| *Chlorella minutissima* | 21.1 | 0.7 | 7.3 | 6.1 | 14.7 | 0.0 | 3.2 | 4.6 | 0.0 | 42.1 | Yongmanitchai & Ward, 1991. |
| *Dunaliella primolecta* | 17.4 | 0.6 | 9.4 | 8.3 | 18.4 | 0.0 | 0.0 | 5.7 | 0.0 | 40.2 | Yongmanitchai & Ward, 1991. |

| Species | 16:0 | 18:0 | 18:1 | 18:2 | 18:3 | 18:4 | 20:4 | 20:5 | 22:6 | Others | Reference |
|---|---|---|---|---|---|---|---|---|---|---|---|
| *Chaetomorpha linum* | 22.1 | 1.6 | 8.8 | 20.5 | 1.1 | 0.4 | 1.0 | 1.9 | 0.4 | 42.2 | Khotimchenka, 1993. |
| *Ulva rigida* | 16.9 | 1.9 | 14.7 | 0.5 | 14.9 | 10.1 | 2.1 | 0.0 | 1.3 | 37.6 | Dembitsky *et al.*, 1991. |
| *Urospora penicilliformis* | 18.4 | 2.3 | 18.1 | 2.4 | 5.6 | 7.4 | 5.6 | 0.0 | 0.8 | 39.4 | Dembitsky *et al.*, 1991. |
| *Entocladia viridis* | 19.6 | 4.2 | 13.7 | 5.1 | 12.1 | 10.7 | 2.0 | 0.0 | 1.0 | 31.6 | Dembitsky *et al.*, 1991. |
| *Enteromorpha prolifera* | 24.3 | 2.0 | 7.9 | 1.3 | 26.7 | 13.1 | 1.0 | 0.0 | 1.0 | 22.7 | Dembitsky *et al.*, 1991. |
| *E. clathrata* | 25.3 | 1.2 | 5.6 | 10.7 | 21.1 | 7.8 | 1.0 | 1.2 | 0.0 | 26.1 | Aknin *et al.*, 1992. |
| *Codium dichotomum* | 28.4 | 1.1 | 12.2 | 7.3 | 18.1 | 1.8 | 7.0 | 2.1 | 0.2 | 21.8 | Aknin *et al.*, 1992. |
| **Phaeophyceae** | | | | | | | | | | | |
| *Fucus vesiculosus* | 17.3 | 2.7 | 17.2 | 9.9 | 7.8 | 8.7 | 15.4 | 13.1 | - | 10.9 | Jones and Harwood, 1992. |
| *Ascophyllum nodosum* | 14.9 | 2.1 | 16.0 | 7.1 | 6.5 | 11.0 | 17.9 | 11.3 | - | 13.2 | Jones and Harwood, 1992. |

success of algal biotechnology. Today, it is being cultivated in various countries on a commercial scale with a productivity of 2000 tonnes/annum. GLA was shown to have many theraputic uses. It is approximately 170 fold more effective in reducing plasma cholesterol level than linolenic acid (Huang *et al.*, 1984), the major constituent of most PUFAs. It was claimed to exert beneficial effects in conditions such as heart diseases, Parkinson disease and multiple sclerosis in humans.

## BACILLARIOPHYCEAE

Microalgae from the class Bacillariophyceae (diatoms) are important in marine food chains. Since the primary storage material in diatoms is lipid, they are of potential value to biotechnology industry for the production of lipids and in particular, polyunsaturated fatty acids. Relative to large number of species within this group (> 1300) (Sournia *et al.*, 1991), very few species have been examined for their lipid class and fatty acid composition and only a few comparative studies have been performed using modern analytical techniques (Kates 1989).

There is no significant difference in lipid class composition among the members of Bacillariophyceae (Table 2). Total cellular lipids are generally dependent on cell size and species known to have larger cells have more lipids. e.g., *Coscinodiscus* sp.) and the accumulation of storage lipid. Polar lipids (mostly phospholipid and glycolipid, and in diatoms sulpholipid (Kates 1989) are major components of cellular membranes and typically the abundant lipid class in microalgal cells. Significant amounts of tryacylglycerols are also evident in some species (0.3-48.0%) as reported by Dunstan *et al.*, (1994). In logarithmic phase cultures of microalgae, the triacylglycerols are present in low amounts (Dunstan *et al.*, 1992) but they are known to accumulate as storage lipid in stationary phase cells (Dunstan *et al.*, 1993).

Diatoms generally contain high concentrations of (n-3) PUFA (20:5) (Orcutt and Patterson 1975) and smaller amounts of 20:6 (n-3) (Ackman *et al.*, 1968). The fatty acid profile of various diatoms is given in Table1. 14:0, 16:1 (n-7) and 16:0 are the most abundant in all species. 24:0 is present in four species at low concentrations though its presence is not often reported in literature. All species contain 20:5 (n-3) with the highest proportion in *P. tricornutum* (12.5%). This species has the characteristic distribution of Bacillariophyceae with a predominance of 16:1 (n-7), 16:0, 22:5 (n-3) and 14:0.

**Table 2. Total cellular lipid content (pg/cell) and percent composition of individual lipid class in diatoms**

| Species | Total lipids | PL | DG+ST | FFA | TG | HC |
|---|---|---|---|---|---|---|
| **Pennales** | | | | | | |
| *Thalassothrix heteromorpha* | 45 | 59.4 | 3.6 | 25.9 | 5.9 | 2.0 |
| *Navicula* sp. | 11 | 50.1 | 0.1 | - | 47.7 | 2.0 |
| *Haslea ostrearia* | 67 | 53.9 | 1.7 | 14.7 | 14.5 | 10.1 |
| *Amphora* sp. | 20 | 92.4 | 0.9 | 2.1 | 3.1 | 1.5 |
| *Amphipora hyalina* | 16 | 81.4 | 1.4 | 6.8 | 2.1 | 2.0 |
| **Centrales** | | | | | | |
| *Thalassiosira stellaris* | 30 | 84.7 | 2.1 | 2.1 | 9.3 | 1.8 |
| *Skeletonema costatum* | 10 | 63.6 | 0.7 | 17.4 | 4.5 | 0.3 |
| *Rizosolenia stigera* | 114 | 72.6 | 3.1 | 9.4 | 7.1 | 3.9 |
| *Coscinodiscus* sp. | 1196 | 84.3 | 2.3 | - | 11.9 | 1.5 |

PL—polar lipids and chlorophylls: DG+ST— diaglycerols and sterols: FFA—free fatty acids: TG—tryacylglycerols: HC—hydrocarbons
(Ref. Dunstan *et al.*, 1994)

Diatoms generally contain high concentrations of (n-3) PUFA (20:5) (Orcutt and Patterson 1975) and smaller amounts of 20:6 (n-3) (Ackman *et al.*, 1968). The fatty acid profile of various diatoms is given in Table1. 14:0, 16:1 (n-7) and 16:0 are the most abundant in all species. 24:0 is present in four species at low concentrations though its presence is not often reported in literature. All species contain 20:5 (n-3) with the highest proportion in *P. tricornutum* (12.5%). This species has the characteristic distribution of Bacillariophyceae with a predominance of 16:1 (n-7), 16:0, 22:5 (n-3) and 14:0.

A decrease in 14:0 with a concomitant increase in 20:4 (n-3) has been reported to occur in aged cultures (Chu and Dupuy 1980). The relative proportions of many fatty acids have shown to be significantly correlated with growth rates controlled by variations in photon flux density (Thompson *et al.*, 1990), an increase of the relative amount of 20:4 (n-3) in marine phytoplankton including Bacillariophyceae occurs when the photon flux density decreases.

Microalgae containing elevated levels of 20:4 (n-3) may prove to be a valuable source of this PUFA. Lipid content and composition can be manipulated by changing the culture conditions (Seto *et al.*, 1984), as a result that the favourable PUFA content of diatoms may be further enhanced. The decarboxylase enzyme which is thought to form the n-C 21:6 hydrocarbons from 22:6 (n-3) PUFA in microalgae (Lee and Loeblich 1971) may reduce the amounts of this PUFA. In spite of the reduced proportions of 22:6 (n-3) due to decarboxylation, the diatoms still contained very high proportions of the other essential PUFA, 20:5 (n-3). Many species have been shown to be valuable feeds for larval animals (Volkman *et al.*, 1993). Some diatoms may also be a ready source of free (non-esterified) fatty acids and monoglycerols.

## CHLOROPHYCEAE

Studies on chlorophycean fatty acids have not so far been systematic, but general peculiarities of fatty acids have been reported (Ackman and Mc Lachlan 1977; Khotimchenko and Svetsashev 1983; Kaneniwa *et al.*, 1987; Stefanov *et al.*, 1988; Khotimchenko 1993). However, the reports on fatty acids in different species in this phylum are contradictory. Stefanov *et al.*, (1988) have found that 16:3 is a major C16 PUFA in *Ulva rigida*, whereas, Dembitsky *et al.*, (1990) have reported 16:4 as the dominant fatty acid of the same species. Klenk *et al.*, (1963) and Sato (1975) believe that 16:3 is the characteristic fatty acid of *Codium* sp. whereas, others does not even report the presence of this fatty acid in two species of *Codium* (Kato and Ariga 1983). Discrepancies in data on fatty acid composition of different species might be due to different environmental conditions to which these algae belong. It has been reported for various algae that fatty acid composition is influenced by environmental factors such as temperature, light and nutrient concentrations (Johns *et al.*, 1979; Pohl and Zurheide 1982). However, by contrast, it was shown for algae of *Gracilaria* and *Sargassum* genera that various algal species of the same genus differ in their pattern and ratio of fatty acids (Araki *et al.*, 1990; Khotimchenko 1991; Khotimchenko *et al.*, 1991). It is therefore, possible that the distinction of fatty acid composition in same species may reflect genetic characteristics of the species rather than being related to environmental factors.

Typically, lipids of Chlorophyceae contain high amounts of C18 and C16 PUFAs. Johns *et al.*, (1979) investigated the fatty acids of ten marine algae including green, red and brown representatives, and concluded that hexaadecatetraenoic acid (16:4 n-3) could be taxonomically distinguished for species of Chlorophyta. Green algae are also able to synthesize not only C20 {20:4 (n-6), 20:5 (n-3)} but also C22 {22:4 (n-6) and 22:6 (n-3)} fatty acids. *Blindiga minima, Ulva fenestrata, Enteromorpha liza* and *E. flexuosa* have similar fatty acid compositions (Khotimchenko 1993). Major components among fatty acids of various species of Chlorophyceae is given in the Table 1.

Information on C20 and C22 polyunsaturated fatty acids in green algae is very important. Chlorophyta capable of synthesizing 20:4 (n-6), 20:5 (n-3), 22:5 (n-3) and 22:6 (n-3) are known, but the amounts present are much lower than Phaeophyceae and Rhodophyceae (Pohl and Zurhheide 1979; Khotimchenko and Svetashev 1987; Kayama *et al.*, 1989). However, there is an unusually high content of C20 PUFAs in green algae from the Black Sea: the concentration of 20:5 makes up 18.3% of all fatty acids in *Ulva rigida* and 8.3% in *Cladophora sericae* (Dembitsky *et al.*, 1990).

Barring few exeptions, unicellular and filamentous green algae are not suitable for the commercial production of C20—polyunsaturated fatty acids (Cojocaru *et al.*, 1988; Cranwell *et al.*, 1990). According to Attavian *et al.*, (1977), green algae that contained considerable amounts of these fatty acids were fila-mentous and more closely resembled the bryophytes. However, algae such as *Pyramimonas virginica, Chlorella* sp. and *Pseudoisochrysis paradoxa* (Chu and Dupuy, 1980) and *Skletonema costatum, Chaetocaros gracilis* and *Tetraselmis chuii* (Su *et al.*, 1988), which are suitable for rearing aquatic animals, contain considerable proportions of C20 and C22 polyunsaturated fatty acids. Some *Chlorella* sp., e.g., *Chlorella minutissima,* may produce upto 45% of eicosapentaenoic acid (20:5) of the total fatty acids (Seto *et al.*, 1984).

## RHODOPHYCEAE

The red algae grow mainly in tropical, sub-tropical and temperate seas and a few species are also found in sub-tropical regions. The morphologies of thallii of different species vary: unicellular, filamentous or cylindrical or circular flat forms and more complex pseudoparenchymatous and parenchymatous. There are 56 genera of red algae: 344 species are of economic value as food for humans and animals, as fertilizers, for medicinal or pharmaceutical purposes.

Red algae (Rhodophyceae) are known to contain appreciable amounts of C20 polyunsaturated fatty acids, mainly 20:4 (n-6) and 20:5 (n-3), with relatively small amounts of C16 and C18 polyunsaturated compounds (Pohl and Zurheide 1979; Khotimchenko and

Svestakov 1984). Stefanov *et al.*, (1988) have reported the same pattern in two red algal species from Black Sea, except that 20:5 (n-3) was the most abundant fatty acid in *Corallina granifera*, while 20:4 (n-6) (more often associated with animal lipids as a precursor of the prostaglandins) was the abundant component of *Phyllophora nervosa*. Ackman (1981) has reported that the amount of 20:5 fatty acid in *Palmaria palmata* is 70%. Another red alga, *Holosaccion ramentaceum*, contains 63.7% of this acid (Ackman and McLachlan 1977). The amount of 20:5 (n-3) is much more in lipids of *Corallina pilulifera* from the Sea of Japan (Christie *et al.*, 1988), *C. officianalis* from the Firth of Clyde (Scotland) (Jamieson and Reid 1972) and *C. tenella* from Red Sea (El-Sayed 1983). Gerwick *et al.*, (1988) reported on the isolation of several new bioactive fatty acids from red algae. These fatty acids (icosanoids) are natural products with potential biomedical applications.

## PHAEOPHYCEAE

The phaeophyceae (brown algae) are the major flora of many temperate littoral zones. The detailed lipid composition of brown algae, however, is not very well documented. The major lipids of brown algae are monogalactosyldiacylglycerol (MGDG), digalactosyldiacylglycerol (DGDG), sulphoquinovosyldiacylglycerol (SQDG), phosphatidylglycerol, phosphatidylethanolamine and tryacylglycerol with phosphatidylcholine, phosphatydyleethanolamine, phosphatidylserine, diphosphatidyleglycerol and unesterified fatty acids (Pham-Quang and Laur 1976). DGDG was the major lipid in *Fucus serratus* (75% of total lipids) along with an unidentified lipid (8%) (Smith and Harwood 1984).

Jones and Harwood (1992) studied the lipid composition of *Fucus vesiculosus* and *Ascophyllum nodosum* and found that the lipid profile of these two algae were very similar. MGDG was the most unsaturated lipid with high proportions of eicosapentaenoic and octadecatetraenoic acid. Relatively high amounts of the essential fatty acid aracidonic acid are also present in brown algae.

## FACTORS AFFECTING THE PRODUCTION OF FATTY ACIDS

Fatty acid composition of microorganisms, both qualitatively and quantitatively, is influenced by environmental conditions. Parameters such as media composition, aeration, light intensity and light-dark cycle, temperature and culture age play a major role in the biosynthesis and accumulation of polyunsaturated fatty acids in algae. Nitrogen content of the medium has been reported to affect the proportion of saturated to unsaturated fatty acids in green algae (Shaw 1966). Under nitrogen stress *Botryococcus braunii, Dunaliella bardawil* and *D. salina* produced higher percentage of EPA (Ben-Amtoz *et al.*, 1985). In contrast, the proportion of PUFA in fresh water algae *Scenedesmus* and *Chlorella*, increased at higher concentrations of nitrate (Piorreck *et al.*, 1984).

Several metal ions promoted the synthesis of lipid and fatty acids in algae (Seto *et al.*, 1984; Enright *et al.*, 1986). As it has been clearly shown that microorganisms required molecular oxygen for the desaturation mechanism in the biosynthetic pathway of PUFAs, and the availability of oxygen will naturally determine the degree of unsaturation of fatty acids produced. Increased oxygen tension in the medium enhanced the unsaturated fatty acid content in the alga *Gyronidium cohinii* (Harington and Holz 1966) and *Euglena gracilis* (Miyatake *et al.*, 1985). Low light intensity enhanced formation and accumulation of PUFAs in many diatoms such as *Cyclotella meneghiniana* (Sicko-Goad *et al.* 1984) and *Nitzschella closterium* (Orcutt and Patterson 1974). However, for the green alga, *Chlorella minutissima* (Seto *et al.*, 1984) and for the red algà, *Porphyridium cruentum* (Rezanka *et al.*, 1987) the effect of light intensity was reversed. Increased synthesis of unsaturated fatty acids at low temperatures has been observed in certain blue-green algae (Holton *et al.*, 1984). Seto *et al.*, (1984) observed this phenomenon in *Chlorella minutissima* also. Brown and Rose (1962) postulated that, because of increased solubility of oxygen at low temperature, a greater amount of extracellular molecular oxygen is available, to meet the requirements of by oxygen dependent enzymes catalysing the desaturation of long-chain saturated fatty acids. On the other hand it has been suggested that the temperature associated production of PUFAs in microalgae is a means of adaptation to environment. It is apparent that much more research on this subject is still needed to study other interesting organisms and various parameters. It is known that the nitrogen and phosphorus levels in the medium, control the switching of metabolism between protein and lipid synthesis. But, nothing is known about the effect of nitrogen and phosphorus contents of the medium on the fatty acid composition of algae.

## BIOTECHNOLOGICAL POTENTIAL

Micro and macroalgae possess the whole complement of enzymes capable of catalysing de-novo synthesis of C20-polyunsaturated fatty acids. Research efforts aiming at implementing some of these organisms for biotechnological production of these essential fatty acids have already started. There are fairly simple techniques for the production of microalgal biomass on a large scale in outdoor ponds (Venkataraman and Becker 1984). Considering the economy of adjusting light and temperature, the cheapest systems would be ponds in tropical or semi-tropical or arid regions (Litchfield 1989). Many such ponds are operational in different countries such as Mexico, Israel, India, Thailand, China, Japan and Australia. Since the conditions for optimal production of biomass are usually not identical with those for optimal C20-PUFA production, a biphasic production system may be advisable. For example in the first phase, the biomass might be produced at a relatively high temperature and in the second phase the harvested cells might be chilled to increase the PUFA content.

**Table 3. C20 PUFA content of some promising algal species that could be used for commercial exploitation**

| Species | 20:3 | 20:4 | 20:5 | Reference |
|---|---|---|---|---|
| **Microalgae** | | | | |
| *Chlorella minutissima* | | | 45.0 | Seto *et al.*, 1984. |
| *Porphyridium cruentum* | 2.0 | 36.0 | 17.0 | Nichols & Appleby, 1969. |
| *Skeletonema costatum* | 2.0 | 2.0 | 30.0 | Chuecas & Riley, 1969. |
| *Lauderia borealis* | | 1.0 | 30.0 | Chuecas & Riley, 1969. |
| *Navicula pelliculosa* | | | 26.0 | Kates & Volcani, 1966 |
| *Dunaliella tertiolecta* | | 4.0 | 10.0 | Chuecas & Riley, 1969. |
| **Macroalgae** | | | | |
| *Gelidium latifolium* | | 10.1 | 24.6 | Radwan *et al.*, 1988. |
| *Polysiphonia lanosa* | | 10.2 | 38.6 | Pettitt *et al.*, 1989. |
| *Chondrus crispus* | | 25.0 | 26.9 | Pettitt *et al.*, 1989. |
| *Sargassum boveanum* | | 12.6 | 13.3 | Al-Hasan *et al.*, 1991. |

Recent interest in PUFA and limited availability of GLA and EPA prompted scientists to look for genetic means to overproduce these fatty acids in algae. Cohen *et al.*, (1992) were the first to develop fatty acid overproducing strains of algae. They developed the cell lines of *Spirulina platensis* and *Porphyridium cruentum*, which are resistant to the growth inhibiting herbicide Sandoz 9785 through the serial transfer of these cells to increasing concentrations of herbicide. They have shown that herbicide resistant cells of *Spirulina* overproduced GLA and that of *Porphyridium* overproduced EPA. These resistant cells were stable upto 50 generations. Attempts are now on to produce other important fatty acids also through genetic and physiological manipulation of the cells.

Green algae have an advantage over other algae in their ability to grow faster (Borowitzka *et al.*, 1988). Most of them are microscopic, unicellular or filamentous and usually found in fresh water. They generally contain less PUFA. However, a marine unicellular alga, *Chlorella minutissima*, contained very high EPA content (about 45% of total lipids) at its optimum culture conditions. This particular species has been shown to be a promising source of EPA for commercial exploitation. Moreover, EPA obtained from this algae is claimed to be superior than that of fish oil because it has no unfavourable fish flavour. The red algae are mainly macroscopic. They contain high levels of arachidonic acid and EPA. Due to their size, it is almost impossible to culture these algae under controlled conditions (Araki *et al.*, 1987, 1990). However, experience gained from culturing edible macroalgae in the marine environment in some Asian countries can certainly be implemented for propagating those macroalgae rich in C20-PUFA.

## CONCLUSION

Some polyunsaturated fatty acids (PUFA) are considered to be essential fatty acids (EFA) since they cannot be synthesised in the human body. Absence of EFA can cause schizophrenia, thrombosis, arthritis and cerebro and cardiovascular diseases. EFA includes, C18-22 unsaturated fatty acids viz., linoleic acid (18:2, n-6), linolenic acid (GLA: 18:3, n-6), arachidonic acid (ARA; 20:4, n-6), eicosapentaenoic acid (EPA; 20:5, n-3) and docosahexaenoic acid (DHA; 22:6, n-3). In view of its therapeutic uses, EFA has gained considerable commercial interest. The feasibility of producing EFA from algae have shown by Cyanotech, Woodwille, USA. The major barriers to the commercialization of algal fatty acids are the fact that no single species is distinctly a better producer than the another, although *Phaeodactylum* and *Porphyridium* are the major likely ones (Venkataraman, 1996).

It is recently reported in print media that Scotia Pharmaceuticals, a British company and the French Heliosynthese Research Company have proposed to develop a new pharmaceutical and nutritional product, based on medically important PUFAs produced from oil-rich micro-algae grown in closed photobioreactors. However, the species of algae to be used has not been announced and *Spirulina* is likely to be one of them. Genetic engineering techniques to enhance essential fatty acids and producing transgenic algae will be of great interest for the future. India has a vast flora of algae, both fresh water and marine forms, in view of its different climatic zones and vast coast line. The full biochemical and biotechnological potential of many forms has not yet been explored. It is important that these valuable flora are preserved, protected and further exploited using innovative biotechnological methods to produce novel nutritional products in an economical way for the benefits of mankind.

## ACKNOWLEDGEMENT

TAK thanks the Council of Scientific and Industrial Research, New Delhi for financial assistance.

## REFERENCES

Ackman, R.G. 1981. In: *New Sourcess of Fats and Oils*, AOCS Champaign. pp. 189-225.

Ackman, R.G. and McLachlan, J. 1977. *Proc. N. Scot. Inst. Sci.* **28**: 47.

Ackman, R.G., Tocher, C.S. and McLachlan, J. 1968. *J. Fish Res. Bd.* Canada. **25**: 1603.

Adam, O. 1990. *Fat Sci. Technol.* **92**: 423.

Ahmadjian, V. 1967. *Phycologia.* **6**: 127.

Ahren, T.J., Katoh, S. and Sada, E. 1983. *Biotechnol. Bioeng,* **25**: 1057.

Aknin, M., Mollet-Nzaou, R., Cisse, E., Kornprobst, J.M., Gaydou, E.M., Samb, A. and Miralles, J. 1992. *Phytochem.* **31(8)**: 2739.

Albers, D., Reisser, W. and Wiessner. 1982. *Plant Sci. Lett.* **25**: 85.

Al-Hasan, R.H., Hantash, F.M. and Radwan, S.S. 1991. *Appl. Microbiol. Biotechnol.* **35**: 530.

Araki, S., Sakurai, T., Kawaguchi, A. and Murata, N. 1987. *Plant Cell Physiol.* **28**: 761.

Araki, S., Sakurai, T., Oohusa, T., Kayama, M. and Nirizawa, K. 1990. *Hydrobiologia.* **204/205**: 513.

Attavian, B.N., Floyd, G.L. and Fairbrothers, D.E. 1977. *Biochem. Syst. Ecol.* **5**: 65.

Bauld, J. and Brock, T.D. 1974. *J. Phycol.* **10**: 101.

Baus, M., Abo-Elnaga, N. and Elmadfa, I. 1990. *Fat Sci. Technol.* **92**: 430.

Ben-Amtoz, A., Tornabene, T.G. and Thomas, W.H. 1985. *J. Phycol.* **21**: 72.

Ben-Amtoz, A., Fishler, R. and Schneller, A. 1987. *Mar. Biol.* **95**: 31.

Borowitzka, M.A. 1988. In: *Microalgal Biotechnology* (eds., Borowitzka, M.A. and Borowitzka, L.J.) Cambridge University Press, Cambridge. pp. 257-287.

Brown, C.M. and Rose, A.H. 1962. *J. Bacteriol.* **84**: 1260.

Chapman, R.L. and Good, B.H. 1982. In: *Algal Symbiosis: A Continuum of Interaction Strategies* (Goff, L.F. ed.) Cambridge University Press, New York.

Chu, F.L.E. and Dupuy, J.L. 1980. *Lipids.* **15**: 356.

Christie, W.W., Brechany, E.Y. and Stefanov, V. 1988. *Chem. Phys. Lipids.* **46**: 127.

Cohen, Z., Vonshak, A. and Richmond, A. 1987. *Phytochem.* **26(8)**: 2255.

Cohen, Z., Vonshak, A. and Richmond, A. 1987. *J. Phycol.* **24**: 328.

Cohen, Z., Didi, S. and Heimer, Y.M. 1992. *Plant Physiol.* **98**: 569.

Cojocaru, M., Shlosberg, M., Dubinsky, Z. and Finkel, A. 1988. *Biomed. Environ. Mass Spectrom.* **16**: 477.

Cooke, W.J. 1975. *Phycologia.* **14**: 35.

Cranwell, P.A., Jaworsky, G.H.M. and Bickley, H.M. 1990. *Phytochem.* **29**: 145.

Das, U.N., Begin, M.E., Huang, Y.S. and Horrobin, D.F. 1987. *Biochem. Biophys. Res. Commun.* **145**: 15.

Dembitsky, V.M., Pechenkina-Shubina, E.E. and Rozentsvet, O.A. 1991. *Phytochem.* **30(7)**: 2279.

Dunstan, G.A., Volkman, J.K., Jeffrey, S.W. and Barrett, S.M. 1992. *J. Exp. Mar. Biol. Ecol.* **161**: .

Dunstan, G.A., Volkman, J.K., Barret, S.M. and Garland, C.D. 1993. *J. Appl. Phycol.* **5**: 71.

Dunstan, G.A., Volkman, J.K., Barret, M.S., Marie Leroi, J. and Jeffrey, S.W. 1994. *Phytochem.* **35(1)**: 155.

Dyerberg, J. 1986. *Nutr. Rev.* **44**: 125.

El-Sayed, M.M. 1983. *J. Fac. Mar. Sci.* (King Abdulaziz University). **3**: 141.

Enright, C.T., Newkirk, G.F., Craigie, J.S. and Castell. 1986. *J. Exp. Mar. Biol. Ecol.* **96**: 1.

Fraser, A.J., Sargent, J.R., Gamble, J.C. and Seaton, D.D. 1989. *Mar. Chem.* **27**: 1.

Friedman, I. 1971. *Phycologia.* **10**: 411.

Friedman, E.I. 1982. *Science.* **215**: 1045.

Gillan, F.Y., Mc Fadden, G.I., Wetherbee, R. and Johns, R.B. 1981. *Phytochem.* **20**: 1935.

Graham, L.E., Mac Entee, F.J. and Bold, H.C. 1981. *Texas J. Sci.* **33**: 13.

Harwood, J.L. and Russel, N.J. 1984. In: *Lipids in Plants and Microbes*, George Allen and Unwin, London. pp. 162.

Henderson, R.J., Leftley, J.W. and Sargent, J.R. 1988. *Phytochem.* **27(6)**: 1679.

Holton, R.W., Blecker, H.H. and Onore, M. 1964. *Phytochem.* **3(5)**: 595.

Horrobin, D.F. and Huang, Y.S. 1987. *J. Cardiol.* **17**: 173.

Huang, Y.S., Nanku, M.S. and Horrobin, D.F. 1984. *Lipids.* **19**: 664.

Jamieson, G.R. and Reid, E.H. 1972. *Phytochem.* **11**: 1423.

Johns, R.B., Nichols, P.D. and Perry, G.J. 1979. *Phytochem.* **18**: 799.

Jones, A.L. and Harwood, J.L. 1992. *Phytochem.* **31(10)**: 3397.

Kaneniwa, M., Itabashi, Y. and Takagi, T. 1987. Nippon Suisan Gakkaishi. **53**: 861.

Kates, M. 1989. In: *Marine Biogenic Lipids, Fats and Oils* (Ackman, R.G., ed.) Vol. 1, pp. 389.

Kato, M. and Ariga, N. 1983. Kyoyobu K. Hakoku (Gifu Daigaku). **18**: 93.

Kayama, M., Iijima, N., Sado, T., Mankura, M., Araki, S. and Sakurai, T. 1986. Nippon Suisan Gakkaishi. **52(3)**: 575.

Kenyon, C.N., Rippka, R. and Stanier, R.Y. 1972. *Arch. Microbiol.* **83**: 216.

Khotimchenko, S.V. 1991. *Phytochem.* **30**: 2639.

Khotimchenko, S.V. 1993. *Phytochem.* **32(5)**: 1203.

Khotimchenko, S.V. and Svtashev, V.I. 1983. *Biol. Morya* (Vladivostak). **6**: 3.

Khotimchenko, S.V. and Svetashev, V. 1984. *Biol. Morya* (Vladivostok). **5**: 45.

Khotimchenko, S.V. and Svetashev, V.I. 1987. *Biol. Morya* (Vladivostok). **6**: 3.

Khotimchenko, S.V., Vaskovsky, V.E. and Przemenetskaya, V.F. 1991. *Phytochem.* **30**: 207.

Klenk, E., Knipprath, W., Eberhagen, D. and Koof, H.D. 1963. *Z. Phys. Chem.* **334**: 44.

Korn, E.D., Greenblatt, C.I. and Lees, A.M. 1965. *J. Lipid Res.* **6**: 43.

Kyle, D. 1992. *Lipid Technol.* **May/June**: 59.

Lee, R.F. and Leoblich, A.R. III. 1971. *Phytochem.* **10**: 593.

Litchfield, J.H. 1989. In: A Revolution in Biotechnology (Marx, J.L., ed.) Cambridge University Press, Cambridge, p. 71.

Marchant, H.J. 1982. *Phycologia.* **21**: 178.

Metzner, C., Luder, W., Machholz, R., Jung, K. and Bleyl, D. 1990. *Fat Sci. Technol.* **92**: 431.

Miyatake, K., Minamigawa, M., Nakano, Y. and Kitaoka, S. 1987. Nippon Eiyo, Shokuryo Gakkaishi. **38(2)**: 117.

Nichols, B.W. and Appleby, R.S. 1966. *Phytochem.* **8**: 1907.

Noda, H., Amano, H., Kato, S. and Ohta, F. 1983. Nippon Suisan Gakkaishi. **49**: 1583.

Oren, A., Fattom, A., Padan, E. and Tietz, A. 1985. *Arch. Microbiol.* **14**: 319.

Orcutt, D.M. and Patterson, G.W. 1975. *Comp. Biochem. Physiol.* **50B**: 579.

Pettitt, T.R., Jones, A.L. and Harwood, J.L. 1989. *Phytochem.* **28**: 399.

Pham-Quang, L. and Laur, M.H. 1976. *Biochimie.* **58**: 1367.

Piorreck, M., Baasch, K. and Pohl, P. 1984. *Phytochem.* **23**: 207.

Pohl, P. and Zurheide, F. 1979. In: *Marine algae in pharmaceutical science* (Hoppe, H.A., Lerving, T. and Tanaka, Y., eds.) Walter de Gruyter, Berlin, p. 473.

Pohl, P., Wagner, H. and Passing, T. 1968. *Phytochem.* **7**: 1565.

Radwan, S.S. 1991. *Appl. Microbiol. Biotechnol.* **35**: 421.

Reisser, W. 1981. *Ber. Dtsch. Bot. Ges.* **94**: 557.

Rezanka, T., Doucha, J., Mares, P. and Podojil, M.J. 1987. *Basic Microbiol.* **27(5)**: 275.

Sato, S. 1975. Nippon Suisan Gakkaishi. **41**: 1117.

Schopf, J.W. 1976. *Origins Life.* **7**: 19.

Seto, A., Wang, H.L. and Hasseltine, C.W. 1984. *J. Am. Oil Chem. Soc.* **61**: 892.

Shaw, R. 1966. *Adv. Lipid Res.* **4**: 107.

Sicko-Goad, L., Simmons, M.S., Lazinsky, D. and Hall, J. 1988. *J. Phycol.* **24**: 1.

Singer, P. 1990. *Fat Sci. Technol.* **92**: 430.

Smith, K.L. and Harwood, J.L. 1984. *Phytochem.* **23**: 2469.

Sournia, A., Chretiennot-Denet, M.J. and Riccardi 1991. *J. Plankton. Res.* **13**: 1093.

Stefanov, K., Konaklieva, M. Brechany, E.Y. and Christie, W.W. 1988. *Phytochem.* **27(11)**: 3495.

Stroh, S., Elmadfa, I., Schlotzer, E. and Weidler, B. 1990. *Fat Sci. Technol.* **92**: 433.

Su, H.M., Lei, C.H. and Liao, I.C. 1988. *J. Fish Soc.* Taiwan. **15**: 21.

Sugano, M., Ishida, T., Yoshida, K, Tanaka, K., Miwa, M., Arima, M. and Morita, A. 1986. *Agric. Biol. Chem.* **50**: 2483.

Thompson, P.A., Harrison, P.J. and Whyte, J.C.N. 1990. *J. Phycol.* **26**: 278.

Venkataraman, L.V. 1996. In: Proceedings of ITIT Symposium on Microalgal Biotechnology—Basics and Applications, Osaka National Research Institute, Osaka, Japan, pp. 49.

Venkataraman, L.V. and Becker, E.W. 1984. *Biomass.* **4**: 105.

Venkataraman, L.V. and Becker, E.W. 1986. Biotechnology and Utilization of Algae—The Indian Experience. DST-CFTRI Publication, Mysore, India. p. 256.

Viso, A.C. and Marty, J.C. 1993. *Phytochem.* **34(6)**: 1521.

Volkman, J.K., Brown, M.R., Dunstan, G.A. and Jeffret, S.W. 1993. *J. Phycol.* **29**: 69.

Yongmanitchai, W. and Ward, O.P. 1991. *Phytochem.* **30(9)**: 2963.

Recent Trends in Algal Taxonomy
Vol. I. Taxonomical Issues
Editors : Vidyavati and A.K. Mahato
APC Publications Pvt. Ltd., New Delhi, 1999, pp. 21-50

# Nuclear Cytology of Algae—A Promising Adjunct to Taxonomy

S.B. Agrawal and B.R. Chaudhary
Centre of Advanced Study in Botany, Banaras Hindu University
Varanasi-221 005, India

The ultimate goal of taxonomy is the recognition of relationships between individuals by comparing the manifest characters. Among all, morphology is the most accessible and utilized characteristic. Cytological features being more stable and consistent can provide more reliable information for taxonomy (cf. Magne, 1978).

Cytotaxonomy of flowering plants has already proved highly rewarding in elucidating the conflicting problems of systematics (Davis & Heywood, 1963). Although much literature has gathered on the various aspects of algal cytology and cytotaxonomy (Sarma, 1983; Sarma *et al.*, 1990) studies in these areas are still in their infancy as compared with those of higher plants. Furthermore, while rapid progress was noticed in the studies on algal systematics (Lund, 1962), those on the cytology of algae and their cytotaxonomy are still moving at a slow pace. However, on the basis of informations available so far it has now been possible to correlate cytological features with the taxonomy and phylogeny of some of the individual algal classes and utilize these criteria in determining the systematic position of some genera of controversial nature (Sarma, 1978; Agrawal, 1996; Chaudhary & Agrawal, 1996). It is always possible to study the detailed fine structure of organelles by electron microscope even at molecular level, light microscopy has nevertheless been used frequently in the detection and description of cellular details of organelles. In other words, light microscopy has also played a crucial role in providing the valuable data for algal taxonomy as discussed in this paper.

## KARYOLOGICAL FEATURES CONSIDERED FOR CYTOTAXONOMY

Several karyological features generally taken into consideration while making taxonomic decisions are: chromosome number, size of chromosomes, ratio of chromosome arms, nature and position of centromere, number and size of satellites and distribution of eu—and heterochromatin (banding patterns) in the chromosomes of the complement (Stebbins, 1950; Davis & Heywood, 1963, Sarma, 1978). Bhatnagar *et al.* (1996) pointed out that like higher plants and animals Giemsa—C banding could be highly profitable in the taxonomy and phylogeny of algae, which otherwise are not so well explicable on the basis of conventional staining of chromosomes.

Although chromosome number is a fixed feature of taxon several workers have already indicated that the chromosome numbers alone cannot be relied upon as the criterion for taxonomic delineation (Smith, 1933; Smith-White, 1954). Instead, an overall assessment of the nuclear features should be taken into consideration while attempting karyotypic comparisons for taxonomical purposes. This fact has now been well documented by recent studies more specifically with distant taxonomic groups as summarized below (cf. Sarma, 1978, 1983).

## CYTOTAXONOMY AT HIGHER TAXONOMIC LEVELS (CLASS/ORDER)

### Oedogoniales

Oedogoniales based on morphological characteristics have been variously classified. Most of the algal taxonomists treat Oedogoniales as an order of Chlorophyceae (Fritsch, 1935), while Fott (1959) included the members of Oedogoniales (*Oedogonium, Oedocladium* and *Bulbochaete* species) in the order Ulotrichales. Round (1963, 1971), on the other hand, considering morphological features elevated the order Oedogoniales to a new class Oedogoniophyceae. Cytologically, all the three genera of Oedogoniales are characterized by fairly large size of chromosomes and distinct centromeres even when the chromosomes are markedly small (Srivastava & Sarma, 1979; Singh & Chaudhary, 1990; Gupta, 1992; Agrawal & Chaudhary, 1996a) and have little in common with Ulotrichales which possess small to minute chromosomes, normally with undeciferable centromeres. Thus, the merger of Oedogoniales with the Ulotrichales done by Fott (1959) was found highly untenable on the basis of karyological features. The odd number of chromosomes (n=9, 11, 13, 15, 17, 18, 19, 21, 25, 27, 37, 41 and 49) and the distinctive cytological features (Hoffman, 1961, 1967; Henningsen, 1963) coupled with other unique morphological characteristics, on the other hand, have strongly favoured the view of Round (1963, 1971) to the separation of Oedogoniales from Chlorophyceae and its elevation to the status of an independent class Oedogoniophyceae (cf. Sarma, 1983; Singh & Chaudhary, 1990). On the ultrastructural basis, however, the Oedogoniales have been retained to Chlorophyceae (Pickett-Heaps, 1975; Mattox & Stewart, 1984).

**Conjugales**

The researches on the karyology of *Spirogyra*, *Zygnema*, *Zygnemopsis*, *Mougeotia*, *Sirocladium*, *Sirogonium* and desmids (cf. Sarma, 1983; Hoshaw *et al.*, 1988) brought some distinctive karyological features in vast majority of Conjugales: (i) interphase nuclei characterized by chromocentres, (ii) complex nucleolus with internal differentiation of nucleolar tracts, (iii) Geitler's nucleolar substance and its behaviour during nuclear division and (iv) polycentric or diffuse centric chromosomes exhibiting parallel disjunction at metaphase and anaphase are so distintive and unique that the separation of Conjugales from Chlorophyceae and the elevation to a new class Conjugatophyceae as propounded by Fott (1959) and Round (1963, 1971), on morphological basis, seems justifiable also from cytological view points. This class almost corresponds with the Conjugatae of Wille (1909) and Zygophycees of Chadefaud and Emberger (1960).

**Cladophorales**

Round (1963) rearranged the members of Cladophorales and Siphonales of Fritsch (1935) and created a separate class Bryopsidophyceae. Cytologically, however, amalgamation of these orders does not seem to be very sound. In morphology and the size of the chromosomes both the orders stand apart. The chromosomes of Siphonales are very small, dot or short rod-like (Sarma, 1983), while the chromosomes of the Cladophoralean members, in general, are appreciably larger in length (Sinha, 1963, 1965, 1968; Chaudhary, 1978, cf. Sarma, 1983) with well defined centromeric locations as compared to Siphonales. Cladophorales constitute well defined group with established polyploidy and a definite relation between the number of nucleoli and level of ploidy (Sinha 1958, 1963; Balakrishnan, 1961; cf. Sarma, 1964, 1978, 1983).

### Ulotrichales and Chaetophorales

Fritsch (1935) treated Ulotrichales and Chaertophorales as two separate orders under the class Chlorophyceae. Abbas and Godward (1964) and Sarma (1978) while discussing relationship of the two orders (except the genus *Coleochaete*) on the basis of their morphology and cytology supported their merger into a single order owing to close affinity betweem them. While Printz (1964) preferred the inclusion of both the groups under Chaetophorales. Fott (1959) and Christensen (1962), on cytological grounds, supported the treatment of the members of Chaetophorales under Ulotrichales.

### Charophyceae

Charophytes being morphologically distinctly different from Chlorophyceae, variations in the vegetative morphology amongst Charophyta are much less pronounced, and this makes taxonomic identifications quite difficult. While discussing the status of Charophyta it has been delimited as a separate class from Chlorophyceae (Smith, 1955; Fott, 1959; Chapman, 1962), subphylum (Round, 1963) or as phylum (Desikachary & Sundaralingam, 1962; Silva, 1962). The studies on the nuclear cytology of this group so far accomplished (long thick chromosomes exhibiting pronounced definite euploidy, but without discernible centromeres) also show disimilarities in karyological features (Sinha, 1958, 1963.,cf. Sarma, 1983; Khan & Sarma, 1984) and seem to favour the separation of this group from Chlorophyceae. However, the suggestion of a probable Chaetophoralean origin of Charophytes (cf. Sarma, 1978) does not find support on the basis of distinctive differences in cytology of Charophytes and the members of Chaetophorales. Nuclear data presented by various workers on the Charophyta have proved to be of immense value in the cytotaxonomic, cytogeographic and phylogenetic assessments of the world Charophytes (Khan & Sarma, 1984). Based on chromosome data the distribution of world Charophyta, with Bihar and Uttar Pradesh (India) as the primary centres of origin of Charophytes has been suggested.

### Dinophyceae

The members of Dinophyceae, commonly referred to as dinoflagellates that were included under protozoa by zoologists and algae by botanists have been shown to be highly interesting and unique in several aspects of karyology (Dodge, 1966; Shyam & Sarma 1978, cf. Sarma, 1983; Agrawal *et al.*, 1995). Dinophyceae are marked by several unusual nuclear and chromosomal characteristics and significant departures from normal mitotic pattern—"idiosyncrotic mitosis" (Taylor, 1980; Sarma, 1983). Unlike the relaxed state as in other eukaryotic algae, in the interphase nucleus of Dinophyceae the chromosomes remain exceptionally in condensed state, comparable to advanced phases of the cell cycle (Dodge, 1966; Shyam & Sarma, 1978). These chromosomes are devoid of typical histones (Rizzo, 1981) and contain substantial amounts of hydroxymethyluracil (Rae, 1976), and in metaphase chromosomes lack the differential heterochromatic cross banding. The presence of these unusual features has led many investigators to suggest that dinoflagellates are intermediate between prokaryotes and the eukaryotes (Dodge, 1966; Allen et al., 1975; Hamakalo & Rattner, 1977). Dodge (1966) coined the term "Mesokaryota" to include the members of this group. However, the explanations given by various workers to the obvious questions as to how dinokaryotic chromosomes assume compaction in the absence of histone proteins and when the DNA synthesis takes place in the permanently condensed state of dinoflagellate chromosomes are still a subject of debate and enquiries. The other unique feature of dinoflagellates is the enormous amount of DNA in the nucleus (3.2 pg/cell in *Amphidinium* to 200 pg/cell *Gonyaulax*) (Holm-Hansen, 1969; Rizzo, 1987) as compared to eukaryotic algae i.e 0.1 pg/cell in *Porphyridium* to 2.1 pg/cell in *Euglena* (*Charles,* 1977). The fact remains that the number of chromosomes in dinoflagellate cells is comparatively very high (Dodge, 1963; Loeblich, 1976), but as to why there has been exorbitant DNA amplification in the dinoflagellate individual chromosomes remaind unexplicable on evolutionary grounds.

## Euglenophyceae

Karyology of euglenoid flagellates (Leedale, 1966; cf. Sarma, 1982, 1983; Chaudhary, 1991; Gavrila, 1996) showed a number of interesting features viz., (i) persistance of endosome (= nucleolus) throughout mitosis and its division being independent of chromosomes, (ii) the apparent absence of spindle and centromeres, (iii) the arrangement of chromosomes parallel to the division axis at metaphase and the absence of true equatorial plate, (iv) staggering of chromatid separation, segregation and migration at anaphase prolonging anaphase and (v) the apparent retention of nuclear envelope throughout division cycle (intranuclear type of division). Cytological features of the Euglenophyceae largely resemble with those of Dinophyceae, particularly in the lack of any spindle or centromeric organisation. However, in the members of Dinophyceae unlike Euglenophyceae the nuclear membrane disappears during mitosis, and the chromosomes of Dinophyceae contain only DNA without any association of histones characterstic of DNA-histone complexes of the eukaryotes. The members of Euglenophyceae also show a distinctive feature i.e. dumb-bell-shaped appearance of the nucleolus at anaphase. Thus, cytological features provide a basis for classifying and distinguishing the members of Dinophyceae and Euglenophyceae. On evolutionary scale the euglenoid flagellates seem to be more advanced than dinoflagellates, occupying a place between Dinophyceae and the typical eukaryotic algae.

## CYTOTAXONOMY AT LOWER TAXONOMIC LEVELS (GENUS/SPECIES)

### Generic level

The study of nuclear cytology, in certain instances, has also been helpful in clarifying the disputed systematic positions at generic level. For example, the genus *Sphaeroplea* variously classified as under Cladophorinae (Smith, 1933), Siphoneae (Bohlin, 1901), Siphonales (Blackman & Tansley, 1902), Siphonocladales (Fott, 1933) and Ulotrichales (Fritsch, 1935; Chapman, 1962) was finally included in the order Ulotrichales on the basis of cytological features. Separation of *Sphaeroplea* into a separate monogeneric order Sphaeropleales (Round, 1963) was also discarded on the basis of cytological grounds as it showed much affinity with the Ulotrichales (Sarma, 1962; Chaudhary, 1983). The position of the genus *Schizomeris* was also disputed due to differing opinions. Some workers favoured the inclusion of *Schizomeris* under Ulvales, while others preferred its retention under Ulotrichales. Campbell & Sarafis (1972), on the other hand, considered *Schizomeris* as an ecophene of *Stigeoclonium tenue* under Chaetophorales. Studies on the karyology of *Schizomeris leibleinii* showed that the chromosomes of this genus in their overall size and general morphology appear more like those of *Uronema* and

*Hormidium* (Sarma, 1963; Sarma & Chaudhary, 1975) than those of *Ulva*. The view expressed by Campbell and Sarafis (1972) to merge *Stigeoclonium tenue* with *Schizomeris leibleinii* under Chaetophorales did not get support on the basis of chromosome studies made by Sarma & Chaudhary (1975), concluding the closeness of *Schizomeris* with the Ulotrichales. The karyology of Ulvales is still meagre, but similarities exist in the chromosome complements of *Ulva*, *Enteromorpha* and *Ulothrix* (Sarma, 1964), suggesting the inclusion of Ulvaceae under Ulotrichales (Fritsch, 1935). Likewise, on the basis of cytological studies Chowdary (1959, 1963) did not favour Flint's (1959) view of considering *Physolinum monilia* as *Trentepohlia monilia*.

**Species level**

A convincing basis for cytotaxonomy has been shown in a number of Chlorophycean genera as revealed by the chromosome numbers recorded for different species belonging to the same genus (cf. Sarma, 1964; 1978, 1983). Based on cytotaxonomic study of five species of *Stigeoclonium*, Abbas & Godward (1964) recognized only three valid species viz., *S. amoenum* (n=12), *S. tenue* (n=13) and *S. farctum* (n=8), suggesting the merger of *S. nanum* (n=13) with *S. tenue* and *S. helveticum* var. *minus* (n=12) with *S. amoenum*. Karyological features have been successfully used in conjunction with morphological characters in the taxonomic descriptions of a number of species of *Spirogyra* (Godward, 1956; Abhayavardhani & Sarma, 1983). Cave & Pocock (1951) stressed on the taxonomic value of nuclear characters in the members of Volvocales. Nuclear studies of Charophyta yielded a wealth of information for better understanding of the systematics in the group. Wood & Imahori (1965) in their monograph "A revision of the Characeae" have made drastic cut in the number of species, either by relegating several existing species to intraspecific level or by merging several species into one (Sarma, 1968), but it did not find any support from cytological data of which a few are given below.

(i) *Chara* : *C. wallichii* Br. has been reduced to a variety of *C. corallina* and renamed as *C. corallina* var. *wallichii*. Both, on the basis of chromosome number and karyotype organisation Khan & Sarma (1967a) have shown that the combination of the above two taxa is not justified, and advocated for the retention of *C. wallichii* as independent species.

*C. wallichii* (dioecious) n=14, (3m 5sm 6t)S
*C. corallina* (monoecious) n= 42, (1m 3sm 1st 3t)M
+ (4m 2sm 26t)S + (2t)Mi

Similarly, the merger of *Nitella burmanica* with *N. furcata* var. *furcata* f. *furcata* as was done by Wood & Imahori (1965) does not seem to be justified (cf. Khan & Sarma, 1967b). Contrarily, cytological studies have supported some of the combinations effected by Wood & Imahori (1965) e.g. combination of *N. mucronata* and *N. flagellifera* with *N. furcata* as subspecies of the latter taxon. Bhatnagar *et al.* (1996) provided an account of C-banding studies in various taxa of Charophytes and brought out the utility of C-band patterns in the taxonomic and phylogenetic considerations. Based on banding patterns, Bhatnagar *et al.* (1996) also presented similarities in *Chara corallina* (n=28) and *C. wallichii* (n=14) on the basis of st and t chromosomes. Likewise, the banding patterns in karyotypes of *Chara vulgaris* and *C. contaria* showed great resemblances.

**Karyotypes in taxonomic diversification**

For chromosome numbers in major groups of algae reference may be made to Godward (1966), Sarma (1983), Chaudhary (1992), Agrawal (1996), Chaudhary & Agrawal (1996), Agrawal & Chaudhary (1996a, b) and Gavrila & Chaudhary (1996), besides origional publications in the subject. Perusal of chromosome numbers recorded for various groups of algae representing different chromosomal situations such as polyploid, aneuploid and cytological races (Tables 1, 2, 3) and cytogenetic effects related to chromosomal abberations employing various physical (cf. Sarma, 1981; Vidyavati & Sathaiah, (1984) and chemical agents (cf. Sarma & Abhayavardhani, 1987) clearly indicate that the cytogenetic processes, like non-disjunction of chromosomes, chromosome duplication, breakage (agmatoploidy), structural rearrangements, hybridization in sexually reproducing forms and genic changes (mutations), both in the somatic and reproductive cells participating in the regeneration process, have resulted into this whole gamut of taxonomic divergence that we so astonishingly and curiously witness to-day. While, polploidy is well documented in certain groups of algae viz., Cladophorales, Charales etc. (Table 1), aneuploidy seems to be wide spread in most of the groups of algae (Table 2), indicating their decisive role in the speciation and evolution of this highly heterogenous group of plants. More than one chromosome number (cytotypes/cytological or chromosomal races) in one and the same species (Table 3) also exist, suggesting the importance of the vitality for the organism of particular chromosome(s) that alter the numerical status of chromosomes rather than a change merely in the magnitude of chromosome number. This also applies to chromosome structural as well as genic variations (Table 4) responsible for speciation and evolution. In essence, a small vital change is more meaningful than a major alteration as in the latter case even if vital the organisms often do not survive/lead normal life/compete with the parental genotypes in drastically changed karyotypic situations. And, it is appropriate to conceive that minor changes occurring from time to time at very slow pace at (i) chromosome structural, (ii) chromosome numerical and (iii) genic levels have really built up this astounding hetergeneity through adaptability in algal biota in the vast span of nearly three billion years ever since the simplest photoautotrophic form of plants originated on earth. This presumption also satisfied as to why evolution that is essentially a cytogenetic process is a long and at times a slow process.

**Table 1. Euploid chromosome numbers illustrating speciation in representative genera of algae***

| Name of the genus & species | Chromosome number | Reference |
|---|---|---|
| **Scenedesmus** | | |
| *S. platydiscus* | n = 4 | Mathew & Chowdary, 1982 |
| *S. dimorphus* | n = 8 | " |
| *S. quadricauda* | n = 12 | Sulek, 1969 |
| **Oedogonium** | | |
| *O. howardii* | n = 9 | Srivastava & Sarma, 1979 |
| *O. pringsheimii* | n = 18 | Henningsen, 1963 |
| *O. silvaticum* | n = 27 | Srivastava & Sarma, 1979 |
| *O. capilliforme* var. *diversum* | n = 36 | " |
| **Cladophora** | | |
| *C. uberrima* | n = 6, 2n = 12 | Sinha & Ahmad, 1971 |
| *C. flexuosa* | n = 12, 2n = 24 | Sinha, 1963 |
| *C. alpina* | n = 24, 2n = 48 | " 1958 |
| *C. glomerata* | n = 48, 2n = 96 | " 1968 |
| **Rhizoclonium** | | |
| *R. implexum* | n = 12 | Verma, 1986b |

| Name of the genus & species | Chromosome number | Reference |
|---|---|---|
| *R. hookeri* | n = 18 | Sinha, 1958 |
| *R. tortuosum* | n = 24 | " |
| *R. polymorpha* | n = 36 | Verma, 1979 |
| *R. sulfuratum* | n = 48 | Sinha, 1958 |
| **Spirogyra** | | |
| *S. weberi* | n = 2 | Tatuno & Iiyama, 1971 |
| *S. hunanensis* | n = 4 | Harada & Yamagishi, 1972 |
| *S. ghosei* | n = 6 | Abhyavardhani & Sarma, 1983 |
| *S. paraguayensis* | n = 8 | Doraiswami, 1946 |
| *S. britannica* | n = 10 | Newnham, 1962 |
| *S. crassa* | n = 12 | " |
| *S. kundaensis* | n = 14 | Srivastava, 1993 |
| *S. punctulata* | n = 16 | Godward & Newnham, 1965 |
| *S. hyalina* | n = 18 | Abhyavardhani & Sarma, 1983 |
| S. VBG 79 | n = 22 | " |
| *S. fluviatilis* | n = 24 | Tatuno & Iiyama, 1971 |
| *S. majuscula* | n = 32 | Godward & Newnham, 1965 |
| *S. sumargaritata* | n = 38 | " |
| S. G. 73-1 | n = 46 | Vedajanani & Sarma, 1978a |
| *S. porticalis* | n = 50 | Godward & Newnham, 1965 |
| *S. ellipsospora* | n = 52 | " |
| S. VEC 79-2 | n = 72 ± 2 | Abhyavardhani & Sarma, 1983 |
| S. G. 73-3 | n = 76 | Vedajanani & Sarma, 1978a |
| S. sp. | n = 84 | Godward, 1966 |
| S. VNS-2 | n = 90 | Srivastava, 1993 |
| S. VNS-7 | n = 108 | " |
| **Cosmarium** | | |
| *C. meneghinii* | n = 6 | Chaudhary & Srivastava, 1994 |
| *C. subtumidum* | n = 12 | " |
| *C. taxichondriforme* | n = 18 | Chowdhury & Sarma, 1984 |
| *C. contractum* | n = 24 | Nizam, 1974 |
| *C. tumidum* | n = 30 | Chowdhury & Sarma, 1984 |
| *C. granatum* var. *elongatum* | n = 36 | Chaudhary & Srivastava, 1994 |
| *C. paradalis* | n = 42 | " |
| **Chara** | | |
| *C. vulgaris* | n = 14 | Peshkov *et al.*, 1974 |

| Name of the genus & species | Chromosome number | Reference |
|---|---|---|
| *C. fragilis* | n = 28 | Sarma & Khan, 1965 |
| *C. inconstans* | n = 42 | " |
| *C. cannescens* | n = 56 | Hotchkiss, 1965 |
| *C. zeylanica* | n = 70 | Hotchkiss. 1963 |
| **Nitella** | | |
| *N. opaca* | n = 6 | Sarma & Khan, 1964 |
| *N. flagelliformis* | n = 9 | " |
| *N. flexilis* | n = 12 | Sawa, 1965 |
| *N. stuartii* | n = 15 | Sarma & Khan, 1965 |
| *N. acuminata* | n = 18 | " |
| *N. psedoflabellata* | n = 27 | Ramjee & Sarma, 1971 |
| *N. stabilis* | n = 21 | Noor & Mukherjee, 1978 |
| *N. furcata* | n = 24 | Imahori & Kato, 1961 |
| *N. megacarpa* | n = 30 | Hotchkiss, 1963 |
| *N. trnuissima* | n = 36 | " |
| **Englena** | | |
| *E. gracilis* | n = 4 | O' Donnell, 1965 |
| *E. limnophila* | n = 16 | Sarma & Shashikala, 1980 |
| *E. vagans* | n = 24 | Baker, 1926 |
| *E. intermedia* | n = 36 | Sarma & Shashikala, 1980 |
| **Nitzschia** | | |
| *N. pseudofonticola* | n = 6 | Giri, 1989 |
| *N. heufleriana* | n = 12 | " |
| *N. subrostrata* | n = 18 | " |
| *N. acicularis* | n = 24 | " |
| *N. obtusa* | n = 48 | " |

* For original references consult Sarma (1964, 1978, 1982); Chaudhary & Agrawal (1966); Agrawal & Chaudhary (1996a, b).

**Table 2. Aneuploid chromosome numbers illustrating speciation in some representative genera of algae.***

| Name of the genus & species | Chromosome number | Reference |
|---|---|---|
| **Chlamydomonas** | | |
| *C. monoice* | n = 4 | Singh, 1981. |
| *C. psedolunata* | n = 5 | " |
| *C. gracilis* | n = 6 | " |

| Name of the genus & species | Chromosome number | Reference |
|---|---|---|
| *Z. cyanosporum* | n = 51 | Srivastava, 1993. |
| **Volvax** | | |
| *V. globator* | n = 5 | Cave & Pocock, 1951. |
| *V. tertius* | n = 13 | " |
| *V. weismannia* | n = 14 | " |
| *V. powersii* | n = 15 | " |
| *V. gigas* | n = 16 | " |
| **Stigeoclonium** | | |
| *S. variabile* | n = 6 | Chowdary, 1967. |
| *S. fasciculare* | n = 7 | " |
| *S. stagnatil* | n = 8 | " |
| *S. aestivale* | n = 9 | " |
| *S. elongatum* | n = 10 | " |
| *S. flagelliferum* | n = 11 | Abbas & Godward, 1964. |
| *S. tenue* | n = 13 | " |
| *S.* sp. | n = 14 | " |
| *S. variable* | n = 16 | " |
| **Oedogonium** | | |
| *O. howardii* | n = 9 | Srivastava & Sarma, 1979. |
| *O. warmingianum* | n = 11 | " |
| *O. grande* | n = 13 | Ohashi, 1930. |
| *O. pachyandrium* | n = 15 | Kretschmer, 1930. |
| *O. plagiostomum* | n = 16 | Henningsen, 1963. |
| *O. terrestris* | n = 17 | Chowdary, 1964. |
| *O. capilliforme* | n = 19 | Hoffman, 1967. |
| *O. bohemicum* | n = 20 | Srivastava & Sarma, 1979. |
| *O. intermedium* | n = 21 | Singh & Chaudhary, 1990. |
| *O. areolatum* | n = 22 | Srivastava & Sarma, 1979. |
| *O. armigerum* | n = 24 | Gupta, 1992. |
| *O. irregulare* | n = 25 | " |
| *O. silvaticum* | n = 27 | Srivastava & Sarma, 1979. |
| *O. calcareum* | n = 32 | Singh & Chaudhary, 1990. |
| *O. tropicum* | n = 37 | " |
| *O.* sp. | n = 38 ± 1 | Hoffman, 1976c. |
| *O.* sp. | n = Ca.41 | Sinha, 1963. |
| *O. glabarum* f. *maharastrense* | n = 49 | Gupta, 1992. |
| *C. moniliforme* | n = 46 | Chowdhury & Sarma, 1984. |
| **Cladophora** | | |
| *Cl. clavuligera* | n = 11 | Balakrishnan, 1961. |

| Name of the genus & species | Chromosome number | Reference |
|---|---|---|
| *Cl. fracta* | n = 12 | Verma, 1985b. |
| *Cl. gracilis* | n = 13 | Foyn, 1929. |
| *Cl. pellucida* | n = ± 18 | Sinha, 1958. |
| **Spirogyra** | | |
| *S. cylindrica* | n = 2 | Tatuno & Iiyama, 1971. |
| *S. dubia* | n = 5 | Merimann, 1916. |
| *S. pratensis* | n = 11-15 | Abhayavardhani & Sarma, 1983. |
| *S. crassoidea* | n = 14 | " |
| S. VS 79-4 | n = 46 ± 1 | Abhayavardhani & Sarma, 1983. |
| S. VS 75-4 | n = 49 ± 1 | Vedajanani & Sarma, 1978a. |
| S. VBG 72-1 | n = 53 ± 1 | " |
| S. VBG 72-2 | n = 57 ± 1 | " |
| S. VS 75-2 | n = 61 ± 1 | " |
| **Zygmema** | | |
| *Z. pectinatum* | n = 14 | Sinha & Noor, 1967. |
| *Z. circumcarinatum* | n = 19 | Prasad & Godward, 1966. |
| Z. VNS | n = 23 | Srivastava, 1993. |
| Z. WA-4N | n = 25 | Prasad, 1958. |
| *Z. cyaneum* | n = 33 | Srivastava, 1993. |
| Z. VB 73-2 | n = 33 ± 1 | Vedajanani & Sarma, 1978b. |
| Z. VB 75-2 | n = 35 ± 1 | " |
| Z. VB 73-1 | n = 35-38 | " |
| *Z. subcylindrospermum* | n = 37 | Prasad & Godward, 1966. |
| Z. VEC 73 | n = 37 ± 1 | Vedajanani & Sarma, 1978b. |
| Z. M 73 | n = 38 | " |
| *Z. verrucopunctatum* | n = 39 | Prasad & Godward, 1966. |
| *Z. subtenue* | n = 40 | " |
| *Z. cylindrospermum* | n = 42 | " |
| *Z. crusiatum* | n = 43 | " |
| *Z. insigne* St. B | n = 43 ± 1 | Vedajanani & Sarma, 1978b. |
| *Z. B 10 Ogwen* | n = 44 | Prasad, 1958. |
| *Z. normanii* | n = 46 | Harada & Yamagishi, 1980. |
| *Z. insigne* St A. | n = 48 | Vedajanani & Sarma, 1978b. |
| *M. fimbriata* M 123 | n = 97 | " |
| *Z. cylindrium* | n = 70 | Prasad & Godward, 1966. |
| *Z. SP Malham* | n = 72 | " |
| Z. 25 *Millport* | n = 82 | " |
| **Staurastrum** | | |
| *S. pilosum* | n = 6 | Chowdury & Sarma, 1984. |

| Name of the genus & species | Chromosome number | Reference |
|---|---|---|
| *S. cerastes* | n = 7 | Chowdury & Sarma, 1984. |
| *S. alternans* | n = 8 | Chaudhary & Srivastava, 1994. |
| **Cosmarium** | | |
| *C. venustum* f. *minor* | n = 5 | Chaudhary & Srivastava, 1994. |
| *C. meneghinii* | n = 6 | " |
| *C. quadrum* St. B | n = 7 | Chowdhury & Sarma, 1984. |
| *C.* VNS-1 | n = 9 | Chaudhary & Srivastava, 1994. |
| *C. laeve* | n = 10 | " |
| *C. humile* var. *substriatum* | n = 11 | " |
| *C. turpini* St. 612/143 | n = 12 | Nizam, 1974. |
| *C. undulatum* St. A | n = 13 | Chowdhury & Sarma, 1984. |
| *C. sphaelerostichum* | n = 14 | Chaudhary & Srivastava, 1994. |
| *C. remiforme* | n = 15 | Chowdhury & Sarma, 1984. |
| *C. lundellii* | n = 16 | " |
| *C. subimpressulum* | n = 17 | Chaudhary & Srivastava, 1994. |
| *C. taxichondriforme* | n = 18 | Chowdhury & Sarma, 1984. |
| *C. auriculatum* var. *bogoriense* | n = 19 | Vidyavati & Sathaiah, 1983. |
| *C.* VNS-2 | n = 21 | Chaudhary & Srivastava, 1994. |
| *C. protractum* | n = 22 | " |
| *C. contractum* | n = 24 | Nizam, 1974. |
| *C. tenue* | n = 26 | Chowdhury & Sarma, 1984. |
| *C. sphaeroidium* | n = 27 | Chowdhury & Sarma, 1984. |
| *C. subturgidum* f. *minor* | n = 28 | " |
| *C. tumidum* | n = 30 | Abhayavardhani & Sarma, 1982. |
| *C. pseudonitidutum* | n = 30 ± 1 | Chowdhury & Sarma, 1984. |
| *C. moniliforme* f. *punctata* | n = 34 | " |
| *C. difficile* | n = C.38 | Brandham, 1965a. |
| *C. tetrapthalum* | n = 39 | Chowdhury & Sarma, 1987. |
| *C. paradilis* | n = 42 | Chaudhary & Srivastava, 1994. |
| *C. cucumis* | n = 44 | King, 1960. |
| *E. ehrenbergii* | n = 205± 5 | " |
| *C. hammeri* | n = 48 ± 1 | Abhyavardhani & Sarma, 1982. |
| *C. granulatum* var. *ocellatum* | n = 57 | Chaudhary & Srivastava, 1994. |
| **Euastrum** | | |
| *E. crenulatum* | n = 9 | Chowdhury & Sarma, 1984. |
| *E. intermium* var. *burmese* | n = 12 | " |
| *E. binale* | n = 13 | " |
| *E. cornubiense* | n = 14 | Vedajanani & Sarma, 1978d. |
| *E. sculptum* St. C | n = 15 | Chowdhury & Sarma, 1984. |

| Name of the genus & species | Chromosome number | Reference |
|---|---|---|
| *E. sculptum* St. B | n = 16 | Chowdhury & Sarma, 1984. |
| *E. sculpyum* St. A | n = 21 | " |
| *E. cornubiense* f. *minor* | n = 22 | Vedajanani & Sarma, 1978d. |
| **Micrasterias** | | |
| *M. rotata* | n = 19 | Agrawal & Chaudhary, 1993. |
| *M. zeylanica* | n = 20 | " |
| *M. radiosa* | n = 24 | Kasprik, 1973. |
| *M. swainei* | n = 26 | " |
| *M. thomsiana* | n = 34-37 | Waris, 1950. |
| *M. crux-melitensis* | n = 36 | Chowdhury & Sarma, 1984. |
| *M. incisa* | n = 38 | " |
| *M. thomsiana* var. *notata* | n = 39 | Kasprik, 1973. |
| *M. muricata* | n = 41 | " |
| *M. tetraptera* var. *taylori* | n = 48 | " |
| *M. papilifera* var. K 678 | n = 49 | " |
| *M* 67 | n = 50 | " |
| *M* 71 | n = 51 | " |
| *M* 211 | n = 52 | " |
| *Micrasterias* sp. | n = 55 | Mix & Kasprik, 1977. |
| *M. angulosa* | n = 60-64 | " |
| *M. fimbriata* var. *spinosa* M 178 | n = 79 | Kasprik, 1973. |
| *M. denticulata* var. *angulosa* | n = 82 | " |
| *M. torreyi* | n = 85 | " |
| *M. acerocama* var. *boldtii* | n = 88 | " |
| *M. americana* K 1483 | n = 93 | " |
| *M. fimbriata* | n = 96 | " |
| *M. furcata* var. *gracillima* | n = c. 106 | " |
| *M. furcata* M 162 | n = 108 | " |
| *M. laticeps* M 126 | n = c. 114 | " |
| *M. americana* M 139 | n = c. 135 | " |
| *M. rotata* K 829 | n = 159 | " |
| M. 137 | n = 172 | King, 1960. |
| | n = c. 200 | Waris, 1950. |
| *M. americana* M 116 | n = c. 205 | Kasprik, 1973. |
| *M. rotata* M 26 | n = c. 226 | " |
| *M* 1 | n = 229 | " |
| *M* 212 | n = c. 235 | " |
| *M* 53 | n = c. 238 | " |

| Name of the genus & species | Chromosome number | Reference |
|---|---|---|
| **Euglena** | | |
| *E. pisciformis* | n = 15 | Chaudhary, 1992. |
| *E. limnophila* | n = 16 | Sarma & Shashikala, 1980. |
| *E. acutissima* | n = 21 | Chaudhary, 1992. |
| *E. vagans* | n = 24 | " |
| *E. limnophila* | n = 25 | " |
| *E. acus* St. A | n = 29 | " |
| St. B | n = 37 ± 1 | |
| *E. pusilla* var. *longa* | n = 39 ± 1 | " |
| *E. gracilis* | n = c. 45 | Leedale, 1957. |
| *E. tripteris* | n = 51 ± 3 | Chaudhary, 1992. |
| *E. charkowiensis* | n = 52 | " |
| *E. proxima* | n = 56± 2 | " |
| *E. caudata* | n = 60± 2 | " |
| *E. spirogyra* | n = 60± 3 | " |
| *E. geniculata* | n = 80± 3 | " |
| *E. deses* | n = 84± 1 | " |
| *E. granulata* | n = 93± 5 | " |
| *E. gracilis* | n = 106± 5 | " |
| *E. clara* | n = 115± 7 | " |
| *E. sociabilis* | n = 118± 6 | " |
| *E. sanguinea* | n = 166± 6 | " |
| *E. velata* | n = 200±10 | " |
| *E. polymorpha* | n = 204± 8 | " |
| **Phacus** | | |
| *P. cylindricus* | n = 10 | Chaudhary, 1992. |
| *P. pusillus* | n = 12 | " |
| *P. agilis* | n = 13 | " |
| *P. acuminatus* St. A | n = 19 | " |
| *P. segretii* | n = 22 | " |
| *P. curvicauda* | n = 35 ± 2 | " |
| *P. brevicaudatus* | n = 40 | " |
| *P. pseudonordstedii* | n = 40 ± 3 | " |
| *P. angulatus* | n = 43 ± 3 | " |
| *P. pyrum* | n = 50 ± 3 | " |
| *P. trigneter* | n = 108 ± 2 | " |
| *P. pleuronectes* | n = 114 ± 5 | " |
| *P. indicus* | n = 135 ± 6 | " |
| *P. circumflexus* | n = 152 ± 7 | " |

| Name of the genus & species | Chromosome number | Reference |
|---|---|---|
| *P. longicauda* | n = 226 ± 8 | Chaudhary, 1992. |
| **Gymnodinium** | | |
| *G. vitilago* | n = c. 44 | Dodge, 1964. |
| *G. wawrikae* | n = 52 ± 2 | Agrawal *et al.*, 1995. |
| *G.* sp. PBS 69 | n = 84-96 | Shyam & Sarma, 1978. |
| *G.* sp. DLW 69 | n = c. 90 | Shyam & Sarma, 1975a. |
| *G. lalitae* | n = 86-94 | Sarma & Shyam, 1978. |
| *G. uberrimum* | n = 96-100 | Shyam & Sarma, 1978. |
| *G. aeruginosum* | n = 97 ± 3 | Agrawal *et al.*, 1995. |
| *G. paradoxum* | n = 100-108 | Shyam & Sarma, 1978. |
| *G. dodgei* | n = 190-200 | Sarma & Shyam, 1974b. |
| *G. inversum* var. *elongatum* | n = 196-212 | Shyam & Sarma, 1978. |
| *G. inversum* var. *inversum* | n = 200 ± 4 | Agrawal *et al.*, 1995. |
| *G. bohemicum* | n = 254 ± 4 | " |
| *Gymnodinium* sp. NBG 76 | n = 514 ± 10 | Chaudhary & Sarma, 1979. |
| *G.* sp. OBG 70 | n = c. 600 | Shyam & Sarma, 1975b. |
| **Vaucheria** | | |
| *V. geminata* | n = 5 | Mundie, 1929. |
| | n = 7 | Gross, 1937. |
| *V. aversa* | n = 8 | Sharma, 1992. |
| *V. geminata Cyto.* A | n = 9 | " |
| *V. sessilis* f. *clavata* | n = 10 | " |
| *V. terrestris* | n = 11 | " |
| *V. sessilis* var. *major* | n = 12 | " |
| *V. hamata* | n = 13 | " |
| *V. orientalis* | n = 14 | " |
| *V. debaryana* | n = 15 | " |
| **Botrydium** | | |
| *B. granalutum* var. *kalkwitzianum* | n = 8 | Sharma, 1992. |
| *B. wallrothii* | n = 11 | " |
| *B. granulatum* | n = 13 | " |
| *B. cystosum* | n = 15 | " |
| *B. tuborosum* | n = 16 | " |
| *B. milleri* | n = 17 | " |
| *B. lateralis* | n = 16-20 | " |

* For original references consult Sarma (1964, 1978, 1982); Chaudhary (1992); Agrawal *et al.* (1995); Chaudhary & Agrawal (1996); Agrawal & Chaudhary (1996a, b).

**Table 3. Chromosome numbers illustrating the existence of cytological races within representative species of algae.***

| Name of the species | Chromosome number (n) | Reference |
|---|---|---|
| *Chlamydomonas eugametos* | 8 | Buffaloe, 1958. |
| | 10 | Moewus, 1936, 1938. |
| | 16 | Wetherall & Krauss, 1957. |
| | 38 ± 4 | Schaechter & De Lamater, 1954. |
| *C. reinhardii* | 8 | Levine & Folsome, 1959. |
| | 11 ± 1 | Abbas & Godward, 1965. |
| | 16 | Wetherall & Krauss, 1957. |
| *Eudorina elegans* | 8 | Chaudhary & Agrawal, 1980. |
| | 10 | Hartmann, 1921. |
| | 12 | Cave & Pocock, 1951. |
| | 16 | Rayns & Godward, 1961. |
| *E. illinoisensis* | 10 | Hovasse, 1937. |
| | 12 | Merton, 1908. |
| | 14 | Goldstein, 1964. |
| *Cosmarium subtumidum* | 10 | Nizam, 1974. |
| | 12 | Chaudhary & Srivastava, 1994. |
| | 20 | Nizam, 1974. |
| | 24 | King, 1960. |
| *Euastrum sculptum* | 9 | Chowdhury & Sarma, 1984. |
| | 15 | " |
| | 16 | " |
| | 21 | " |
| *Micrasterias papilifera* | 49 | Kasprik, 1973. |
| | 50 | " |
| | 51 | " |
| | 52 | " |
| | 76 | " |
| *Shizomeris libleinii* | 14 | Sarma & Chaudhary, 1975a. |
| | 15 | Chowdary, 1967a. |
| | 28 | Prasad & Srivastava, 1963. |
| | 30 | Patel, 1967. |
| *Trentepohlia aurea* | 16 | Suematu, 1960. |
| | 18 | Abbas & Godward, 1964. |
| | 56 | Jose & Chowdary, 1978. |

| Name of the species | Chromosome number (n) | Reference |
|---|---|---|
| *Chaetophora elegans* | 7, 9 | Sarma & Shashikala, 1980. |
| | 8 | Sinha & Das, 1965. |
| | 12 | Abbas & Godward, 1965. |
| *Cephaleuros virescens* | 18 | Jose & Chowdary, 1977. |
| | 24 | " |
| | 27 | " |
| | 32 | " |
| *Rhizoclonium hieroglyphicum* | 12, 2n = 22 | Sarma & Bala, 1983. |
| | 24 | " |
| | 32 | " |
| | 34 | " |
| | 36 | " |

* For original references consult Sarma (1964, 1978, 1982); Agrawal (1996); Patel (1996); Chaudhary & Agrawal (1996a, b).

**Table 4. Same chromosome numbers in different species of some representative genera illustrating speciation due to chromosomal rearrangements/genic changes.***

| Name of the genus & species | Chromosome number | Reference |
|---|---|---|
| **Chiamylomonus** | | |
| *C. proboscigera* | 5 | Singh, 1981. |
| *C. pseudolunata* | 5 | " |
| *C. matwienkoi* | 5 | " |
| *C. glocopara* | 6 | " |
| *C. gracilis* | 6 | " |
| *C. indica* | 6 | " |
| *C. nasuta* | 8 | Karter, 1929. |
| *C. chlamydogama* | 8 | Buffaloe, 1958. |
| *C. gloeogama* | 8 | Singh, 1981. |
| *C. tranulans* | 8 | " |
| *C. pumilio* var. *ovoides* | 8 | " |
| *C. elliptica* var. *britannica* | 8 | " |
| *C. pachychlamys* | 8 | " |
| *C. sordida* | 8 | " |
| *C. noctigama* var. *minor* | 8 | " |

| Name of the species | Chromosome number (n) | Reference |
|---|---|---|
| *C. reinhardii* | 8 | Buffaloe, 1958. |
| *C. moewusii* | 8 | " |
| *C. eugametos* | 8 | " |
| *C. eugametos* | 10 | Levine & Ebersold, 1960. |
| *C. variabilis* | 10 | Dangeard, 1899. |
| *C. dilli* | 10 | " |
| *C. debaryana* var. *cristata* | 10 | Singh, 1981. |
| *C. bourrellyi* | 10 | " |
| *C. baca* | 10 | " |
| **Volvox** | | |
| *V. globator* | 5 | Cave & Pocock, 1951. |
| *V. barberi* | 5 | " |
| *V. rousseletii* | 5 | " |
| *V. merrilli* | 5 | " |
| *V. prolificus* | 5 | Patel, 1978. |
| *V. tertius* | 13 | Cave & Pocock, 1951. |
| *V. africanus* | 13 | " |
| *V. spermatosphaera* | 13 | " |
| *V.* sp. | 13 | " |
| **Scenedesmus** | | |
| *S. quadricauda* var. *longispina* | 8 | Sasikala, 1980. |
| *S. bijugatus* var. *graeventzii* | 8 | " |
| *S. abundans* | 8 | " |
| *S. acutiformis* | 8 | " |
| *S. dimorphus* | 8 | Mathew & Chowdary, 1982. |
| **Cladophora Freshwater** | | |
| *C. glomerata* var. *fasciculata* f. *fasciculata* | 12 | Sinha & Noor, 1967. |
| *C. callicoma* | 12 | Chaudhary, 1978. |
| *C. speciosa* | 12 | Masaki, 1978. |
| *C. fracta* var. *normalis* | 12 | Geitler, 1936. |
| **Marine** | | |
| *C. flexuosa* | 12 | Patel, 1961. |
| *C. rupestris* var. *distorta* | 12 | " |
| *C. gracilis* | 12 | " |
| *C. expansa* | 12 | " |

| Name of the species | Chromosome number (n) | Reference |
|---|---|---|
| *C. hutchinsiae* | 12 | Patel, 1961. |
| **Oedogonium** | | |
| *O. grande* | 13 | Ohashi, 1930. |
| *O.* spp. (5 species) | 13 | Tschermak, 1943. |
| *O. echinospermum* | 13 | Hoffman, 1967. |
| *O. idiocandrosporum* | 13 | " |
| *O. amplum* | 13 | Sinha & Noor, 1967. |
| *O. magnusii* | 13 | Das, 1971 |
| *O. petri* | 13 | Srivastava & Sarma, 1979. |
| *O. gracilius* var. *globosum* | 13 | Singh & Chaudhary, 1990. |
| *O. plagiostomum* | 16 | Henningsen, 1963. |
| *O. vaucheri* | 16 | " |
| *O. capillare* | 16 | " |
| *O. spectabile* | 16 | " |
| *O. foveolatum* | 16 | Hoffman, 1967. |
| *O. setigenum* | 16 | " |
| *O. welwitschii* | 16 | Srivastava & Sarma, 1979. |
| *O.* spp. (3 species) | 17 | Tschermak, 1943. |
| *O. plagiostomum* | 17 | " |
| *O. terrestris* | 17 | Chowdary, 1964. |
| *O. angustistomum* | 17 | Hoffman, 1967. |
| *O. calliandrum* | 17 | " |
| *O. princeps* | 17 | " |
| *O. vingens* | 17 | " |
| *O. autumnale* | 17 | Das, 1971. |
| *O. acmandrium* | 17 | Sarma & Tripathi, 1973. |
| *O. wabashense* | 17 | Srivastava & Sarma, 1979. |
| *O. gunnii* | 17 | " |
| *O. hindustanense* | 17 | Singh & Chaudhary, 1990. |
| *O. cyathigerum* | 19 | van Wisselingh, 1908. |
| *O.* sp. | 19 | " 1921. |
| *O. pluviale* | 19 | Tschermak, 1943. |
| *O.* sp. | 19 | " |
| *O. capilliforme* | 19 | Howard & Horsley, 1960. |
| *O. cardiacum* | 19 | Hasitschka-Janska, 1960. |
| *O. flavescens* | 19 | Das, 1971. |

| Name of the genus & species | Chromosome number | Reference |
|---|---|---|
| *O. geniculatum* var. *indicum* | 19 | Srivastava & Sarma, 1979. |
| *O. lemmermannii* | 19 | Singh & Chaudhary, 1990. |
| **Spirogyra** | | |
| *S. jugalis* | 6 | Karsten, 1908. |
| *S. triformis* | 6 | van Wisselingh, 1900. |
| *S.* sp. | 6 | Doraiswami, 1946. |
| *S. triformis* | 6 | Godward, 1950. |
| *S. ghosei* | 6 | Abhayavardhani & Sarma, 1983. |
| *S.* VBG 80-1 | 6 | " |
| *S. paraguayensis* | 6 | Doraiswami, 1946. |
| *S. crassa* | 6 | Harada & Yamagishi, 1972. |
| *S. fuellibornei* | 6 | Doraiswami, 1946. |
| *S. neglecta* | 6 | Trondle, 1911. |
| *S. azygospora* | 12 | Abhayavardhani & Sarma, 1983. |
| *S. paradoxa* | 12 | " |
| *S. variformis* | 12 | " |
| *S. bellis* | 12 | Merriman, 1916. |
| **Chara** | | |
| *C. tomentosa* | 14 | Telezynski, 1929. |
| *C. vulgaris* | 14 | Rodrigues, 1945. |
| *C. contraria* | 14 | Hotchkiss, 1965. |
| *C. hispidula* | 14 | Guerlesquin, 1965. |
| *C. crassicaulis* | 14 | " 1963. |
| *C. longibracteata* | 14 | Gonzalves, 1941-42. |
| *C. squamosa* | 14 | Guerlesquin, 1963. |
| *C. conimbrigensis* | 14 | " 1974. |
| *C. oedophylla* | 14 | " 1961. |
| *C. nitelloides* | 14 | Ramjee & Sarma, 1971. |
| *C. imperfecta* | 14 | Guerlesquin, 1963. |
| *C. intermedia* | 14 | " |
| *C. globularis* | 14 | Hotchkiss, 1958. |
| *C. mauretanica* | 14 | Guerlesquin, 1961. |
| *C. connivens* | 14 | " 1963. |
| *C. fragifera* | 14 | Gillet, 1959. |
| *C. virgata* | 14 | Ramjee & Sarma, 1971. |

| Name of the genus & species | Chromosome number | Reference |
|---|---|---|
| *C. delicatula* | 14 | Sarma & Khan, 1965. |
| *C. leiopitys* | 14 | " 1967. |
| *C. fischeri* | 14 | Ahmad & Sinha, 1973. |
| *C. leptosperma* | 14 | " |
| *C. aspera* | 14 | Gillet, 1959. |
| *C. galioides* | 14 | Guerlesquin, 1961. |
| *C. setosa* | 14 | Ramjee & Sarma, 1971. |
| *C. pseudobrachypus* | 14 | " |
| *C. sejuncta* | 14 | Hotchkiss, 1958. |
| *C. braunii* | 14 | Imahori & Kato, 1961. |
| *C. schweinitzii* | 14 | Chatterjee, 1971. |
| *C. coromandeliana* | 14 | " |
| *C. australis* | 14 | Hotchkiss, 1963. |
| *C. wallichii* | 14 | Sarma & Khan, 1965. |
| *C. nuda* | 14 | Sinha & Verma, 1970. |
| *C. fibrosa* | 14 | Ramjee & Sarma, 1971. |
| *C. benthami* | 14 | Sarma & Khan, 1965. |
| *C. flaccida* | 14 | " |
| *C. gymnopitys* | 14 | " |
| *C. curtissii* | 14 | Ramjee & Sarma, 1971. |
| *C. erythrogyna* | 14 | " |
| *C. hydropitys* | 14 | Hotchkiss, 1965. |
| *C. mexicana* | 14 | Ramjee & Sarma, 1971. |
| *C. bharadwajae* | 14 | Sarma & Khan, 1967. |
| *C. leptopitys* | 14 | Hotchkiss, 1964. |
| *C. vulgaris* | 28 | Moutschen & Dahmen, 1956. |
| *C. contraria* | 28 | Hotchkiss, 1958. |
| *C. schaffneri* | 28 | Peshkov *et al.*, 1973. |
| *C. hispidula* | 28 | Guerlesquin, 1967. |
| *C. longibracteata* | 28 | Delay. 1949. |
| *C. nitelloides* | 28 | Chatterjee, 1971. |
| *C. hispida* | 28 | Peshkov *et al.*, 1973. |
| *C. major* | 28 | Guerlesquin, 1961. |
| *C. pedunculata* | 28 | " 1963. |
| *C. baltica* | 28 | " 1961. |
| *C. connescens* | 28 | Tindall & Sawa, 1964. |

| Name of the genus & species | Chromosome number | Reference |
|---|---|---|
| *C. globularis* | 28 | Hotchkiss, 1958. |
| *C. virgata* | 28 | Chatterjee, 1975. |
| *C. delicatula* | 28 | " 1971. |
| *C. chrysospora* | 28 | Ramjee & Sarma, 1971. |
| *C. leptosperma* | 28 | " |
| *C. strigasa* | 28 | Ahmad & Sinha, 1973. |
| *C. setosa* | 28 | Sarma & Khan, 1965. |
| *C. pseudobrachypus* | 28 | " |
| *C. zeylanica* | 28 | Sundaralingam, 1946. |
| *C. haitensis* | 28 | Bharati & Chennaveeraiah, 1974. |
| *C. compressa* | 28 | Sarma & Khan, 1965. |
| *C. elegans* | 28 | Ramjee & Sarma, 1971. |
| *C. filicaulis* | 28 | " |
| *C. diaphana* | 28 | Guerlesquin, 1971. |
| *C. inconstans* | 28 | Bharati & Chennaveeraiah, 1974. |
| *C. vandalurensis* | 28 | Ahmad & Sinha, 1973. |
| *C. schweinitzii* | 28 | Chatterjee, 1971. |
| *C. braunii* | 28 | Verma, 1967. |
| *C. socotrensis* | 28 | Ramjee & Sarma, 1971. |
| *C. nuda* | 28 | Chatterjee, 1971. |
| *C. fibrosa* | 28 | Hotchkiss, 1965. |
| *C. gymnopitys* | 28 | " |
| *C. longicorollata* | 28 | Chatterjee, 1971. |
| *C. hydropitys* | 28 | Ramjee & Sarma, 1971. |
| *C. backellii* | 28 | Tindall *et al.*, 1965. |
| **Nitella** | | |
| *N. opaca* | 6 | Gillet, 1959. |
| *N. hotchkissii* | 6 | Tindall, 1967. |
| *N. mirabilis* | 6 | Sarma & Khan, 1964. |
| *N. inokosiraensis* | 6 | Kanahori, 1971. |
| *N. syncarpa* | 6 | Gillet, 1959. |
| *N. capillaris* | 6 | Guerlesquin, 1964. |
| *N. flexilis* | 12 | Lindenbein, 1927. |
| *N. opaca* | 12 | Strasburger, 1908. |
| *N. laxa* | 12 | Kanahori, 1974. |

| Name of the genus & species | Chromosome number | Reference |
|---|---|---|
| *N. acuminata* | 12 | Guerlesquin, 1972. |
| *N. capitulifera* | 12 | " |
| *N. syncarpa* | 12 | Strasburger, 1908. |
| *N. mucronata* | 12 | Sinha & Noor, 1967. |
| *N. pygmaea* | 12 | Guerlesquin, 1972. |
| *N. hyalina* | 12 | Sato, 1959. |
| *N. axillaris* | 12 | Sinha & Noor, 1971. |
| *N. acuminata* | 18 | Hotchkiss, 1958. |
| *N. subglomerata* | 18 | " |
| *N. capitulifera* | 18 | " |
| *N. obscura* | 18 | " |
| *N. belargeri* | 18 | Tindall, 1970. |
| *N. glomerulifera* | 18 | Ramjee & Sarma, 1971. |
| *N. dilatata* | 18 | Hotchkiss, 1965. |
| *N. pulchella* | 18 | Sasaki, 1961. |
| *N. furcata* | 18 | Sarma & Khan, 1965. |
| *N. burmanica* | 18 | Imahori & Kato, 1961. |
| *N. careana* | 18 | Sinha & Ahmad, 1973. |
| *N. roxburghii* | 18 | " |
| *N. polycarpa* | 18 | Sarma & Khan, 1965. |
| *N. mauritiana* | 18 | Sarma & Ramjee, 1971. |
| *N. microcarpa* | 18 | Wood, 1963. |
| *N. glaziovii* | 18 | Noor & Mukherjee, 1974. |
| *N. mucronata* | 18 | Sarma & Khan, 1965. |
| *N. oligospira* | 18 | " |
| *N. dictyosperma* | 18 | " 1964. |
| *N. pygmaea* | 18 | Ramjee & Sarma, 1971. |
| *N. fascicularis* | 18 | Guerlesquin, 1972. |
| *N. ovigalis* | 18 | Sinha & Ahmad, 1973. |
| *N. flagellifera* | 18 | Sarma & Khan, 1965. |
| *N. tuberculata* | 18 | Sinha & Noor, 1967. |
| *N. graciliformis* | 18 | Sinha & Ahmad, 1973. |
| *N. patula* | 18 | Sarma &Khan, 1965. |
| *N. megacarpa* | 18 | Hotchkiss, 1963. |
| *N. tumulosa* | 18 | Ramjee & Sarma, 1971. |
| *N. tenuissima* | 18 | Sarma & Khan, 1965. |

| Name of the genus & species | Chromosome number | Reference |
|---|---|---|
| *N. transilis* | 18 | Tindall & Sawa, 1964. |
| *N. bengalensis* | 18 | Sarma & Khan, 1965. |
| *N. pseudoflabellata* | 18 | " |
| *N. elegans* | 18 | Sasaki, 1961. |
| *N. sejuncta* | 18 | " |
| *N. leptodactyla* | 18 | Sarma & Khan, 1965. |
| *N. moramangaensis* | 18 | Tindall & Sawa, 1964. |
| *N. wattii* | 18 | Mukherjee & Noor, 1973. |
| *N. pusilla* | 18 | Sinha & Ahmad, 1973. |
| *N. gracilis* | 18 | Guerlesquin, 1961. |
| *N. sphaerocephalla* | 18 | Ramjee & Sarma, 1971. |

* For original references consult Sarma (1964, 1978, 1982); Agrawal (1996); Agrawal & Chaudhary (1996a); Chaudhary & Agrawal (1996); Guerlesquin (1996).

## ACKNOWLEDGEMENTS

Authors thank Dr. M. Agrawal and Mr. S.S. Deepak for computer help rendered by them. SBA is also grateful to C.S.I.R. (Division for Scientific & Technical Personnel), New Delhi for financial assistance.

## REFERENCES

Abbas, A. and Godward, M.B.E. 1964. Cytology in relation to taxonomy in Chaetophorales. *J. Linn. Soc. Lond.* (*Bot.*) **58**: 499-508.

Abhayavardhani, P. and Sarma, Y.S.R.K. 1983. Cytological and cytotaxonomical studies on the genus *Spirogyra* Link (Conjugales, Chlorophyceae). *Cytologia.* **48**: 467-482.

Agrawal, S.B. 1996. Cytogenetical studies in Chlorococcales and Volvocales: a review In: *Cytology, Genetics and Molecular Biology of Algae* (Eds.) B.R. Chaudhary and S.B. Agawral, SPB Academic Publishing, The N herlands (in press).

Agrawal, S.B. and Chaudhary, B.R. 1996a. Cytological observations and evolutionary trends in the genus *Oedogonium* Link (Chlorophyceae). In: *Advances in Phycology: an appraisal* (Eds.) B.N. Verma and A.N. Kargupta, Narendra Publishing House, New Delhi (in press).

Agrawal, S.B. and Chaudhary, B.R. 1996. Chrysophyte cytology: *compilation and evalution.* In: *Cytology, Genetics and Molecular Biology of Algae* (Eds.) B.R. Chaudhary and S.B. Agrawal, SPB Academic Publishing, The Netherlands (in press).

Allen, J.R., Roberts, T.M., Loeblich, A.R. III & Klotz, L.C. 1975. Characterization of the DNA from the dinoflagellate *Crypthecodinium cohnii* and implications for nuclear organization. Cell, **6**: 161-169.

Balakrishnan, M.S. 1961. A cytotaxonomical investigation of Indian members of the Cladophoraceae. *J. Ind. Bot. Soc.* (**Prof. Agharkar Comm. Vol.**): 30-40.

Bhatnagar, S.K., Gupta, P. & Gautam, A. 1996. C-banding and charophycean phylogeny—a new horizon. In: *Cytology, Genetics and Molecular Biology of Algae* (eds.) B.R. Chaudhary and S.B. Agrawal, SPB Academic Publishing, The Netherlands (in press).

Blackman, F.E. and Tansley, A.G. 1902. A revision of the classification of the green algae. *New Phytol.* **1**: 17-24, 67-77, 89-96, 114-120, 133-144, 163-168, 189-192, 213-220, 258-264.

Bohlin, K. 1901. "Utkast till de Grona Algerans och Arkigoniaternas Fylogeni" Almqvist & Wiksell, Stockholm.

Cave, M.S. and Pocock, M.A. 1951. Karyological studies in the Volvocales. *Amer. J. Bot.* **38**: 800-811.

Campbell, E.O. and Sarafis, V. 1972. *Schizomeris*—A growth form of *Stigeoclonium tenue*. *J. Phycol.* **8**: 6-28.

Chadefaud, M. and Emberger, L. 1960. Trite de Botanique Systematique. Tome I-Chadefaud, M. Les Vegetaux non Vasculaires, Paris.

Chapman, V.J. 1962. The Algae, London.

Charles, D. 1977. Isolation and characterization of DNA from unicellular algae. *Plant. Sci. Lett.* **8**: 35-44.

Chaudhary, B.R. 1978. Meiotic chromosome number in *Cladophora callicoma* Kuetzing. *Isr. J. Bot.* **27**: 212-216.

Chaudhary, B.R. 1983. Karyology of an Indian strain of *Spaeroplea annulina* (Roth) *Ag. Bibliotheca Phycologica.* **66**: 101-109.

Chaudhary, B.R. and Agrawal, S.B. 1996. Conjugalean cytology: reassessment. In: *Cytology, Genetics and Molecular Biology of Algae* (Eds.) B.R. Chaudhary and S.B. Agrawal, SPB Academic Publishing, The Netherlands (in press).

Chowdary, Y.B.K. 1959. Cytology of *Trentepohlia* and *Cephaleuros*. *Proc. Symp. Algol.* UNESCO-ICAR, Delhi, pp. 65-69.

Chowdary, Y.B.K. 1963. On the cytology and systematic position of *Physolinum monilia* Printz. The Nucleus, **6**: 43-48.

Christensen, T. 1962. *Alger.* Copenhagen.

Davis, P.M. and Heywood, V.M. 1963. Principles of angiosperm taxonomy. Oliver and Boyd, Edinburgh and London.

Desikachary, T.V. and Sundaralingam, V.S. 1962. Affinities and interrelationship of the Characeae. *Phycologia.* **2**: 9-16.

Dodge, J.D. 1996. The dinophyceae. In: *The chromosomes of the algae* (ed.) M.B.E. Godward, Edward Arnold, London, pp. 96-115.

Flint, E.A. 1959. The occurrence of oospores in *Physolinum* Printz. *New Phytol.* **58**: 267-270.

Fott, B. 1959. Algenkunde, Jena.

Fritscn, F.E. 1935. The structure and reproduction of algae. Vol. 1. Cambridge University Press, Cambridge.

Gavrila, I.L. 1996. Light and electron microscopic studies of euglenoid nuclei. In: *Cytology, Genetics and Molecular Biology of Algae* (Eds.) B.R. Chaudhary and S.B. Agrawal, SPB Academic Publishing, The Netherlands (in press).

Gavrila, I.L. and Chaudhary, B.R. 1996. Structure and behaviour of dinoflagellate nuclei. In: *Cytology, Genetics and Molecular Biology of Algae* (Eds.) B.R. Chaudhary and S.B. Agrawal, SPB Academic Publishing, The Netherlands (in press).

Godward, M.B.E. 1996. Cytotaxonomy of *Spirogyra* I. *S. submargaritata, S. subechinata* and *S. britannica* spp. *Novae. J. Linn. Soc. (Bot.)* **55**: 532-546.

Godward, M.B.E. 1966. The chromosomes of the algae. Edward Arnold, London.

Gupta, A.K. 1992. Morphological and cytological studies of some members of Oedogoniaceae (Chlorophyceae). Ph.D. Thesis. Banaras Hindu University.

Hamakalo, B.A. and Rattner, J.B. 1977. The structure of a mesokaryote chromosome. *Chromosoma (Berl.)*. **60**: 39-47.

Henningsen, K. 1963. Chromosome numbers in five species of *Oedogonium. Phycologia.* **3**: 29-36.

Hoffman, L.R. 1961. Chromosome numbers in the genus *Oedogonium. Phycol. News Bull.* **3**: 29-36.

Hoffman, L.R. 1967. Chromosome numbers in twelve species of *Oedogonium. Am. J. Bot.* **54**: 271-281.

Holm-Hansen, O. 1969. Algae: amounts of DNA and organic carbon in single cells. *Science.* **163**: 87-88.

Hoshaw, R.W. and McCourt, R.M. 1988. The Zygnemataceae (Chlorophyta): a twenty-year/update of research. *Phycologia.* **27**: 511-548.

Khan, M. and Sarma, Y.S.R.K. 1967a. Studies on cytotaxonomy of Indian Charophyta I. *Chara. Phykos.* **6**: 36-47.

Khan, M. and Sarma, Y.S.R.K. 1967b. Studies on cytotaxonomy of Indian Charophyta II. *Nitella. Phykos.* **6**: 48-61.

Khan, M. and Sarma, Y.S.R.K. 1984. Cytogeography and Cytosystyematics of Charophyta. In: *Systematics of the Green Algae* (Eds.) D.E.G. Irvine and D.M. John. Academic Press, London, pp. 303-330.

Leedale, G.F. 1966. The Euglenophyceae. In: *The chromosomes of the Algae* (Ed.) M.B.E. Godward, Edward Arnold, London, pp. 78-95.

Lund, J.W.G. 1962. Classical and modern criteria used in algal taxonomy with special reference to genera of microbial size. *Symp. Soc. Gen. Microbiol.* No. 12. Microbial classification, pp. 68-110.

Magne, F. 1964. La mitose calliblepharidenne de certaines Rhodophycees. *C.R. Acad. Sci. Paris.* **259**: 3811-3812.

Mattox, K.R. and Steward, K.D. 1984. Classification of the green algae: a concept based on comparative cytology. In: *Systematics of the green algae* (Eds.) D.E.G. Irvine and D.M. John. Academic Press, London & Orlando, pp. 29-72.

Pickett-Heaps, J.D. 1975. Green Algae. Sinauer Associates, Inc., Publishers, Sunderland, Massachusetts.

Printz, H. 1964. Die Chaetophoralean der Binnengewasser. *Hydrobiologia.* **24**: 1-456.

Rae, P.M.M. 1976. Hydroxymethyluracil in eukaryotic DNA: a natural feature of Pyrrophyta (Dinoflagellates). *Science.* **194**: 1062-1064.

Rizzo, P.J. 1987. Biochemistry of the dinoflagellate nucleus. In: *The biology of*

*Dinoflagellates*. (Ed.) F.J.R. Taylor, Blackwell Scientific Publications, Oxford, pp. 143-173.

Rizzo, P.J. 1981. Comparative aspects of basic chromatin proteins in dinoflagellates. *Biosystems*. **14**: 433-443.

Round, F.E. 1963. The taxonomy of the Chlorophyta. *Br. Phycol. Bull.* **2**: 224-235.

Round, F.E. 1971. The taxonomy of the Chlorophyta. II. *Br. Phycol. Bull.* **6**:

Sarma, Y.S.R.K. 1962. Nuclear cytology of *Sphaeroplea annulina* (Roth.) Ag., and its bearing on the systematic position of *Sphaeroplea*. *Cytologia*. **27**: 72-78.

Sarma, Y.S.R.K. 1964. Cytology in relation to systematics of algae with particular reference to Chlorophyceae. *The Nucleus*. **7**: 127-136.

Sarma, Y.S.R.K. 1963. Contributions to the karyology of the Ulotrichales. II. *Uronema* Lagh. and *Hormidium* Klebs. *Caryologia*. **16**: 515-519.

Sarma, Y.S.R.K. 1968. Cytology and cytotaxonomy of Indian Charophyta—A resume. *The Nucleus. Proc. Sem. Chr.*, pp. 128-137.

Sarma, Y.S.R.K. and Chaudhary, B.R. 1975. On a new cytological race of *Schizomeris leibleinii* Kuetz. *Hydrobiologia*. **47**: 181-191.

Sarma, Y.S.R.K. 1978. Cytotaxonomy in algae. In: *Glimpses in Plant Research: Modern Trends in Plant Taxonomy* (Ed.) P.K.K. Nair, Vikash Publishing House, New Delhi, pp. 19-43.

Sarma, Y.S.R.K. 1981. Irradiation studies on green algae—A resume. In: *Recent Advances in Cryptogamic Botany* (ed.) D.C. Bharadwaj, The *Palaeobot. Soc.*, Lucknow, pp. 16-32.

Sarma, Y.S.R.K. 1983. Algal karyology and evolutionary trends. In: *Chromosomes in Evolution of Eukaryotic Groups* Vol. I (Eds.) A.K. Sharma and A. Sharma, CRC Press, Florida, pp. 178-206.

Sarma, Y.S.R.K. and Abhayavardhani, P. 1987. Cytological effects of some chemicals on algae—a review. In: *Aspects of Plant Sciences*, (Ed.) S.S. Bir, Today & Tomorrow Printers & Publishers. **9**: 1-33.

Sarma, Y.S.R.K., Chaudhary, B.R. and Chowdary, Y.B.K. 1990. Algal cytology and cytogenetics from Banaras School. In: *Glimpses of Phycology in India* (Eds.) O.N. Srivastava and M. Srivastava, Today and Tomorrow's Printers and Publishers, New Delhi, pp. 1-42.

Shyam, R. and Sarma, Y.S.R.K. 1978. Cytology of Indian Freshwater Dinophyceae. *Bot. J. Linn. Soc.* **76**: 145-159.

Singh, H.V. and Chaudhary, B.R. 1990. Karyological investigations on the genus *Oedogonium* Link (Oedogoniales, Chlorophyceae). *Caryologia*. **43**: 169-181.

Sinha, J.P. 1958. Cytological and cultural studies on some members of Cladophorales and Oedogoniales. Ph.D. Thesis, London University.

Sinha, J.P. 1963. Cytotaxonomical studies on *Cladophora flexuosa* (Griff.) *Harv. a marine* sp. *Cytologia*. **28**: 1-11.

Sinha, J.P. 1965. Cytotaxonomical studies on Harv. and *C. refracta* Aresch, two marine spp. *Cytologia*. **30**: 375-384.

Sinha, J.P. 1968. Cytotaxonomical studies on *Cladophora glomerata*, four freshwater forms. *Cytologia*. **32**: 507-518.

Smith, G.M. 1933. Some aspects of the bearing of cytology *Cladophora hutchinsiae* on taxonomy. *Proc. Linn. Soc. Lond.*, **145**: 151-181.

Smith-White, S. 1954. Chromosome numbers in the Boronieae (Rutaceae) and their bearing on the evolutionary development of the tribe in the Australian flora. *Austral. J. Bot.* **2**: 287-303.

Srivastava, S. and Sarma, Y.S.R.K. 1979. Karyological studies on the genus *Oedogonium* Link (Oedogoniales, Chlorophyceae). *Phycologia*, **18**: 228-236.

Stebbins, Jr. G.L. 1950. Variation and evolution in plants. Columbia Univ. Press, New York.

Taylor, F.J.R. 1980. On dinoflagellate evolution. *Bio Systems*. **13**: 65-108.

Vidyavati and Sathaiah, G. 1984. Radiobiological studies in algae (A review). *Bull. Pure & App. Sci.* **38**: 63-88.

Wille, N. 1909. Chlorophyceae and Conjugatae. In: Engler and Prantl, Die Naturlichen Pflanzenfamilen. Nachtrage Zur. I Teil. Abt. 2. Leipzig.

Wood, R.D. and Imahori, K. Revision of the Characeae. Verlag von. J. Cramer.

Recent Trends in Algal Taxonomy
Vol. I. Taxonomical Issues
Editors : Vidyavati and A.K. Mahato
APC Publications Pvt. Ltd., New Delhi, 1999, pp. 51-63

# *Phycotaxonomy in Tamil Nadu*

V. Sankaran
Dept. Botany, NGM College, Pollachi-642 001

Pioneering work on phycotaxonomy in Tamil Nadu has been carried out by Prof. M.O.P. Iyengar and the illustrious batch of competent algologists viz., Profs. T.V. Desikachary, M.S. Balakrishnan, V. Krishnamurthy, and V.S. Sundaralingam. The contributions of Prof. M.O.P. Iyengar are very rich (See Desikachary 1964). Contributions towards knowledge on Cyanophyta (Anand 1994). Rhodophyta (Krishnamurthy 1994), Phaeophyta Rao (1994) and Chlorophyta and Charophyta (Sundaralingam 1994) of India have been recently reviewed. These reviews do not particularly pertain to Tamil Nadu. The present review covers new taxa and new records of algae not covered by the above reviewers. This review also includes new records and new taxa of Dinophyta and Bacillariophyta pertaining to Tamil Nadu area.

Thajuddin and Subramanian (1991) have surveyed the Indian Ocean coasts of Tamil Nadu for marine Cyanobacteria. Three species of cyanobacteria *Radaisia violacea* (Pleurocapsaceae), *Siphonema polonicum* (Siphonemataceae) and *Dichothrix spiralis* (Rivulariaceae) recorded by them have not been recorded before from the Indian coasts. They come to the conclusion after scanning the pertinent taxonomic literature.

The crustose thallus of *Radaisia violacea* Fremy is differentiated into a prostrate and an erect system. The erect system is unbranched and composed of cells, variable in number and shape, enveloped by a single homogenous, thin, non-stratified sheath. The cells of the prostrate system divide in all planes, whereas, those of the vertical system divide only transversely. The authors comment that this algal differs from earlier descriptions in the absence of branches in the vertical system. Sporangia and endospores are also not recorded in the Indian specimen. The algae has been reported from the Cape Comorin and Gulf of Mannar region.

The thallus of *Siphonema Polonicum* Geitler is at first single celled, undergoes divisions in three planes resulting in many rowed stigonemalean structure. The constituent cells are variously shaped and coloured, enveloped by yellow sheath. Reproduction is reported by constriction of Chroococcoid cells, commenting on the alga, the authors feel that this could be a variant of the species recorded by Geitler, in view of the absence of certain elongated cells. The alga is reported from Tiruchendur and Gulf of Mannar regions.

*Dichothrix spiralis* Fritsch F.E., has false branched filaments with pale yellow or brown, unlamellated sheath, ending in a long hair. Trichomes coiled inside sheaths with basal and intercalary heterocysts. The alga differs from earlier specimen in the presence of elongated hetrocysts and absence of inner sheath layer.

In an earlier report Thajuddin and Subramanian (1990) have given an account of marine cyanobacterial phytoplankton of the Gulf of Mannar Region. They have monitored intensively this planktonic cyanobacterial population, in terms of abundance and diversity. These belong to 9 families, 24 genera and 37 species. These are from the following genera: *Microcystis, Chroococcus, Gloeocapsa, Aphanocapsa, Aphanothece, Synechocystis, Merismopedia, Johannesbaptistia, Stichosiphon, Xenococcus, Hyella, Arthrospira, Spirulina, Oscillatoria, Trichodesmium, Phormidium, Lyngbya, Richelia, Nostoc, Pseudanabaena, Nodularia, Plectonema, Calothrix* & *Mastigocoleus*.

The taxonomic studies in *Spirulina* undertaken by Desikachary and Jeeji Bai (1992) deserves special attention. This study attempts to fill the gap in taxonomic description and identification of *Spirulina*. They recognise *Arthrospira* and *Spirulina* as two district genera. To determine the nomenclatural and taxonomic identify of the Indian forms they have sought to examine *Spirulina* material from different sources in India, Europe and U.S. The type slide of *A. platensis* has also been examined. The samples identified as *S. fusiformis*, Woronichin by Jeeji Bai and isolates from Adayar river estuary, Mariamman temple tank, Madurai, isolates from Lonar lake, Warda, isolates from Sambhar lake, Texas collection, chad isolates, have all been identified as *Arthrospira indica Sp. Nov.* Despite minor variations, these isolates collections are considered to belong to same taxa, described as new species namely *Arthrospira indica*. They characterise this new species as having blue green trichomes, with close or distant coils, showing attenuation at both the ends, with capitate and calyptrate and cells, granulated at the cross walls and general presence of gas vacuoles. Multiplication by fragmentation with the formation of necridial cells.

*Batrachosperdmum desikacharyi* is a new species described by Sankaran (1984) from Valparai, Anaimalais, Tamil Nadu. The alga is pinkish/violet, monoecious mucilaginous, 15-20 cm in diameter profusely branched and typically beaded in appearance. The glomeruli are 700 μm in diameter. The older portions of the thalli are corticated. The solitary/paired spermatangia, terminal on determinate laterals are 6.0–8.5 μm in diameter. The 5–10 celled long (each cell with paired bracts), straight/curved carpogonial branches, arising form the basal cells of the primary determinate laterals—one/two in number, bear an ellipsoid carpogonium 8.5–12.0 μm long and 6.0–8.5 μm broad with a club shaped stalked or rarely sub-sessile trichogyne 40–46.0 μm long × 11.5–14.0 μm broad. The carpogonial branch becomes enveloped after fertilization, forming a stalk to the developing globose gonimoblast 120.0 μm in diameter. The carposporangia are 34.0–47.6 μm long × 24.7 μm broad.

In its colour and general morphology, this species resembles those belonging to the section TURFICOLA. But in this section, the trichogyne is almost sessile and obconical, whereas, in this form it is stalked/subsessile club shaped/cylindrical as in the section VIRIDIA. Thus in trichogyne morphology, this new species is nearer to the species of the section Viridia. Further, the carpogonium terminates a branch from the periaxial cell. This type of carpogonial branch is homologous with a branch of unlimited growth and in this report it is close to *B. vagum* of section Turficola. A remarkable feature of *B. desikacharyi* is the formation of certain thread like cells from the bract cells. These thread like cells grow downwards and envelope the carpogonial branch resulting in a pedicel that protrudes from the primary whorls. This pedicellate carposporophyte found in this species is a very unique feature that distinguishes it from all other species of *Batrachospermum.* (See also Desikachary 1990)

Krishnamurthy *et al.* (1977) reported on the Dinophyta from Porto Novo, Tamil Nadu. They have undertaken a detailed study of Phytoplankton species composition of the Dinophyta. It is the first report of its kind from the east coast of India. They deal with dinoflagellate species composition from the neretic and estuarine biotopes of portonove (parangipattai) based on 5 years of intensive study. They recorded 94 species belonging to 20 genera. *Exuviella, Phalachroma, Pyrophacus, Diplosalis, Diplocelopsis, Goniaulax, Gymnodinium, Goniodoma, Pyrocystis, Ceratium, Peridinium, Dinophysis, Arnithocer-*

*cus, Ceratocorys, Podolampas, Amphisolenia, Prorocenea, Oxytoxum, Noctiluca,* and *Glenodinium*—They have identified and tabulated the species. Of these, 11 species are new to India. These are *Prorocentrum, micans, Phalachroma, palchellum, Peridinium ampalliforme, P. brevipes, P. latispinum, P. Ovum, P. umbonatum, P. africanum, P. rosaceum, P. adeliense, P. tuba,* But this work does not contain any description or sketches.

There have been very few reports from India on Cryptophyceae. Prasad (1983) has described a cryptophycean flagellate, *Chroomonas norstedtii* f *minor* Nygaard, for the first time from Madras. This blue-green fresh water phytoflagellate was collected from a fresh water puddle in Madras. The alga is described as having longish oval cells, dorsoventrally flattened with somewhat convex dorsal side and slightly concave or straight ventral side and with indistinct trichocysts. Anterior end broader than posterior end; with anterior contractile vacuoles, Stigma not present. Gullet and furrow not clear. Chromatophores single, later appearing as two, pyrenoid single, mid posterior, two unequal subapically inserted flagella. During cell division there is mucilage secretion. The author comments that the two Chromatophores present in the older cells may represent a common bilobed condition of the chromatophore. He distinguished the described form *C. acuta* Utermohl and *C. norstedtii* Hansg. He assigns the alga to *C. norstedtii, f. minor* based on cell size, colour and number of chromatophores. The paper includes phase contrast photomicrographs of the alga, giving details of cell shape and flagellation.

Prasad (1982) reports on the occurrence of a packet forming genus of Chlorosarcinales viz., *Chlorosarcinopsis* from Madras. The alga has been isolated in culture from sandy soil samples. The spherical cells are 14–12 μm in diameter are uninucleate with a parietal chromatophore and single pyrenoid. The chromatophore becomes massive cell on maturation. Three dimensional, irregular sarcinoid packets result from vegetative cell division. These packets are surrounded by a common colloidal matrix. Zoosporogenesis takes place on packet dissociation. When transferred to freshmedia, zoospores are produced by successive bipartitioning of the protoplast. Commonly 4–6 Zoospores are produced per cell. The fusiform, naked zoospores are 2.7–4.5 μm broad × 4.5 – 6.3 μm long with a prominent anterior stigma and inconspicuous pyrenoid. Sexual reproduction is not observed. The author discusses the similarities and differences exhibited by the alga under consideration with closely related taxa. Based on these discussions, the alga is assigned to *Chlorosarcinopsis arenicola* Grover and Bold, The paper includes photomicrographs and sketches of the various stages of development of the alga. This investigation is a step towards enhancing the knowledge on soil algae of India and adds validity to the concepts of more recent orders such as the chlorosarcinales.

Inspired by the efforts of R. Subramanian, Desikachary has brought out an atlas of marine diatoms of India. The atlas is in 6 fascicles and has been released in a phased manner during 1986-1989. The atlas includes photomicrographs of diatoms collected by him and also in slides deposited in the British museum, Philadelphia Academy of Natural sciences & Mus. Hist. Nat. Paris.

Fascicle I deals with fossil diatoms of Middle meiocene material. The new taxa of fossil diatoms described by Desikachary (1986) in this fascicle include:

*Actinocyclus srinivasani* sp nov

*Arachnoidiscus chrenbergii* var *nankoorensis* var nov

*Auliscus prunosus* var *nankooransis* var nov

*A. Caelatus* var *nankoorensis* var nov

*Campylodiscus nankooransis* sp nov

*Coscinodiscus nankooransis* sp nov

*Pyxidicula reniformis* sp nov

*Astero lampra nankoorensis* sp nov

Fascicle II (Desikachary 1987) gives photomicrographs of diatoms found in the sediments of Indian ocean region from various sources such as Scripps' Institution of Oceanography, Lamont-Doherty Geological observatory, Antarche Research facility, Florida, National Institute of Oceanography, Goa, India, The new taxa described here include:

*Biddulphia infundibuloides*, sp. nov.

*Chaetoceros neobulbosum* sp. nov.

Fascicle III (1987) includes diatoms from the Bay of Bengal The new taxa described here include:

*Campylodiscus ornatus* var *indicauns*, var nov.

*C. bifurcatus* var *minor* var nov.

*Actinocyclus ingens* var *indica* var nov.

*Nitzschia subconstricta* sp nov.

*N. cornicobarica* sp nov.

*Neofragilaria nicobarica* Desikachary. Prasad and Prema *gen et sp nov.*

The places of collection include Madras city, Mahabalipuram, Pichavaram, Mandapam, Tuticorin, Kovalam, Pamban, Tiruchendur, Rameswaram, Krusadi, Cape comerin etc.

Fascicle IV (1987) includes diatoms of the W. Indian Ocean region from Cochin, Someswar, Karwar, Goa, Malvan, etc. No new taxa are described.

Fascicles V & VI (1988-89) include also the diatoms of the Indian Ocean region based on collections from the British museum and also gut contents of Holothurians of Krusadi & Karwar. The new taxon described is:

*Diploneis biseriata cleve var australis* var nov. (Fasc V) *Rhaphoneis Krusadiensis sp. nov.* Fas VI.

The atlas of diatoms in 6 fascicles represent a very monumental work of high quality and is the product of devoted work spread over several years. The descriptions of the new diatom taxa are given below:

*ACTINOCYCLUS SRINIVASANII* sp. Nov.

Valve broadly elliptical with slightly convex surface, distinct single apical submarginal pseudonodule. Surface punctate, puncta scattered in the middle and few in number near the margin puncta arranged in compact short radial rows. Valve 24.0 × 16.5 μm

*A. INGENS* VAR *INDICA* VAR. Nov.

Valve circular, marginal circlet of spines absent. Central space hyaline often with 3-4 puncta. Areolae round, arranged in regular radial rows, the border crowded, hyaline inter spaces present Valve 40–84.0 μm, areolae 3–6 in 10 μm. This differs in larger areolae. But other differences do not seem to be clear cut.

*ARACHNOIDISCUS EHRENBERGI VAR NANKOORENSIS* var nov.

Valve circular with well marked rays. Distinct primary rays and short secondary rays alternating. Areolae arranged in concentric rings. Innermost circle of areolae slightly elongated valve diameter 136 μm.

*ASTEROLAMPRA NANKOORENSIS* sp. nov.

Valve circular central rosette and areolate area absent. Rays 8, straight, the compartments reaching upto three-fourths of the radius inwards. Small round granules present close to the compartment. Border narrow, hyaline, indistinct, valve 42.0 µm in diameter.

*AULISCUS PRUINOSUS* var *NANKOORENSIS* var nov.

Valve circular surface punctuate, puncta and epiculi connected by delicate lines. Processes two, near the margin, placed opposite to one another. Central space present. Valves 113–157.0 µm in diameter.

*A. CAELATUS VAR NANKOORENSIS* var nov.

Valves with two ocelli. Central area with distinct punctal row. The ring of puncta in the central area distinguishes this from other varieties.

*BIDDULPHIA INFUNDIBOLOIDES* sp. nov.

Valve seen only in side view, chimney shaped with a thick thorn like projection at each corner and two long thick horn in between the two projections. Valve in side view with reticulate areolation, areolae punctate, prominent. Pervalvar axis 69.0 µm including the long spine. Valve 12 µm broad in narrow region, 35.0 µm at the end. Areolae 8 in 10 µm. Thorn at the corner 10.0 µm long Middle spine 35.0 µm long.

*CHAETOCEROS NEOBULBOSUM* sp. nov.

Frustule in girdle view somewhat octagonal valve surface flat with small median inflation, mantle obliquely sloping setae emerging within the edge of the valve and parallel to the pervalvar axis; setae forming short swollen appendices with longitudinal costae and ending to a short attenuate apex. Valves 10–30.0 µm in diam. and 29–40.0 µm in pervalvar axis. The diagnostic feature of this species is the absence of central spine and the presence of spherical terminal portion of the setae. For the reason, the species is separated from the balbous species of *Chaetoceros*.

*CAMPYLODISCUS BIFURCATUS* var *minor* var nov.

Valves sub orbicular with costae, broad hyaline central area and distinct punctate area on either side of the hyaline space towards the margin. Valves 30 × 42.0 µm costae 2–3 in 10 µm. The variety differs from the species in the absence of bifurcating line in the centre, smaller size and closer costae.

*CAMPYLODISCUS NANKOORENSIS* sp. nov.

Valves nearly circular, much bent with two bands of radiating canaliculi, the marginal one narrow, the inner one much broader, the canaliculi distant, with two rows of puncta between them.

Central space hyaline, bordered with a circular row of oblong granules valves 115 × 105 μm.

*CAMPYLODISCUS ORNATUS VAR INDICUS VAR* nov.

Valve saddle-shaped, rays in two, marginal narrow inner distinct with punctae. Median space distinct bordered by a row of puncta near the rays. Valves 56.3 × 57.6 μm diameter. The variety is more saddle-shaped than the species and smaller in size.

*COSCINODISCUS NANKOORENSIS* sp. nov.

Valve ovate, central aereolae 2 in 10 μm, smaller areolae 7 in 10 μm in the outer half. Central area hyaline. Periphery with punctae in radial rows appearing as striae. Valve 44 × 33 μm.

*DIPLONEIS BISERIATA* var *AUSTRALIS DESIK & PREMA* var nov.

Valve elliptical, not constricted; central nodule large, costae parallel; furrows linear, narrow with a row of large granules, ocelli forming double rows, costate, costae alternating with double row of coarse punctae. Valve 51.0 μm × 18.0 μm. Costae 6 in 10 μm. This variety differs from others in smaller cells.

*NEOFRAGILARIA NICOBARICA* Desik Prasad and Prema gen et sp. nov.

Valve linear elliptical, or elliptical lanceolate, tapering towards the apices; apices narrower and somewhat produced, rounded; axial area hyaline, distinct, narrow; valve with transverse rows of areolae, areolae on both sides of the central axial area not aligned in straight and continuing transverse rows, but alternating, the exterior valves with elongated apical pore fields, rinoportulae absent, marginal spines absent, interior valve shows the velum with several stratus attached to the areolar wall. Valve 15-35.0 μm long × 8–10.0 μm broad areolae 6–7 in 10 μm.

*NITZSCHIA CORNICOBARICA* sp. nov.

Valves linear, constricted in the middle, ends subacute, median space broad. Valves 38.0 μm long × 10.0 μm broad, 5.0 μm at the constriction. Carinal puncta 11–12 in 10 μm, transverse striae 22–24 in 10 μm. This species differs from the closely related *N. marginulata* in the presence of median constriction and apiculate apices.

*NITZSCHIA SUBCONSTRICTA* sp. nov.

Valves broadly apical, slightly constricted in the middle and tapering towards the somewhat truncate ends. Keel excentric valve surface striate. Striae coarsely punctate, valve 44–96.0 μm long,

12–18.0 µm broad, striae 10–13 in 10.0 µm.

*PYXIDICULA RENIFORMIS* sp. nov.

Valve reniform, surface covered with irregularly distributed spines, 6 in 10 µm. Valves 44.0 × 24.0 µm.

*RHAPHONEIS KRUSSADIENSIS* Desikachary & Sridharan sp. nov.

Valve broadly rhomboidal, distinctly swollen in the middle, appearing like a rectangle with parallel transapical longer sides and arched shorter sides, as broad as or slightly shorter than long. 40 × 30 µm strial panctae 7 in 10 µm. In holuthurian gut contents.

Subsequently Prema and Desikachary (1989) have investigated and reported some new and interesting marine fossil diatoms from the Indian ocean core samples. These are described in brief.

*QUADRODISCUS* gen nov.

This genus is a coscinodiscoid form with a bipolar symmetry, with distinctive quadrangular valve structure. This genus has been erected on the basis of the quadrate shape with bipolar symmetry. The quadrangular valve has two arms long and two arms short. The arms are tapering with rounded apices. Central area areolate. A single process subterminally placed on the two longer arms. Ocelli and peudonodulus absent. The type species is *Q. indicus* sp. nov. Besides this species the other species described are *Q. minutus* (Greville) the comb. no. and *Q. gombosii* sp. nov.

*PORODISCUS MINOR* sp. nov.

Valve circular slightly depressed in the centre valve 35 µm diameter, central area hyaline with hazy limits, surrounded by radially punctate zone. Outer part of the valve with irregular rows of punctae. Margin with a distinct band of radial striae. Striae punctate, punctae 5 in 10 µm in the valve face.

*PORODISCUS VENKATARAMANII* sp. nov.

Similar to the *P. minor* in valve morphology punctae inconspicuous upto 1/3 of the diameter. There is a zone of decussately arranged puncta—puncta 8 in 10 µm, striae 10 in 10 µm. Submarginal process present. These fossil taxa are from oligocene.

There is a further report of fossil diatoms by Gowthaman and Desikachary (1989) from core samples of Indian Ocean. No new taxa are described, but there are new reports to the Indian Ocean

region. These include:
*Craspedodiscus undulatus, Isthmia squina boli, Strangulonema barbadense, Triceratium banayae var trilobata* & *T. kanayae var quadriloba*
These belong to ocene/oligocene. Prasad and Janaki (1981) have reported *Scotiellicystis oocystiformis* (Lund) Fott for the first time from South India. The alga has been isolated from soil samples of the Konai Falls area. The alga was also subjected to SEM analysis.

The alga was grown in soil water, Bristol, and modified Chu 10. Morphological variations were minimum in these media and autospore formation was present in all the media. They have given a description of the alga on the basis of cumulative observations.

The solitary cells were spindle shaped 8.0–11.2 (–19.2) μm long 6.4–11.2 (–14.4) μm wide, lacked contractile vacuoles and eye spots. Enlarged, ellipsoidal and abnormal cells were also observed. The most prominent feature of the alga was the presence of 10 longitudinal wing-like extensions of the wall, regularly and equally disposed, extending from pole to pole. These are evident in the top-view of the cell, but not clear in lateral view, incomplete flanges were also observed. The flanges were not interconnected, but converged at the polar nodules. Single parietal chromatophore with a single pyrenoid breaking into discoid chromatophores. A "pinching of" division of the cell is reported but not clearly understood. 2, 4 or 8 autospores were formed in a cell. No motile stages were observed. The authors discuss at length the taxonomic distinction and validity of allied forms. They transfer in the light of taxonomic studies based on developmental investigations three species recognised under *Scotiella* to *Scotiellocystis*. Excellent SEM pictures are included in this paper.

Cladophorella—a terrestrial green alga first described by Fritsch in 1944 from Cambridge has been recorded from Pakistan and Bengal. This alga was first recorded and described by Sankaran 1990 from Tamil Nadu. The alga was collected from calcareous rocks and bricks in seewage water from a reservoir near Pollachi in Tamil Nadu. The alga consists of prostrate system of loose cluster of cells from which erect threads and rhizoids arise. The cells of the prastrate system are variously shaped, cylindrical (250.0 μm × 75.0 μm broad) irregular or globular, with diameter upto 150.0 μm. The prostrate system gives rise to long tubular rhizoids reaching upto a length of 700.0 μm. They may be straight or undulate in outline and may end as dilated lobed structures. The cells of the vertical system are upto 500.0 μm long × 19–68.0 μm wide, rarely branched, branches arising from near the upper septum. The cells exhibit a typical cladophoralean structure with reticulate chloroplast with many pyrenoids and are multi nucleate. The cell walls are X. 1.0 μm in thickness. Reproduction by akinetes, commonly produced in the vertical system, terminal, solitary, paired or in chains, oblong or short cylindrical, 34.0–220.0 μm long X 27.2–50.0 μm broad. The akinetes are multi necleate and have dense cell contents. The alga agrees with *C. calcicola Fritsch.*

Premila *et al.* (1990) pointed out the lack of information on the B. Subaerial algae in Tamil Nadu. They have studied the sub-aerial algae—epiphytic and epiphyllous algae from Ootacammand, Tamil Nadu. They have described *Trentepohlia aurea* (Linn) Martius growing on the trunks of *Quercus, Eucalyptus* & *Cinchona, T. Odorata* (Wiggers) Wittrock on the trunks of *Acacia* and barks of some Gymnosperms, *T. abietina* (Flotow) Hansg. on the barks of *Quercus* and *Eucalyptus, Physolinum monilia* (De Wildemann) Printz on the trunks of *Cinchona* and *Anacardium*, and *Phycopeltis epiphyton* Millard on leaves of *Eucalyptus*, leaves of Lauraceae and on bunch palms *P. arundinacea* (Mont) De Toni, on leaves of bunch palm, *Cephaleuros virescens* Kunze on lauraceous leaves, leaves of *Eucalyptus Anacardium* & *Artocarpus; Hormidium Klebsii* Smith on barks of *Quercus, Araucaria* & *Magnolia.*

This survey besides, the above listed taxa, includes a new variety and a new record, This new variety is *Physolinum monilia* (De Wildemann) Printz var. sub *sphaerica* var nov. growing on barks of *Anacardium.* Thallus green or red in colour, branched cells almost spherical, constricted at the septum 10–25 μm long, 15.25 μm broad with sporangia not well developed. The difference, it appears is in the smaller cell size.

The first report is that of *Apatococcus constipatus.* All these forms are described and adequately illustrated.

Studies on chaetophorales have been very few in Tamil Nadu. Nirmala *et al.* (1990) have made a brief report on *Stigeoclonium* from Madras. They reported for the first time 3 species of the genus.

*Stigeoclonium lubricum,* Kutzing epiphytic on aquatic angiosperms from covelong *S. aestivale* (Hazen) Collins from a Madras cement tank and *S. amoenum* Kutzing are thoroughly described by them bringing out the morphological and morphometric characters. Critical diagrams find a place in this paper.

Phycotaxonomic work in Tamil Nadu is making further progress. Algologists here are inspired by the work still be carried out by Prof. T.V. Desikachary and Prof. Sundaralingam at the University of Madras, the centre from which Prof. M.O.P. Iyengar made several monumental contributions. Prof. V. Krishnamurthy Institute of Algology, is another centre that is aiming at consolidating phycotaxonomic research in Tamil Nadu. The Bharathidhasan University, Tiruchi, the group of algologists at Mandapam are concentrating on marine algae. The Pachiyappas College at Madras is also a centre of Phycotaxonomy.

Sankaran has been studying the fresh water algae of the Anamalay ranges, for the past more than 20 years from N.G.M. College, Pollachi. This phycotaxonomic work is being continued currently and is expected to be completed shortly.

## ACKNOWLEDGEMENT

Grateful thanks are due to Prof. T.V. Desikachary and Dr. N. Jeeji Bai for their critical suggestions and to the Principal NGM College for facilities.

## REFERENCES

Anand, N. 1994. Cyanobacteria (Cyanophyta, Blue-green algae) in Botany in India History and Progress 1, B.M. Johri (ed), pp. 165-179.

Desikachary, T.V. 1964. M.O.P. Iyengar (1886-1963) *Phycologia*, **4**(1): 54-61.

Desikachary, T.V. 1986-89. Atlas of Diatoms, Madras Science, Foundation Fasc: I-VI.

Desikachary, T.V. and Jeeji Bai, N. 1992. Taxonomic studies in *Spirulina* in C.V. Seshadri & N. Jeeji Bai ed. *Spirulina ETTA Symposium*, Madras. pp. 12-37.

Desikachary, T.V., Krishnamurthy, V. and Balakrishnan, M.S. 1990. Rhodophyta—Madras, Science Foundation.

Gowthaman, S. and Desikachary, T.V. 1989. Some new and interesting fossil diatoms: *Phykos*, **28**(1 & 2): 132-135.

Krishnamurthi, K. and Jothi, K. 1977. The Dinophyta from Portonovo, S. India *Sea Weed Res. Utiln.* **2**(2): 65-68.

Krishnamurthy, V. 1994. Rhodophyta in "Botany in India History and Progress of Botany"—I B.M. John (ed). pp. 211-222.

Nirmala, R.N., Premila and Bharatnan, S. 1990. On some *Stigeoclonium* from Madras, *J. Indian Bot. Soc.* **69**: 477-478.

Prasad, A.K.S.K. 1982. Notes on soil algae—Chlorosarcinopsis Herndon (Chlorosarcinales) (Chlorophyceae) in India. *Arch. Protistenk.* **126**: 273-282.

Prasad, A.K.S.K. 1983. Notes on Phyto flagellates—I A *Chroomonas Hansg.* from Madras, India. *Phykos*, **22**: 57-60.

Prasad, A.K.S.K. and Janaki, V.C. 1981. *Scotiellocystis* Fott from S. India and Comments on some species of *Scotiella* Fritsch. *Arch. Protistenk.* **124**: 417-429.

Prema, P. and Desikachary, T.V. 1989. New and interesting fossil diatom from the Indian Ocean *Phykos*, **28**(1 & 2): 127-131.

Premila, N.R., Nirmala and Bharathan, S. 1990. Preliminary report on the epiphytic and epiphyllous green algae from the Nilgiris, *Phykos*, **29**(1 & 2): 159-165.

Rao, V.N.R. 1994. Phaeophyta in "Botany in India History and Progress"—I B.M. Johri ed. pp. 201-209.

Sankaran, V. 1984. *Batrachospermum desikachari* sp. nov. from Valparai, Anamalays T.N. *Phykos*, **23**(1 & 2): 163-170.

Sankaran, V. 1990. *Cladophorella Calcicola Fritsch.* from Anamalays *Perspective in Phycology*. V.N. Raja Rao ed. Today & Tomorrows Publishers, New Delhi. pp. 69-71.

Sundaralingam, V.S. 1994. Chlorophyta and Charophyta in "Botany in India—History & Progress of Botany" 1 ed. Johri. pp. 181-199.

Thajuddin, N. and Subramanian, G. 1990. Cyanobacterial phytoplanktons of the Gulf of Mannar region in *National Symposium on Cyanobacterial Nitrogen Fixation IARI*, New Delhi. pp. 457-463.

Thajuddin, N. and Subramanian, G. 1991. New reports of Cyanobacteria from the Southern east coast of India, *Phykos*, **30**: 19-23.

# Section - 2

Recent Trends in Algal Taxonomy
Vol. I. Taxonomical Issues
Editors : Vidyavati and A.K. Mahato
APC Publications Pvt. Ltd., New Delhi, 1999, pp. 67-100

# *Recent Trends in the Taxonomy of Selected Chlorococcalean Algae*

Francis R. Trainor and Eduardo A. Morales
Department of Ecology and Evolutionary Biology
University of Connecticut, Storrs, Connecticut-06268-3042, USA

## INTRODUCTION

During the last decade great advances have been made in our understanding of the evolution of microscopic organisms. Even at the level of kingdom, *e.g.* the Protista, there have been proposals to abandon some systems of classification. As an example, after viewing the evolution of aquatic organisms from the bottom up, we have determined that experimentation with amitochondriate eukaryotic microorganisms will help us understand both the evolution of the various categories of microorganisms, as well as other important evolutionary concepts. Inasmuch as some previous interpretations of the organization of kingdoms were formulated after examination of characteristics of macroscopic organisms exclusively, this approach is refreshing.

In a recent introductory biology text, Minkoff & Baker (1996) discuss the manner in which biologists deal with such topics, as well as how they add to the fund of knowledge, that is, by testing falsifiable hypotheses. New discoveries are constantly being announced, and are continuously accumulated; most are interesting, but not revolutionary. This process is called *normal* science by Minkoff & Baker. Kuhn (1968) talked about this type of activity somewhat differently:

**"Mopping-up operations are what engage most scientists throughout their careers."**

In contrast, some discoveries ultimately contribute to a revolutionary approach in science, to a completely different way in which we consider and interpret the facts before us. "Models of achievement which are sufficiently unprecedented, but at the same time open-ended have been termed paradigms" (Kuhn, 1968). If the system previously was driven by an older paradigm, it may now be clear to many in the field that a significant change is necessary. The new paradigm simply appears better than the old, and although it can employ a different research methodology, one should not expect that it provides all of the answers. It will simply be the focal point for additional research, until some time in the future when a new paradigm shift is clearly indicated to at least some in the field.

In general, one should expect some resistance to a new approach, for old ideas are not too easily surrendered. This resistance is the guarantee that we will not be easily distracted, shifting quickly from paradigm to paradigm, as well as assurance that the new approach is well grounded. Even with such careful preparation, the passage to a new paradigm, formed by new evidence, cannot be made a step at a time. Once the pieces are in place, like a master switch, it is often rapid and complete. Two examples of major paradigm shifts would include Darwinian evolution and the molecular revolution.

At times, even though the need for a new approach might be apparent, the path to take is by no means clearly outlined. In such cases, as evidence for change accumulates, articulation of the new paradigm is not always rapid, especially as rapid as a few might hope. Engrossed in the daily travail of experimentation and its data gathering, a researcher might not clearly see theory, or adequately articulate it, in support of the new paradigm. Thus even with a recognized need, the pronouncement has not yet gelled. Only later might one clearly see many of the important ramifications, and thus the paradigm itself.

## THE GREEN ALGAE

During the last few years there has been considerable interest in numerous members of the Chlorococcales (Comas & Komárek, 1984; Hegewald & Silva, 1988; Komárek, 1987; Kouwets, 1995; Mladenov & Furnadieva, 1995). In this review, rather than briefly outlining the important findings with individual organisms, we have chosen to present data in support of a new paradigm. Those data will be concentrated in two areas: investigations with the zoospore-producing unicell, *Neochloris*, and developmental research with *Scenedesmus*.

The paradigm: **The microalgal species, often difficult to define, is much more complex than previously conceived.**

To understand, classify, utilize and appreciate individual members of the Chlorococcales, organisms must be studied more intensively and in considerably more depth than would have been foreseen even a decade or two ago. When such information is assembled and evaluated, the systematics of the individual Chlorococcalean alga changes dramatically. With supporting data from a variety of sources, we are quickly beginning to understand the biology, relationships, and evolution of individual taxa.

Initial support for our speculation now comes from many organisms in a variety of classes and orders. For example, Costas *et al.* (1995) looked historically at the standards used for defining marine dinoflagellate species and concluded that morphological features, because they are the most accessible, were the predominant criteria. But they discussed the conflict which surfaces when one attempts to use data from immunological techniques or DNA probes; "species" group differently into what they called *morphospecies* [the older concept] and *genospecies*. "Morphological features are the complex expression of the genotype allowing for phenotypic change due to environmental, physiological, and other interactions." Costas *et al.*, 1995.

One of our goals is to present data from experimentation which enables us to provide a better understanding of phylogenetic relationships among organisms. Manhart & McCourt (1992) took into consideration molecular data in their recent summary of algal species concepts. As a result they talked of genetic, paleontological, evolutionary, and biosystematic species. Some might prefer to concentrate solely on the biological species, which considers reproductive isolation of interbreeding natural populations, a concept difficult to interpret when organisms are strictly asexual, or reproduce rarely. Reproductive cohesiveness is a feature of many putative species, but interspecific hybridization among some taxa and the occurrence of strictly asexual taxa are problems which must be addressed when applying the biological species concept. On the other hand we have much more morphological data with which to work when it is known that the organism in question exhibits extensive plasticity. Thus our present concern surfaces, *i.e.* we have often failed to understand the extent of phenotypic plasticity in many microorganisms.

First, in order to supply background, let us examine our understanding of the cytology of higher organisms. Reflecting on mammalian cell diversity, or that of an angiosperm, we note that an individual genotype of the former can produce a multiplicity of cell types, *e.g.* those in the liver, bone, brain, blood, etc. Each cell type has its special characteristics and they differ in both structure and function. In a species of *Pinus* we are not surprised to learn that the meristematic cells of an individual genotype can produce parenchyma cells, cork cells, fibers, sieve tubes and tracheids, a plastic expression from one genotype. Thus the pine needle, the bark and the wood are remarkably different, a fact which we often take for granted.

When we examine a population of *Golenkinia* in a body of water we do not normally consider, or expect, phenotypic plasticity, *i.e.* many *Golenkinia* morphs in the population. However, we are fully aware that the organism, the individual genotype, will find itself in several different environments. With a drop in the pH, would the wall have the same thickness? Does the taxon produce the same spine pattern in both spring and fall? Are these patterns somewhat similar, or remarkably different? Do chemical signals from predators affect morphology? If so, in what particular ways, affecting spine number or spine length? How often, if ever, does one find motile cells present in the life history of that genotype? We simply do not have these answers for *most* Chlorococcalean species, for we often lack the hard data and we realize that these data are often difficult to obtain in the field. Yet we have to make choices using the limited data pool, and we do.

Further influencing our decision making is the absence of adequate discussion or argumentation regarding such phenomena; one might cite an ambiguous statement or the absence of clarity. In one discussion or another, the questions may be posed: Do we know whether morphological differences are genetically determined or are expressions of phenotypic plasticity?

In addressing this topic, we would first declare that morphology is determined by way of the inherited genotype. For the majority of microalgae, especially diatoms, we *assumed* that some features, *e.g.* wall cytology used to delimit a species, were stable. Yes, one would expect some variation, but within a narrow range. Such morphological differences were explained as being genetically determined. But what are the limits, what would be consiered normal variation and what would be called plasticity?

We might first want to consider what is known about related organisms. When data for an individual taxon reflected plasticity, some would initially avoid discussion of the role of the genotype, stating that these were expressions of phenotypic plasticity. But upon reflection, this plasticity is also genetically determined. In the first case the determination simply appears to us to be more direct. In the second example, when we consider extensive plasticity, a genotype may express a feature in one environment, but not in another. These various expressions of morphological stages are also genetically determined, even though the determination might be more indirect. Because some considered plastic organisms the exception, organisms not presently worthy of much further consideration, there has been a delay in acceptance of any concept proposing extensive plasticity. Is there but limited plasticity?

There are also some historical reasons for a limited understanding of phenotypic plasticity. With most species descriptions based on preserved material, *e.g.* the ashed diatom frustule, there was little opportunity to unravel a complex life history, especially one which could so easily be altered by the environment. But had an earlier generation adequately explored an alternate hypothesis, that some organisms routinely exhibit an extraordinary level of phenotypic plasticity? Such taxa produce one display of cytological features in one environment, but a different assemblage of features in another environment.

## GREEN ALGAE IN CULTURE

Historically some became interested in examining organisms in culture, and generated data which were at times in conflict with the established stability dogma. Organisms in culture do not always conform to established theory, especially narrowly focused theory. The new standard, *i.e.* that considerable cultured material actually reflected natural morphological expressions, was not always seriously considered, and some conflict resulted. When occasions arose, in some laboratories emphasis was placed on the examination of very plastic organisms. The record will indicate that expressions of extensive plasticity were often first observed in culture studies. It is valuable to keep in mind the words of H.C. Bold (1950) almost 50 years ago.
"I have reference to the classification of many unicellular and colonial algae, especially Chlorophytan forms, the taxonomy of which is nothing short of chaotic." Since he recognized that intensive studies of organisms give us a broader working base, especially for taxonomic studies, he recommended the use of cultured material for such investigations.

When we examine individual taxa, is it possible to distinguish between (a) normal variation, *i.e.* the narrow range of differences in size or wall thickness or appearance of the plastid, and (b) extensive phenotypic plasticity. Can the *normal* range be precisely defined? We frequently deal with it, *e.g.* young *Chlamydomonas reinhardtii* cells are small, ellipsoidal and flagellated, while older cells are much larger, spherical and sedentary. We might avoid presenting an answer, discussing these differences in another way, as individual components of the life history. But could this not also be an example of limited phenotypic plasticity?

In contrast, some organisms exhibit a phenomenal level of plasticity, a topic discussed later with *Scenedesmus*. We will learn that many in the field have not yet become comfortable with this topic. Inasmuch as the range of variability is different from organism to organism, it can be difficult to initially formulate an hypothesis; this has been the source of much confusion.

In order to distinguish between normal variation and plasticity we should first determine if the variability which we are observing falls under the umbrella of the definition of the species. With our present understanding of individual species it is assumed that most organisms exhibit just slight variability—they have "normal" variation. In this paper we do two things. First, point out that often we do not have the hard data to support this assumption and then show examples of extensive phenotypic plasticity. Such examples include more complete data sets which are available for select organisms, data sets based on intense investigation for more than a decade. In this review we will concentrate on just two. We will demonstrate that there are quite useful stable characters, often unknown, or even overlooked in the past. We also emphasize that we must now give serious consideration to the taxonomic value of expressions of phenotypic plasticity.

## Part I. *NEOCHLORIS*

The evolution of our concepts for the classification of members of the genus *Neochloris*, and organisms formerly called *Neochloris*, is an example of the value of a multidisciplined approach, where we use different experimental data in our attempts to unravel phylogenetic affinities. The genus has been known for over 40 years (Starr, 1955) and more recent information, gathered by means of novel techniques, provides a new opportunity for discussing relationships among these organisms. The original description has been modified as a result of these newly discovered facts. These data point to the need for a new, modern paradigm, for only with the benefit of these multidisciplinary studies will we arrive at a more natural classification system.

The parity between laboratory and field data has been emphasized by several investigators; however, acceptance of such reasoning has been curtailed because of a long held conviction that organisms respond differently under laboratory conditions than they do in the field. This interpretation was somehow justified by the assertion that all culture studies were carried out in "unnatural conditions, *e.g.* culture media possessed far too high nutrient concentrations. However, with the development and refinement of better laboratory protocols, *e.g.* use of dilute media or examination of frequently transferred cultures, researchers were able to simulate natural conditions more easily. Nevertheless, some continued to postulate that organisms in the laboratory respond differently than they do in the field; supporting data for such an assertion are very limited or in most cases wanting.

In attempting to counter such arguments, we might first understand that the genetic constitution of an organism enables it to produce one (or more) phenotypic patterns. Can we then assume that such an organism expresses specific patterns **only** in the laboratory? Are there data to support this assumption? When we utilize the "standard conditions" combinations of physical and chemical parameters in the laboratory, is it reasonable to speculate that organisms will **never** see such combinations in nature? There would be no driving force for such an evolutionary pattern. As Cox (1993) pointed out, it is more probable that we have thus far failed in recognizing those changes in the field.

In contrast to the literature containing protologues of numerous other unicellular algae, most of the data used for the classification of *Neochloris* comes from laboratory investigations, where one can use culture data for determining cytological features (Fig. 1), cell cycles and reproductive mechanisms, RNA and DNA sequencing, etc. Little field data have been accumulated, especially when the original organisms used for the description of the genus or a species, such as *Neochloris aquatica*, were isolated and characterized.

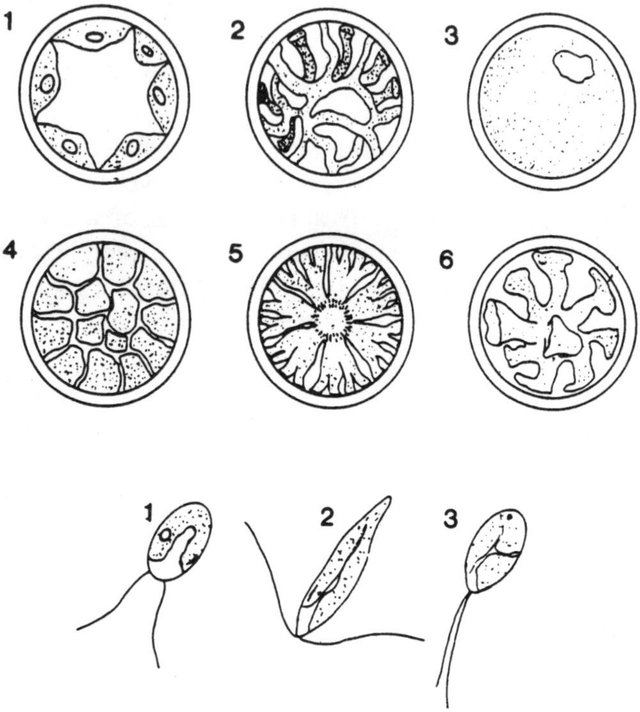


Fig. 1. Types of both vegetative cell chloroplasts and zoospores of selected chlorococcalean algae. **Plastids**: 1. Optical section of complex type composed of several plates, each with a pyrenoid. 2. Net-like plastid. 3. Surface view of a cup-like plastid. 4. Surface view of complex type, with individual units lacking pyrenoids. 5. Asteroid plastid. 6. Axile plastid. **Zoospores**: 1. *Chlamydomonas* type, with a wall and flagella of equal length. 2. *Protosiphon*-type, without a wall but with flagella of equal length. 3. *Bracteacoccus*-type, with a wall and flagella of unequal length.

## A CHRONOLOGICAL SUMMARY

Prior to 1955 there was confusion about the identity of a number of spherical-shaped and zoospore-producing chlorococcalean organisms, especially those found in the soil environment. In most enumerations of such floras at any locality, the chlorococcalean genus *Chlorococcum* was almost always included, in spite of the lack of even the most basic information about the genus itself. This problem lead Starr (1955) to study several genera in the Chlorococcales, with the objective of providing better boundaries among them.

His observations were mainly focused on a few features: the type of zoospores, as well as in the types of chromatophores (chloroplasts) and pyrenoids in both vegetative and reproductive cells. He found that presence or absence of pyrenoids, as well as the cytological details of the chromatophore were stable characters. There were several distinct plastid types, including those which were spherical, axile, asteroid, net-like, or one which was a system composed of separate plates (Fig. 1). When examining the individual zoospores, he considered the presence or absence of a cell wall, as well as the length of the flagella. Thus there were three zoospore types: 1. The *Chlamydomonas* type, with a wall and flagella of equal length. 2. The *Protosiphon* type, which lacked a wall (naked cell) but also had flagella of equal length. 3. The *Bracteacoccus* type, possessing naked zoospores, and flagella, which were not of equal length (Fig. 1).

Based on these features he was able to differentiate ten genera in the Chlorococcales, including two which were new to science. One of these was *Neochloris*, at that time with only one species, *N. aquatica*. The genus included organisms with *Protosiphon* type zoospores and with vegetative cells having a hollow and spherical chromatophore containing a pyrenoid.

After this seminal work by Starr several new species were quickly added to *Neochloris* so that thirteen other species were soon available (Table 1). Watanabe & Floyd (1989) studied nine species, dividing them into three groups based on the old attributes, pyrenoid characteristics and presence or absence of the cell wall on the zoospores, but focused on ultrastructure of the flagellar apparatus. They discovered that flagella basal bodies could be arranged in three different ways: clockwise, counterclockwise and directly opposite to each other (Fig. 2). With the addition of this new stable character, the organisms did not group in the same way and thus some necessary alterations in the classification were immediately apparent. Two groups could remain in the Class Chlorophyceae, one of them as *Neochloris*. One could then add another character to the definition of that genus, organisms with directly opposite basal body orientation. The second member of the Chlorophyceae was a member of a newly created genus, *Chlorococcopsis*. An essential character for this genus was the clockwise orientation of basal bodies. The third group of species, those with a counterclockwise basal body orientation, were classified in another new genus, *Parietochloris*. This genus would be placed in the class Pleurastrophyceae.

**Table 1. List of original *Neochloris* species, along with taxonomic status over time. *N. fusispora* is not included for Watanabe (1983) synonymized it with *Pseudotetracystis terrestris*.**

| Species | Comments |
|---|---|
| *Neochloris alveolaris* (Bold, 1958) | Synonymized with *N. bilobata* (Vinatzer, 1975) by Watanabe (1983). Later transferred to *Parietochloris* as *P. alveolaris* (Watanabe & Floyd, 1989). It is the type species of *Parietochloris*. |
| *N. aquatica* (Starr, 1955) | Type species of *Neochloris*. |
| *N. cohaerens* (Groover & Bold, 1969) | Transferred to *Parietochloris* by Watanabe & Floyd (1989). |
| *N. conjuncta* (Archibald, 1973) | |
| *N. gelatinosa* (Herndon, 1958) | |
| *N. minuta* (Arce & Bold, 1958) | Transferred to *Chlorococcopsis* (Watanabe & Floyd 1989). Later this name was abandoned and *Ettlia* (*E. minuta*) replaced it. Currently known as *Nautococcus minuta* (Kouwets, 1995). |
| *N. oleobundans* (Chantanachat & Bold, 1962) | Placed in Ettlia as *E. oleoabundans* (Komárek, 1989), but returned to *Neochloris* by Kouwets (1995). |
| *N. pseudoalveolaris* (Deason & Bold, 1960) | Transferred to *Parietochloris* by Watanabe & Floyd (1989). |
| *N. pseudostigmatica* (Bischoff & Bold, 1963) | |
| *N. pyrenoidosa* (Arce & Bold, 1958) | |
| *N. terrestris* (Herndon, 1958) | |
| *N. texensis* (Archibald, 1973) | Transferred to *Ettlia* (Komárek, 1989), but returned to *Neochloris* (Deason *et al.*, 1991). |
| *N. vigenis* (Archibald, 1973) | |
| *N. wimmeri* (Archibald & Bold, 1970) | Transferred from *Chlorococcum* (Starr, 1955). Then placed in *Chlorococcopsis* (Watanabe & Floyd, 1989). Komárek (1989) changed it to *Ettlia carotinosa*, the type species of the genus. |

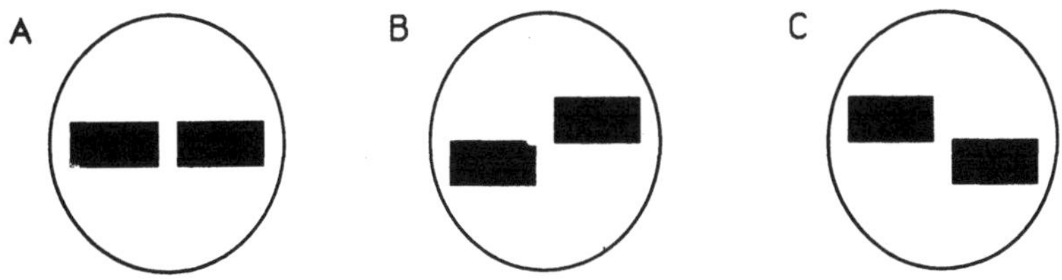


Fig. 2. Diagrammatic representation of basal body orientation in the flagellar apparatus at the apex of flagellated cells. A. Directly opposite. B. Counterclockwise. C. Clockwise.

One reason for some taxonomic confusion was that both Watanabe & Floyd (1989) and Komárek (1989) were working, and publishing, simultaneously on selected coccal green algae in the family Neochloridaceae. Komárek was considering the importance of number of nuclei present, *i.e.* uninucleate or multinucleate vegetative cells. He then proposed a new genus, *Ettlia* to include the uninucleate organisms previously placed in *Neochloris*, leaving in the latter only species with multinucleate cells. In 1991, Deason *et al.* reconciled both positions and synonymized *Ettlia* with *Chlorococcopsis*, but unfortunately misplaced the former in the family Neochloridaceae, a fact later pointed out by Kouwets (1995).

To some readers it may be surprising that very little experimentation concerning phenotypic changes has been published in the "*Neochloris*" complex, even though considerable attention has been devoted to this topic with another green alga, *Scenedesmus* (see later). First, we have to consider that we might not always be open to significant change, and second we should be aware that emphasis on other activities could prove to be temporarily distracting. One should always question whether particular characters, *e.g.* mitotic spindle apparatus and cytology of the apical region are conservative and thus less likely to be affected by environmental factors, or merely assumed to be "stable". One could design experiments to resolve such a concern. The original basic assumptions which had been made with the *Neochloris* complex are now supported by molecular data (Lewis *et al.*, 1992; Wilcox *et al.*, 1992). As a result, the inferred hypotheses based on ultrastructural studies are again supported and emphasis on the elucidation of ultrastructural detail appears to be sound.

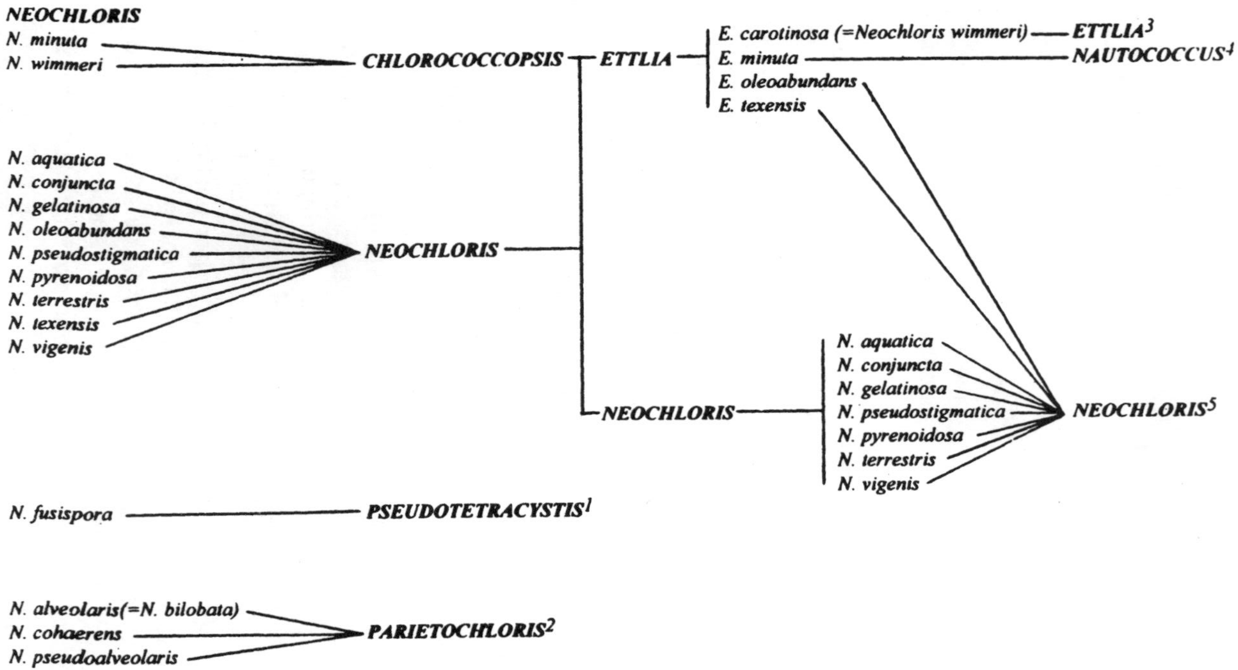


Fig. 3. Evolution of the classification of the "*Neochloris*" complex. The species that once belonged to the genus, are currently placed in five different genera, within three different classes of the Chlorophyta. [1]Borodinellaceae, Chlorococcales, Chlamydophyceae; [2]Pleurastraceae, Pleurastrales, [3]Pleurastrophyceae; and [4]Chlorococcaceae, Chlorococcales, Chlamydophyceae; and [5]Neochloridaceae, Sphaeropleales, Chlorophyceae.

Further ultrastructural data were reported in 1995 by Kouwets, who examined mitotic events along with cytokinesis, as he attempted to understand the cell cycle for some *Neochloris* species. He found that the reproductive and cytological features of *Ettlia minuta* resembled those found in several species of *Nautococcus*, another chlorococcalean genus, and thus the organism should be known as *N. minuta*. Kouwets also proposed that *Ettlia oleoabundans* and *E. texensis* remain in *Neochloris* for the time being, until the details of cytokinesis are understood.

As seen in Fig. 3, the classification of all former species of *Neochloris* is not yet resolved, even though the available information has increased greatly since the original protologue by Starr (1955). Although most authors seem to agree on the best taxonomic characters, as new tools are used and new approaches are taken, the affinities both within the genus and with other taxa remain controversial. Ettl & Komárek (1982), Deason *et al.* (1991), and Kouwets (1995) have proposed placement of species in different orders and even in different classes!

In our opinion, this apparent disarray should not be discouraging; rather the flow of new information supports the need for a new paradigm, and thus new paths are suggested. When we begin to examine a broader assemblage of microorganisms, we learn of proposed changes at a higher taxonomic level, the class, some of which have been presented in the last decade. Several of these recently proposed algal classes are no doubt not universally accepted within the phycological community. Often there is a reluctance to use them, perhaps because specimens within the class may not be frequently encountered. Thus we might ask at times whom do I follow as I attempt to identify an organism? What name should I apply to my organism?

No doubt some of the controversy and confusion is a result of the utilization of a multiplicity of characteristics that, ideally, most taxonomic studies should employ, for there are advantages of a multidisciplinary approach (Fig. 4). But recall that we are also dealing with "algae", organisms which are not monophyletic. The achievable goal is to produce a new scheme reflecting natural relationships among the organisms. Accordingly, when the approach indeed becomes multidisciplinary, new taxonomic arrangements are suggested, often with some frequency. These developments can be confusing to many, and at times disconcerting to

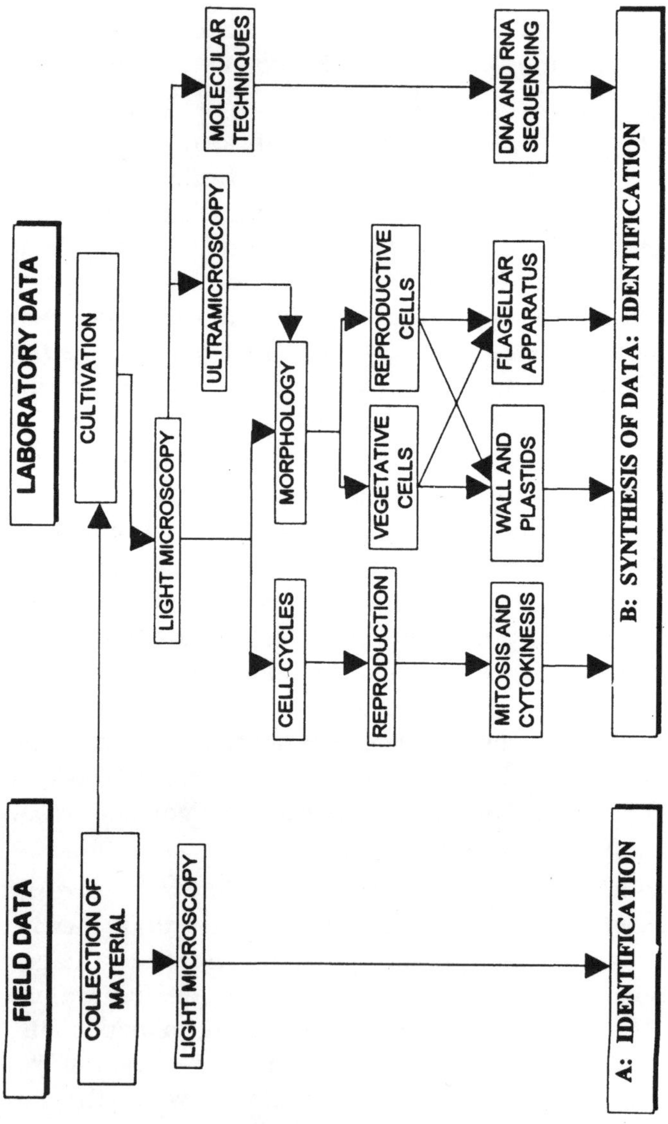


Fig. 4. Two approaches to the classification of *Neochloris*: A. Older, traditional and limited approach. B. Proposed paths to be taken when following a new paradigm, which combines various efforts in a multidisciplinary endeavor.

most phycologists, especially when identifying a particular organism. We simply should ask, are we using a system which is updated, or even accurate?

## Part II. SCENEDESMUS

Earlier in this century Chodat (1926) had initiated the experimentation designed to understand *Scenedesmus* development, while in the last decade we have expanded this data base considerably, extending our grasp of the way in which phenotypic plasticity is expressed. At the same time, every few months we now learn of a new account of the flora in a given region, *e.g.* the study of the occurrence of several green algal taxa in Argentine waters (Echenique & Arena 1994). The latter reported *Scenedesmus bourrellyi* as a new record for Argentina and incorporated diagrams of 13 *Scenedesmus* taxa, including 10 species. Also presented were the unicellular genera *Franceia* and *Lagerheimia*, but the latter might be components of the life history of one or more species of *Scenedesmus*. Unfortunately, in the absence of considerably more data or available cultures we will not be able to determine which of the above are discrete taxa and which are morphs of just a few species [see later].

Can we reconcile such published information with a recent Hindák (1990) statement regarding the genus *Scenedesmus*? "Diagnostic features are demonstrated in well growing cultures, but their quantitative manifestations may change depending on culture conditions." **That is, new morphs of one taxon may appear as the conditions of culture are altered.** He continued, "the ability of strains to form, *e.g.* additional spines, teeth, ribs, granules, bicaudated or spineless coenobia, mucilage, etc. manifests itself mostly in cultures." **Are there data to support this belief, or is Hindák merely expressing a bias for an assumed stability of form in nature?** Hindák continued "In spite of this we assume that the variation of morphological features found in cultures represents, in comparison with natural conditions where there is an incomparably greater amount of combinations of physical, chemical and biological factions, only a definite portion of the overall morphological plasticity of an individual species."

**Should we not then conclude that culture studies do provide valuable information for ecologists and systematists? And also that presence or absence of certain specimens at one site might be a useful environmental indicator?**

Trainor (1991) recently commented on the compilation, and subsequent publication, of all *Scenedesmus* protologues in a volume along with additional taxonomic data (Hegewald & Silva, 1988). With this annotated catalogue the stage is set for the generation of a new comprehensive monograph. Although initially there might seem to be little connection between history, systematics and developmental studies, the plasticity data will provide the conceptual framework needed for this effort. With an understanding of this information concerning morphological variability, relationships among species as well as evolution in the genus are more clearly understood.

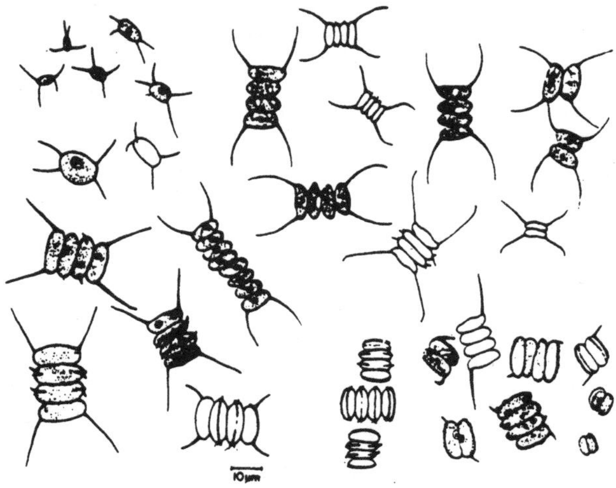

Fig. 5. The numerous ecomorphs which are produced in a single genotype [strain] of *S. armatus*. In culture over time one sees a sequence of development (Table 2).

As a result of experimentation in several laboratories we know that many species in the genus, if not all, may or may not join cells side by side, cementing cells at specific sites (Pickett-Heaps, 1975; Pickett-Heaps & Staehelin, 1975). Consequently taxa have the capacity to produce either single cells or coenobia. In addition, cells may or may not activate specific sites for spine production (Trainor, 1993); an individual genotype may form several coenobial types, each with a different spine pattern. A very useful organism for such experimentation is the common species, *S. armatus*. In batch cultures of several strains of that organism, it became clear that spine development was suppressed as a culture aged (Trainor, 1966). As a result, organisms with long spine thrived in young cultures; older material had morphs with reduced spination. In time, after numerous publications dealing with control mechanisms for individual ecomorphs, these data were presented as an **ordered sequence of ecomorph development** for *S. armatus* (Egan & Trainor, 1989). With this information in place, an expanded description of that species would appear to be needed. The major ecomorph types which can be produced by that species are diagrammed in Fig. 5.

In spite of these findings, close examination of these data still presents a confusing picture. Recently, Hegewald *et al.*, (1990) commented on this variability within *S. armatus* var. *spinosus* stating, "The taxon is characterized by the variability of spination. Besides normalspined coenobia, also very short-spines coenobia always occur. These are often bicaudate. The percentage of long—and short-spined taxa varies between strains and depends on culture conditions." Thus they too were able to detect considerable plasticity, even within what they determined to be one variety of the species.

Once sufficient experimental data had been accumulated, using *S. armatus* and several other spine-bearing species (Trainor & Egan, 1990; Trainor, 1992b, 1993, 1995), two significant conclusions could be reached: (a) The ecomorphs for each taxon developed in an orderly sequence (Table 2), and (b) The overall pattern of ecomorph development, *i.e.* unicells, multispined specimens, quadricaudates, bicaudates and acaudates (Fig. 6), was similar from taxon to taxon.

Consider some of the ramifications of this knowledge. In any limnological experimentation, the field biologist faces enormous problems, not only when our knowledge of this plasticity is limited, but also if one attempts to apply concepts of plasticity. Consider an attempt to identify the organisms present in a drop of water collected in June, when one could encounter a population of unicells, each with a number of elongate spines precisely placed at the cell poles. We might "confidently" place it in the genus *Lagerheimia*, and even call it *L. quadriseta*. Then, if one returned to the same site in July, the unicells might no longer be dominant, or even present, and we might assume that they were either grazed or died. Thus, in the July collection one could have a population of four-celled colonies with quadricaudate spination. Then, in August the population could have many colonies with **just two** precisely placed elongate spines, each diagonally positioned on the terminal cells. An experienced field investigator might feel con-

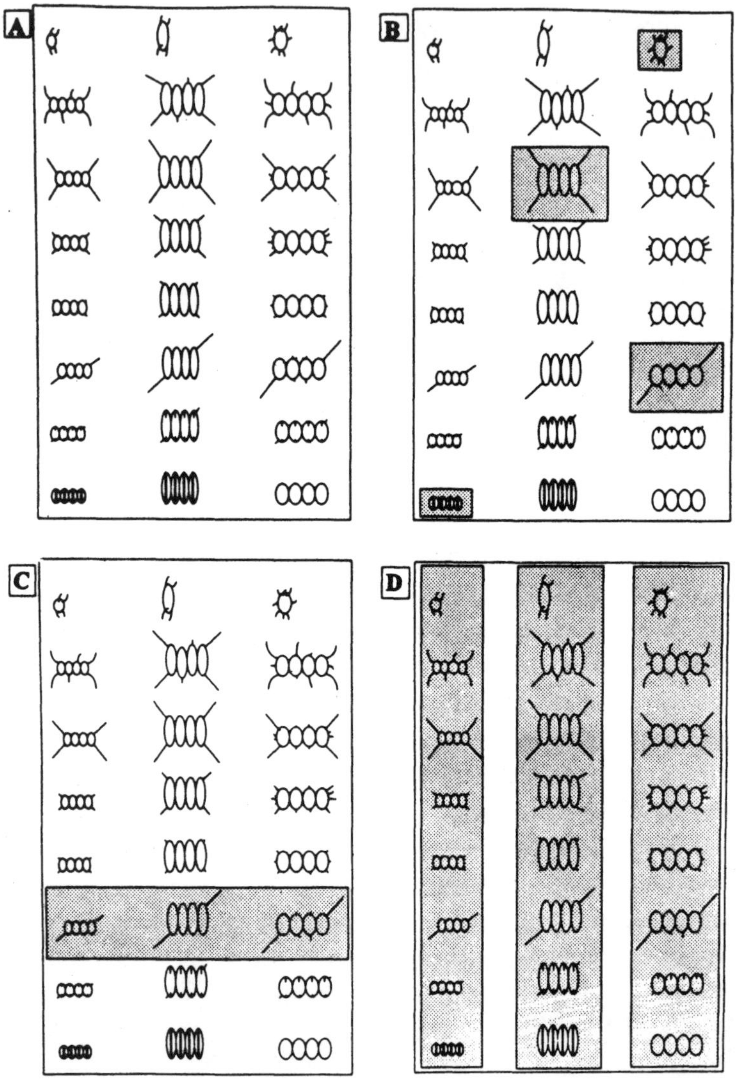

Fig. 6. Theoretical display of ecomorphs in three "species" of *Scenedesmus*, showing unicells and coenobia. Formerly the coenobia of various species were distinguished by appearance of individuals. In A, from top to bottom, the individuals have been arranged as a developmental sequence. **A.** In an older classification the three unicell types could have been considered species in one genus, while the coenobia were in *Scenedesmus*. **B.** Without recognition of plasticity, each individual (shaded box) would have been called a taxon. Because of differences in size, shape, number of spines and ornamentation, some might recognize close to 21 taxa! **C.** In some systems of classification, similar morphs (shaded box) would have been called varieties within just one species, such as *S. aymmetricus*. **D.** The similar pattern of ecomorphs development, in an orderly sequence in three plastic species of *Scenedesmus* (shaded boxes). These would be the species following the new paradigm.

fident calling this organism *S. asymmetricus*. But unless one witnesses stages in reproduction, *i.e.* release of juvenile colonies from the parent, how can one be aware that these types of green algae were **not** three taxa, but rather stages in a sequence of development of **one** taxon (Fig. 6)?

The critical question here has been: can one be certain of the number of genotypes present in that pond, genotypes which represent both the unicell populations and the two colony types? If one of the organisms had been a *Synura* and the other a *Navicula*, there would be no reason to doubt your species determinations. In addition, there would be no reason to believe that there was a *Navicula-Synura* connection.

But in an older, narrower approach to classification, especially one based on observations of preserved material, unicells and coenobia were obviously different genera. Even with *Scenedesmus* coenobia, number of spines on individual cells was formerly considered a stable character. We assumed that in each specific genotype there was stability, *e.g.* typical quadricaudate colonies had two spines on both terminal cells and two interior cells without any spines. But now it is clear that, in *Scenedesmus* and perhaps related genera, the production of each of these ecomorphs is under environmental control. After an examination of synchro-nized cultures of some spiny taxa, Steenbergen (1975) concluded that both day length and nutrition were involved in ecomorphs formation. We would support such a statement and additionally report that temperature influences both cell size, sometimes markedly (Komárek & Růžička, 1969), and display of spines (Trainor, 1992a). Can we be certain that other environmental factors are not involved in expressions of *Scenedesmus* plasticity?

**Phenotypic plasticity**

The *Lagerheimia-Scenedesmus* relationship is simply an example of a single genotype capable of expressing different morphologies, because the genotype has the capacity to take cues from a changing environment. The microalgae which are capable of this type of development are said to express **phenotypic plasticity**. In the past, phycologists have been unaware of many expressions of phenotypic plasticity because of infrequent sampling, *e.g.* a failure to sample from one locality week after week, or an inability to make connections. Thus a possible sequence of development for a specific genotype was not known. But over time evidence was accumulated (Table 2).

Now we know that, in the future, expressions of phenotypic plasticity must be melded into the existing rather rigid classifiation. As this process is initiated we become aware that the older classification has served as a barrier, with the result that new insights often were slighted. Even when some of the many weaknesses in the old approach were exposed, there was still a reluctance to change. On the other hand, we can be much more optimistic about the future.

**Table 2. The percentages of S. *armatus* coenobia which develop over a period of eight weeks when grown in Bristol's medium at low light intensity [50 μmol $m^{-2}s^{-1}$] are given. Experiment initiated with a 100% unicell population. Similar results were obtained at higher light intensities when there was more rapid growth. However, when the light intensity was increased, the population of unicells matured much more rapidly.**

| | PERCENTAGE ON DAY | | | | | | |
|---|---|---|---|---|---|---|---|
| **ECOMORPH** | 0 | 10 | 14 | 21 | 26 | 39 | 56 |
| UNICELLS | 100 | 44 | 15 | | | | |
| LARGE UNICELLS | 0 | 22 | 30 | 21 | 16 | 1 | |
| TWO CELLED COENOBIA | | 5 | 17 | 31 | 28 | 24 | |
| LONG SPINED COENOBIA | | 29 | 38 | 45 | 41 | 29 | |
| SHORT SPINED COENOBIA | | | | 3 | 15 | 46 | 60 |
| ACAUDATE COENOBIA | | | | | | | 40 |

After Trainor (1993).

## The ordered sequence of ecomorph development

We have introduced briefly the supporting data for extensive phenotypic plasticity in the genus. Our first encounter with the sequence, many years ago, was with culture UTEX 1236 [available at the University of Texas Culture Collection (Starr & Zeikus, 1993)], a prolific unicell producer which did not form many colonies (Trainor & Hilton, 1963). Later when we examined this problem in detail we determined the mechanism for controlling colony formation, detected additional ecomorphs of UTEX 1236 and first saw evidence of a sequence of development.

Over time cultures of the well known species, *S. armatus*, have yielded much information regarding the developmental sequence. In batch cultures, when a low *S. armatus* cell density is used as the inoculum, unicells appear in early log growth and remain as long as the cell density is minimal and the nutrient supply is adequate. The maximum spine number on *S. armatus* unicells, as a matter of fact on any *S. armatus* cell, whether a unicell or component of a colony, is four. As the culture of unicells ages, many colonial ecomorphs begin to appear. This display of colonies is quite striking, for the ecomorphs which are produced clearly change over time. Colonies with very long spines appear earlier in development than do colonies with much shorter, or fewer, spines or colonies without spines. And as we now have come to expect, colonies without spines formed when nutrients were depleted. If one manipulates the nutrient environment, such as that of a dilute medium, spineless colonies emerge more quickly (Trainor & Shubert, 1974). In contrast to the production of a reduced number of spines with aging, long ridges appear as spine development is curtailed. The ridge is the ornamentation running lengthwise on the cell surface (Fig. 7). As we picture this display of ecomorphs, it becomes apparent that one rarely observes true unicells once the population of coenobia consists of organisms bearing ridges.

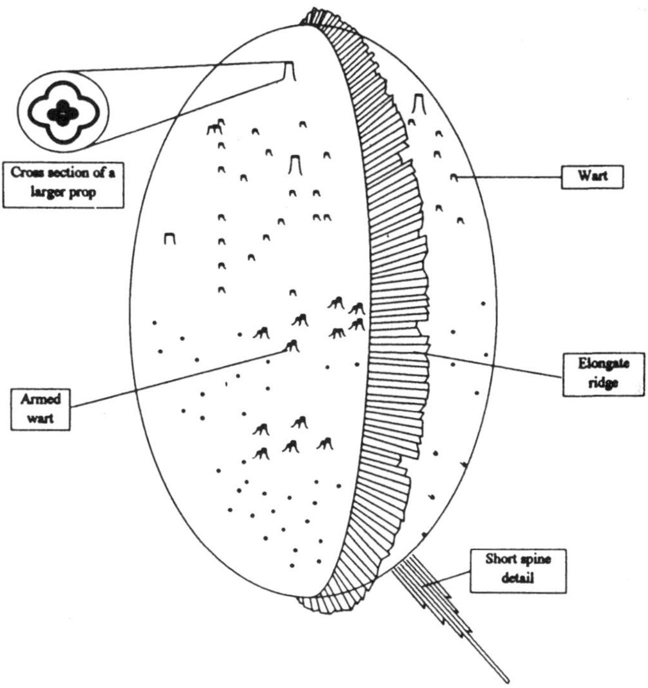


Fig. 7. Selected ultrastructural features, diagrammatic presentation, of the cell wall of a theoretical Scenedesmus species. The ridge, not present in all species, is composed of a linear series of props. These may run the length of the cell, or only shortened portions may be present at each pole. Props may appear individually, but when they are shorter, as well as more numerous, they may be termed warts. Armed warts are aggregations of two, three or more short props, which can give the appearance of a buttressed wart. Individual spines are composed of a series of elongate "props". Tapering of a spine is achieved when the number of props diminishes from base to apex. Spines may be quite short or upto 2-2.5 times the cell length.

Occasionally in batch culture the organisms will stop growing, because of a decreasing nutrient supply, before the spineless ecomorph develops. If a transfer is made, perhaps just one volume of medium to one volume of cells, growth is resumed, but that population will not revert to the long-spined stage. Rather a morph with very short spines, or a spineless morph develops; this is a cell density/nutrient availability response. Additionally in batch cultures, it is possible to reverse the developmental sequence at almost any stage. Note however that the reversal might sometimes itself be sequential, e.g. spineless colonies develop initially into a spiny colony before unicell production can be expected (Egan & Trainor, 1989).

**Cytological features. Light microscope.**

Some of the best examples for recognizing that either spine display or spine length can be altered are observed in the unicell/colony transformation. Although *S. armatus* unicells display two spines at each pole, in a 4-celled colony we do not see 16 spines!! Thus colony formation suppresses spine formation; many sites simply are not activated. Environmental factors apparently influence steps in spine production, such as utilization of an individual spine site (Trainor, 1993). Inconsistent data are sometimes recorded since spines do not always unfold away from the cell of the surface of the coenobium. Therefore it might be difficult to detect spine presence.

Plasticity in the expression of the ridge can be observed in any individual clone of *S. armatus* var. *armatus*. One can detect individual coenobia without ridges, with ridges not at all robustly expressed, as well as cells with incomplete or complete ridges. Hegewald (1989) pointed out that the ridge is the most variable feature, and certainly EM surveys would not be in contradiction.

Surrounding the colonies of this and other species, varying amounts of mucilage have been reported. Using the axenic *S. armatus* strains UTEX 1588, 1589 and 2193, Hegewald (1989) reported that they "could produce small amounts of mucilage (maximal 2 μm thickness), although this was usually lacking." Mucilage is seldom reported from field collections or from xenic cultures, no doubt because it is substrate for non-photosynthetic microorganisms. Even in axenic cultures production seems to be suppressed in certain environments.

Also, since evidence points to the fact that any transformation of unicells to colonies, or in the spring from colonies to unicells, is under nutritional control (Steenbergen, 1975), the sequence will not occur with the same intensity as conditions change from day to day, [*e.g.* cell density], from culture to culture, laboratory to laboratory or from one field site to another. The sequence itself persists, but precision in the actual display of ecomorphs may be lacking. A complete reversal of the sequence may be brought about by the addition of new nutrients; soon the unicellular population will be formed once again. This apparent absence of uniformity has made it difficult to formulate ecological theory.

In the laboratory, when an organisn is maintained in log growth, and the cell density is low, a high percentage of unicells will persist (Table 2). When colonies form after log growth they are long spined. These two growth forms, unicells and quadricaudate coenobia, develop only when growth conditions are ideal, e.g. when there are abundant nutrients. Colonies with shorter spines will be found in older cultures, in stationary growth.

Bristles, easily detected on desiccated preparations of laboratory and field samples, are present on many organisms, especially when examined in the laboratory. Bristles do form in nature, but often are not displayed as well as in axenic cultures. Grazing, competition and the more rigorous field environment are no doubt involved in the survival of bristles. Bristles are readily formed on log phase and early stationary phase cultures, but our knowledge of the pattern of distribution is still too sketchy to enable us to include this information in a developmental sequence.

**Cytological features Ultrastructure**

As we have learned with several other microalgal genera, at the ultrastructural level the *Scenedesmus* cell wall is composed of several layers, including structures such as elaborate spines fashioned of individual soda-straw-like units, rosettes, props and ridges or ribs. Although spines and ridges are sufficiently robust that they can easily be detected with the light microscope, there is cytological detail uncovered when the electron microscope is used. Using ultrastructural observations of the developmental stages of ridges, perhaps we will soon understand the sequence of formation in a population over time. There appears to be a marvelous diversity (Comas & Komárek, 1984; Hegewald, 1989; Komárek & Fott, 1983) in both the cytological detail and the arrangement of these individual units.

On a small number of spiny species a hexagonal network may be present, but most species are limited to a few or numerous projections, called tubercles or warts. Excellent detail of these features can be seen in Fig. 3.50 of Pickett-Heaps (1975). The majority of studies from numerous laboratories have not shown the detail which Pickett-Heaps presented. In some cases warts are simple, but they often are more complicated, with three or four arms or buttresses (Fig. 7). Such buttressed warts have been termed armed warts (Hegewald *et al.*, 1990). At times structures with as many as 7 arms, probably a melding of adjacent simpler warts, have also been detected. On the wall of most species a limited number of openings or rosettes, the site of bristle development, are also present.

Initially, after examining the sizable ultrastructural data base available, one is overwhelmed. But as we synthesize these data, often with different ornamentation patterns on individual coenobia of a species, we realize that we must sometimes incorporate this information in our systematic ventures. On the other hand, over a decade ago Comas & Komárek (1984) cautioned that there was no reason to erect new species simply because of the quantity of ultrastructural ornamentation, a valuable suggestion. Unquestionably we will find these data useful, but we have yet to determine how much ultrastructural information is needed either to define the taxon or to make a precise identification.

Some time ago it seemed that the EM data would be the key to understanding the limits among individuals, for with additional characters the task should have been easier. Perhaps some even thought that we would be able to describe numerous additional taxa. However, as John (1995) recently reported "Still largely untested experimentally is any variation exhibited by morphological traits as revealed by electron microscopy." In view of the present chaotic condition of *Scenedesmus* systematics, information concerning ultrastructural plasticity could prove to be very valuable. The detailed information gathered at this level, although still incomplete, will later be found to be essential information in the synthesis of systematic data.

### An hypothesis

**There is an ordered sequence in the development of ultrastructural features**. The uncovering of this sequence is almost a certainty. Initially one might speculate that young, actively growing coenobia would have one type of ornamentation, perhaps a simpler pattern, and that mature coenobia would have a quite different wall ultrastructure. As we learn more about the steps in development, it will be important to correlate the light microscope and ultrastructural cytological sequence data. Even now, if we reexamine published data (Hegewald *et al.*, 1989), as well as additional observation with an organism such as *S. armatus* var. *spinous* [see earlier], we should be able to support this hypothesis, reinforcing what is known for organisms at the light microscope level. Gathering the data will not be difficult, but it will be time consuming, for surely the production of ultrastructural features will also be affected both by environmental conditions and the presence of other organisms. Again we will learn that the developmental stage of the coenobium reveals considerable information.

At present the hypothesis itself is based on the following observations. With every *S. armatus* examined to date it is known that ridge development is more vigorous in older cultures. On the other hand unicells seldom, if ever, bear ridges and long spined coenobial ecomorphs almost without exception lack ridges. Some published photographs show the best development of armed warts on coenobia which are bearing ridges, whereas, the simple warts are more abundant on cells without ridges, or on the walls of unicells. Both the warts and ridges should prove to be fundamental units in the EM sequence.

Also, since the light microscope features can easily be associated with growth data, *e.g.* whether one is examining log phase or stationary phase material, one would expect a similar sequence of development at the ultrastructural level. After all smaller units, invisible at the light microscope level, are assembled to fabricate a ridge or a spine. Considerable effort must be employed to amass all the information needed to support this hypothesis, as well as to expand it to include several species. A thorough survey of published material will not be sufficient, but such information can point us in the proper direction for additional experimentation. While assembling the essential data, one might consider four points:

**First**, there is a need for examining material over time, either in an actively growing culture, or when collected from a natural habitat. One would then be able to observe a variety of ecomorphs, each produced in sequence. **Second**, the typical random sampling of specimens, an approach which thus far has provided both new information and excellent background data, must now be avoided. At this time we should be aware of the conditions of culture, or field conditions, as well as the age of the material under observation. This change in sampling procedures is needed to reconfirm a developmental sequence. **Third**, since we recognize that all individual light microscope ecomorphs, or even the majority, do not form precisely at one time, there will be a similar overlap of ultrastructural features. After an examination of the data in Table 2 one can see the overlap among the ecomorphs, such as long spined morphs which first appear on day 10, peak at 3 weeks and are last seen after 6 weeks. Each morphs brackets a specific time span, a fact which initially made it difficult to unravel the sequence at the light microscope level. **Fourth**, one must recognize that much of these data will be obtained from observations of discarded walls, the best material for transmission electron microscopy. It would be essential to know that on day 42 a wall selected for observation was recently discarded and not one present in the culture during the early stages of growth. By careful preparation and manipulation of the inoculum, we can have that assurance.

**Local aquatic floras: phenotypic plasticity**

In spite of numerous recent developments, it is still extremely difficult to determine the cytological limits of most *Scenedesmus* taxa, including *S. armatus*. Even after more than three decades of experience in this field we do not feel comfortable identifying the majority of specimens we encounter. Recent research has sometimes served to further intensify this problem, rather than enabling us to clarify issues. On the other hand, there seems to be some consensus on at least one species, *S. armatus*, but not at the level of the *S. armatus* variety. At this time additional dialogue and data would be quite helpful in this difficult task.

Inasmuch as we now know that each genotype can produce not only its own kind, but also a variety of morphs (Fig. 5), this information should be available to those interested in the identification of field specimens. Such knowledge now has become an essential part of our concept of the species of *Scenedesmus*. Ideally, this information would have been included in the species description, but over time we have accumulated an enormous data bank concerning the display of useful features, other than those recognized when the protologue was written. Even at the light microscope level we have compiled valuable information on the precise positioning of the spines, and the display of the elongate bristles.

**Amending species descriptions**

Perhaps the description of *S. armatus* would be the place to begin, for we know a great deal about its plasticity. Although there is often a lack of consensus for the proper name for many species, that is not true for *S. armatus*. However, the present protologue for *S. armatus* does not give any evidence that the genotype is capable of producing several ecomorphs, and there is not a hint of a developmental sequence. Realistically, we can not suggest that individuals who want the appropriate name for this 4-celled green alga, often the dominant organism in the plankton, examine the voluminous literature. Essential summary data must be assembled in one place, and an amended protologue would appear to be ideal. In the process of amending descriptions, determining which species names have priority, and/or are the most appropriate, will be difficult. But when this is accomplished, we will have many excellent descriptive terms available to us (Trainor & Egan, 1991).

## SYNTHESIS

Earlier we pointed out that over time a series of discoveries may eventually contribute to the need for a more diversified approach to experimentation, to a completely different consideration and interpretation of the facts before us. Kuhn (1968) has termed these unprecedented shifts paradigms. With this in mind we have proposed that experimental microalgal research in the genera *Neochloris* and *Scenedesmus*, especially that conducted for taxonomic purposes, be significantly altered, even though the new methods themselves will not necessarily provide all of the answers. The new data gathered will enable us to understand problems more completely, and then concentrate on new focal points in our research. At some time in the future another paradigm shift will again be clearly articulated.

In this case, the path to take appears, at least at present, to be rather clearly outlined. Inasmuch as both the *Neochloris* complex, and individual species of *Scenedesmus* can now be visualized more clearly, we know that we can understand the complexity of these microalgal species only if we follow a multidisciplinary approach (Fig. 4). The limited light microscope approach of Starr (1955), a giant step forward at that time, has now been exposed. Some former *Neochloris* species are now known to be unrelated, a conclusion arrived at only after examination of additional cytological data, ultrastructural detail and some biochemical information along with confirmation by way of molecular systematics.

Where will the endorsement of a phenomenal level of *Scenedesmus* phenotypic plasticity eventually take us? As yet it is still difficult to speculate concerning the impact of the recognition of plasticity on the classification of these or other microorganisms. With our present system, in many cases stability of form is always assumed and we maintain the *status quo*. In the future we might examine an alternate approach where the *species* is originally defined as being very plastic. As we then scrutinize the similarities among ecomorphs in individual strains (Fig. 6), a thorough reconsideration of all data is suggested and this would lead to considerable synonymy. With *Scenedesmus* we have yet to detemine the number of species, but a conservative might speculate that there were not over 300, but perhaps as few as 30! However, with a need for considerably more solid data in many areas, it might be wise for a time to avoid the temptation to consider finalizing our decisions at this time.

Although a developmental sequence has now been solidly confirmed only for *S. armatus*, *S. communis* and *S. subspicatus*, a conservative premise would be that a similar developmental sequence will be detected for each *Scenedesmus* species. This hypothesis might first seem to be premature, but one might ask, where is there supporting evidence that any taxon within the genus has a stable morphology? For decades we **assumed** stability of form, but there were no supporting data. Without recognition of plasticity, will we not continue to expose many conflicts and inconsistencies?

The historical record, even recent history, has provided the background needed so that we are able to understand precisely how we arrived at our present state, often with an inadequate, perhaps even fragile, taxonomic base. The causes have been many. With reliance on a foundation based on light microscope observations, many new algae were discovered and described. Any attempt at recognition of the enormous diversity was overwhelming; in haste, brief descriptions were the norm. Very soon an astonishing number of species were described, and relationships were rapidly resolved based on their habitat, shape, and even their pigmentation. The artificial schemes became more and more powerful, but as the field developed, researchers began to consider more natural systems of classification. At that time we were overwhelmed with many quite similar entities. Detailed experimentation with individual taxa seemed out of the question, and new species continued to be published, often without the discussion and argumentation central to excellent taxonomic work (Hegewald & Silva, 1988).

In some areas or centers for phycological studies, especially when a physiological approach was limited or quite difficult to execute, each new flora was considered a definitive work. When we consider the impact within the research hierarchy of both authority and respect in some cultures where the data were assembled and synthesized, such conclusions were inevitable. Discussions of alternate approaches were limited, discouraged or even wanting. Specialists might now consider following the example of individuals working with selected members of the marine ultraplankton. Information gathered from the light microscope, EM, analysis of pigments and finally, gene sequences have been used to describe a species of *Nannochloropsis*, which has a cell diameter between 2-4 μm (Karlson *et al.*, 1996). In this cooperative effort, four individuals from two laboratories were involved, perhaps a portent of the new taxonomy.

The practicality of the artificial systems, which apparently is useful for some organisms such as diatoms, is without doubt another reason it is difficult to establish a more natural approach today. We should note that many merely want a name for the organisms which they encounter. Organisms such as diatoms are readily identified from preserved material, but in contrast cultured material is essential for zoospore-producing microalgae. When examining diatom systematics, consider the difficulty we would encounter in identifying these organisms (ashed or acid clean material), if it was not for the classification based on frustule (wall) morphology. Inasmuch as certain ecological studies are enormously time consuming, it would be very difficult at present to convince investigators that both cultured material and ultrastructural investigation were now needed, especially for identification of organisms which they frequently encounter in lakes and rivers. Although this practical approach remains in place, should we remain complacent or rather search for a better, a multidisciplinary approach?

Merging disparate approaches or arriving at a consensus, even within a rather small group of scientists, is a very complicated process. When the proponents of different methodologies have been molded by cultural and idiomatic differences, there are many barriers to free and open exchange. Indeed there may be a hidden agenda. Each culture possesses its own postulates which influence how one understands and transmits factual information. Thus both how we individually see information, and then communicate this to others are products of our education, culture and background; we would also include acquisition of skills and research training.

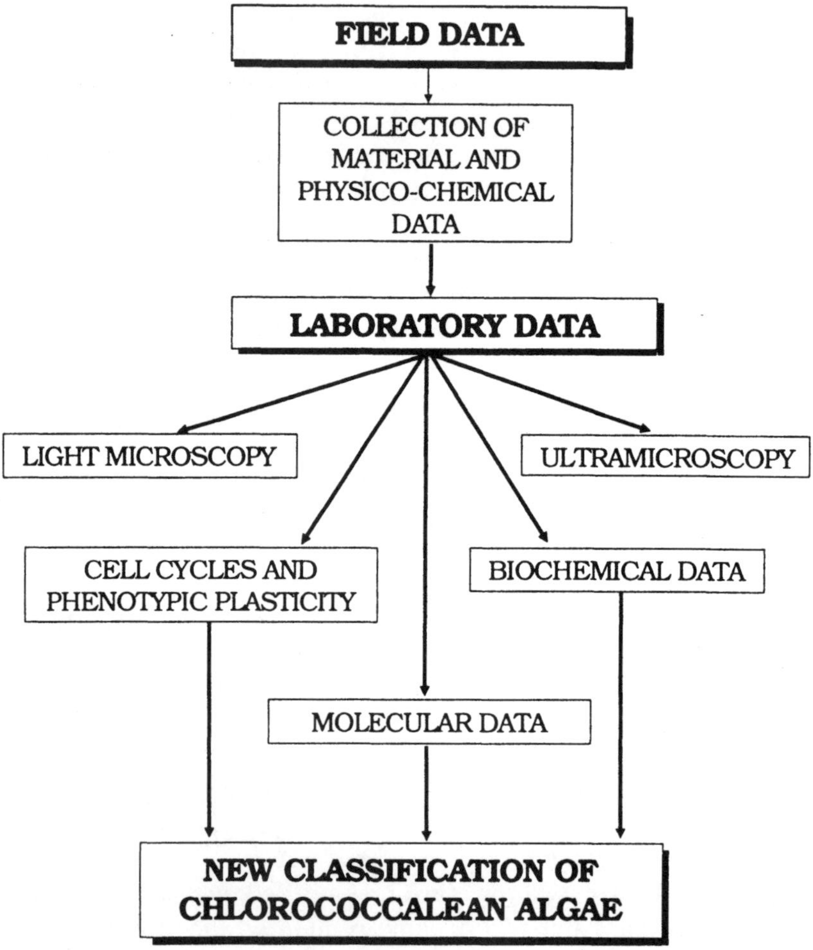


Fig. 8. In some investigations physico-chemical data are collected along with a field sample. Organisms would then be studied extensively in culture., using a variety of techniques and ideally involving more than one laboratory. The new paradigm emphasizes a multidisciplinary approach.

Each publication of a particular *research school* thus reflects the thinking of that individual school, whose basic philosophy is not necessarily shared by others. Is it no wonder then that we sometimes either fail to communicate on the one hand, or that others might not really understand our central message? In countries often called *third world*, *n.b.* by others, **access** to essential literature clearly affects free **choice** and thus taxonomic decisions. In both ecological investigations and phycological studies this results in different approaches to the identification of members of the same group of algae.

The existing approach to green microalgal identification utilized by field practitioners is practical, but limited. The new paradigm (Fig. 8) would point to the need for a detailed laboratory study at least for *Neochloris* and *Scenedesmus*, and this presents a real obstacle for many. But might we not reflect on the words of Chodat (1926), freely translated: Isn't it immoral to lead the young to believe that one can identify microalgae simply with the aid of a microscope? In some cases we might then honestly pronounce that a particular organism, seen just as a mature vegetative cell, is *Neochloris aquatica*-like. We thus communicate at least selected essential information, but we do not make a definitive pronouncement concerning the evolution of the organism in question.

In summary, at present we believe that for identification of members of either *Neochloris* or *Scenedesmus* there are three possible avenues. Let us assume that one wishes to put an name on a 20 μm spherical green alga. First, after consulting the only available general key to the fresh water algae, the investigator might conclude it was *Chlorococcum* sp. Some of us might prefer that it be determined to be *Chlorococcum*-like. Second, aware of recent research, but lacking several critical publications, (s)he might attempt a thorough literature search; a colleague in another country or at another institute might not have the resources allowing the initiation of such a literature search. Even so, the organism might now be determined to be *Neochloris* sp., even *Neochloris*-like. Third, perhaps but a *few* investigators would be able to follow what *most* would agree to be the ideal avenue: Use a multidisciplinary approach before attempting to identify a member of either of these genera to species, and utilize information obtained from a study of material in culture. One might then confidently say *Neochloris aquatica*.

## REFERENCES

Arce, G. and Bold, H. 1958. Some Chlorophyceae from Cuban soils. *Amer. J. Bot.* **45**: 492-503.

Archibald, P. 1973. The genus *Neochloris* STARR (Chlorophyceae, Chlorococcales). *Phycologia*. **12**: 187-193.

Archibald, P. and Bold, H. 1970. Reclassification of three unicellular green algae. *Phytomorphology*. **20**: 383-389.

Bischoff, H. and Bold, H. 1963. Phycological Studies 4. Some soil algae from Enchanted Rock and related algal species. Univ. of Texas Publication No. 6318. pp. 32-36.

Bold, H. 1950. Some problems in the cultivation of algae. In: Brunel, J., Prescott, G. and Tiffany, L. [Ed.] *The Culturing of Algae*. Kettering Foundation. Antioch Press. pp. 11-17.

Bold, H. 1958. Three new chlorophycean algae. *Amer. J. Bot.* **45**: 737-743.

Chantanachat, S. and Bold, H. 1962. Phycological Studies 2. Some algae from arid soils. Univ. of Texas Publ. No. 6212. pp. 1-75.

Chodat, R. 1926. *Scenedesmus*. Étude de génétique, de systématique experimentale et d'hydrobiologie. *Revue Hydrobiol.* **3**: 71-258.

Comas, A. and Komárek, J. 1984. Taxonomy and nomenclature of several species of *Scenedesmus* (Chlorellales). *Arch. Hydrobiol.* **67**: 135-157.

Costas, E., Zardoya, R., Bautista, J., Garrido, A., Rojo, C. and Lopez-Rodas, V. 1995. Morphospecies vs. genospecies in toxic marine dinoflagellates: An analysis of *Gymnodinium catenatum/Gyrodinium impudicum* and *Alexandrium minutum/A. lusitanicum* using antibodies, lectins, and gene sequences. *J. Phycol.* **31**: 801-807.

Cox, E. 1993. Diatom systematics—a review of past and present practice and a personal vision for future development. Nova Hedwigia, *Beiheft.* **106**: 1-20.

Deason, T. and Bold, H. 1960. Phycological Studies 1. Exploratory Studies of Texas soil algae. Univ. of Texas Publ. No. 6022. pp. 28-30.

Deason, T., Silva, P., Watanabe, S. and Floyd, G. 1991. Taxonomic status of the species of the green algal genus *Neochloris. Pt. Syst. Evol.* **177**: 213-219.

Echenique, R. and Arenas, P. 1994. Fitoplancton de la Laguna Vitel, Argentina I. Gayana *Botanica.* **51**: 89-104.

Egan, P. & Trainor, F. 1989. Low cell density: the unifying principle for unicell development in *Scenedesmus* (Chlorophyceae). *Br. Phycol. J.* **24**: 271-283.

Ettl, H. and Komárek, J. 1982. Was versteht man unter dem Begriff "coccale Grünalgen"? *Algological Studies.* **29**: 345-374.

Groover, R. and Bold, H. 1969. Phycological Studies 8. The taxonomy and comparative physiology of the Chlorosarcinales and certain edaphic algae. Univ. of Texas Publ. No. 6907. pp. 45-47.

Hegewald, E. 1989. The *Scenedesmus* strains of the Culture Collection of the University of Texas at Austin (UTEX). Arch. Hydrobiol. Suppl. 60, *Algol. Stud.* **29**: 375-406.

Hegewald, E., Hindák, F. and Schnepf, E. 1990. Studies on the genus *Scenedesmus* Meyen (Chlorophyceae, Chlorococcales) from South India, with special reference to the cell wall ultrastructure. Beih. *Nova Hedw.* **99**: 1-75+135 Plates.

Hegewald, E. and Silva, P. 1988. An annotated catalogue of *Scenedesmus* and nomenclaturally related genera including original descriptions and Figures. *Bibl. Phycol.* **80**: 1-587.

Herndon, W. 1958. Some new species of chlorococcalean algae. *Amer. J. Bot.* **45**: 308-323.

Hindák, F. 1990. Studies on the Chlorococcal algae (Chlorophyceae). V. Biologicke Prace. VEDA, Bratislava. pp. 225.

John, D. 1995. Species usage, concepts and problems in Green Algae. The *Phycologist.* **40**: 22.

Karlson, B., Potter, D., Kuylenstierna, M. and Anderson, R. 1996. Ultrastructure, pigment composition, and 18S rRNA gene sequence for *Nannochloropsis granulata* sp. nov. (Monodopsidaceae, Eustigmatophyceae), a marine ultraplankter isolated from the Skagerrak, northeast Atlantic Ocean. *J. Phycol.* **35**: 253-260.

Komárek, J. 1987. Species concept in coccal green algae. *Arch. Hydrobiol. Suppl.* **45**: 437-471.

Komárek, J. 1989. Polynuclearity of vegetative cells in coccal green algae from the family Neochloridaceae. *Arch. Protistenk.* **137**: 255-273.

Komárek, J. & Fott, B. 1983. Chlorophyceae (Grünalgen) Ordnung: Chlorococcales. In Huber-Pestalozzi, G. [Ed.] *Das Phytoplankton des Süsswassers. Systematik und Biologie.* 7. Teil, 1. Hälfte. Schweizerbart'sche, Stuttgart. pp. 813-941.

Komárek, J. & Růžička, J. 1969. Effect of temperature on the growth and variability of *Scenedesmus quadricauda* (Turp.) Breb. *In Studies in Phycology* (Fott, B., editor) E. Schweizerbarrt'sche, Stuttgart. pp. 262-292.

Kouwets, F. 1995. Comparative ultrastructure of sporulation in six species of *Neochloris* (Chlorophyta). *Phycologia.* **34**: 486-500.

Kuhn, T. 1968. *The Structure of Scientific Revolutions.* The University of Chicago Press, Chicago, Illinois. p. 210.

Lewis, L., Wilcox, L., Fuerst, P. & Floyd, P. 1992. Concordance of molecular and ultrastructural data in the study of zoosporic chlorococcalean green algae. *J. Phycol.* **28**: 375-380.

Manhart, J. & McCourt, R. 1992. Molecular data and species concepts in the algae. *J. Phycol.* **28**: 730-737.

Minkoff, E. & Baker, P. 1996. *Biology Today: An Issues Approach.* The McGraw-Hill Companies, Inc. p. 533.

Mladenov, R. & Furnadžieva, S. 1995. Morphologial variability of the clonal cultures of *Scenedesmus falcaltus* Chod. and *Scenedesmus acuminatus* (Lagerh.) Chod. Arch. Hydrobiol., Suppl. Algological Studies. **79**: 93-107.

Pickett-Heaps, J. 1975. *Green algae. Structure, reproduction and evolution in selected genera.* Sinauer, Sunderland, MS. p. 606.

Pickett-Heaps, J. & Staehelin, L. 1975. The ultrastructure of *Scenedesmus* (Chlorophyceae). II. Cell division and colony formation. *J. Phycol.* **11**: 186-202.

Starr, R. 1955. *A comparative study of Chlorococcum Meneghini and other zoospore-producing genera of the Chlorococcales.* Indiana University Publications Science Series No. 20, Indiana University Press, Bloomington. p. 111.

Starr, R. & Zeikus, J. 1993. The culture collection of algae at the University of Texas at Austin. *J. Phycol.* **29 (Suppl.)**: 1-106.

Steenbergen, C. 1975. Light-dependent morphogenesis of unicellular stages in synchronized cultures of *Scenedesmius quadricauda* (Turp.) Breb. (Chlorophyceae). *Acta. Bot. Neerl.* **24**: 391-396.

Trainor, F. 1966. A study of wall ornamentation in cultures of *Scenedesmus*. *J. Bot.* **53**: 995-1000.

Trainor, F. 1991. *Scenedesmus* plasticity: Facts and hypotheses. *J. Phycol.* **27**: 555-556.

Trainor, F. 1992a. Cyclomorphosis in *Scenedesmus armatus* (Chlorophyta): An ordered sequence of ecomorph development. *J. Phycol.* **28**: 553-558.

Trainor, F. 1992b. Cyclomorphosis in *Scenedesmus communis* Hegew. Ecomorph expression at low temperature. *Br. Phycol. J.* **27**: 75-81.

Trainor, F. 1993. Cyclomorphosis in *Scenedesmus subspicatus* R. Chod. Cell behavior during the unicell-colony transformation of a phenotypicly plastic organism. *Arch. Protistenk.* **143**: 55-61.

Trainor, F. 1995. The sequence of ecomorph formation in a phenotypicly plastic, multispined *Scenedesmus* species (Chlorophyceae). *Arch. Hydrobiol.* **133**: 161-171.

Trainor, F. & Egan, P. 1990. Phenotypic plasticity in *Scenedesmus* (Chlorophyta) with special reference to *Scenedesmus armatus* unicells. *Phycologia.* **29**: 461-469.

Trainor, F. & Egan, P. 1991. Discovering the various ecomorphs of *Scenedesmus*: The end of a taxonomic era. *Arch. Protistenk.* **139**: 125-132.

Trainor, F. & Hilton, R. 1963. Culture of *Scenedesmus longus*. *Bull. Torrey Bot. Club.* **90**: 407-412.

Trainor, F. & Shubert, L. 1974. *Scenedesmus* morphogenesis. Colony control in dilute media. *J. Phycol.* **10**: 28-30.

Vinatzer, G. 1975. Neue Bodealgen aus den Dolomiten. *Plant Syst. Evol.* **123**: 213-235.

Watanabe, S. 1983. New and interesting green algae from soils of some Asian and Oceanian regions. *Arch. Protistenk.* **127**: 223-270.

Watanabe, S. & Floyd, G. 1989. Comparative ultrastructure of the zoospores of nine species of *Neochloris* (Chlorophyta). *Plant Systematics & Evolution.* **168**: 195-219.

Wilcox, L., Lewis, L., Fuerst, P. & Floyd, G. 1992. Addressing the relationships of autosporic and zoosporic chlorococcalean green algae with 18S rDNA sequence data. *J. Phycol.* **28**: 381-386.

Recent Trends in Algal Taxonomy
Vol. I. Taxonomical Issues
Editors : Vidyavati and A.K. Mahato
APC Publications Pvt. Ltd., New Delhi, 1999, pp. 101-126

# *Cladophorales: An Overview Over Morphotaxonomy vis-a-vis Cytotaxonomy*

B.N. Verma and Vidyavati
Professor, Univ. Department of Botany, B.R.A. Bihar University
Muzaffarpur (Bihar)

Professor, Department of Botany, Kakatiya University
Warangal (A.P.)

**ABSTRACT**

Cladophorales forms a small but interesting group of green algae. Its long coenocytes organized in unbranched or branched thallus have always posed a serious controversy amongst the phycologists in regard to its systematic position and phylogenetic relationship. Prevalence of euploidy with long polyploid series is another feature of this order which has no parallel in any other group of algae except Charophyta. Good size of Cladophoralean chromosomes having distinct localized centromere provide an ample opportunity of undertaking investigations into comparative chromosome morphology, chiasma frequency and hybridization to resolve the prevailing problems of taxonomical nature. In the present endeavour, karyological investigations made so far in the order Cladophorales have been critically reviewed and the karyological vis-a-vis morphological data have been discussed in relation to its systematic position and phylogenetic relationships. Stability of the order has been discussed on the basis of the basic chromosome number and the frequency of bivalent formation at meiosis. An attempt has also been made to discuss the evolution and interrelationships within the order at generic level in the light of karyotypic variations.

## INTRODUCTION

The order Cladophorales (Fritsch, 1935) forms an interesting small group of green algae especially from taxonomical point of view. Broadly it can be characterized by the thalli consisting of branched or unbranched uniseriate filaments whose cells are multinucleate but uninucleate cells are not uncommon. It possesses numerous discoid angular chloroplast united in a parietal reticulum or a more or less continuous layer with numerous pyrenoids. Motile cells are biflagellate or quadriflagellate zoospores and biflagellate isogamete. The life cycle generally involves an alternation between isomorphic generations. It is represented by 10 living genera and more than 419 species enjoying world wide distribution both in fresh water and marine habitats.

## PHYLOGENY

Diversities in the cellular as well as thallus organizations among the cladophoralean members (as conceived by Fritsch, 1935) have always provided sufficient materials for confusions among the phycologists. Probably this leads to the postulation of three divergent views regarding the origin of this order:

(i) from Chlorococcales (Papenfuss, 1957).

(ii) as an offshoot of Ulotrichales (Fritsch, 1937; Smith, 1938, 1955; Chapman, 1954; Nizamuddin 1964).

(iii) from Siphonales (Faridi vide Nizamuddin, 1964).

Chlorococcalean origin explains how the coenocytic segments of Cladophorales might have been derived along with Siphonales through gradual elaboration of the thallus seen in Characium (Chlorococcales). This hypothesis, however, does not explain further phylogenetic sequences in the order. Also the morphological and cytological evidences do not support it. Therefore, the origin of Cladophorales is invariably regarded either from the Ulotrichalean stock or from the Siphonales. Morphological and cytological data have been proved useful in resolving this controversial issue.

Cladophorales is commonly characterized by the cells possessing only one parietal chloroplast showing variable degree of reticulation and numerous pyrenoids (Carter, 1919; Chadefaud, 1936; Schmitz, 1883). Subsequently it becomes divided into a large number of irregular pieces as seen in marine Cladophorales (Feldmann, 1938) which are quite different from the polygonal pieces of septate Siphonae (Boergessen, 1914). According to Oltmann (1923), reticulate parietal chloroplast of Cladophorales was derived from single girdle shaped chloroplast of *Ulothrix* through its elaboration following increase in the dimension of cells. Reticulation of chloroplasts in *Schizomeris* (Korochikoff, 1927) and in *Microspora* (West, 1916) lend support to this hypothesis. The chloroplast of *Urospora* is taken as the link between the chloroplasts of two orders.

The branched forms of Cladophorales (*Cladophora*, *Pithophora* and *Spongomorpha*) with apical or intercalary vegetative growth and in showing specialized division of labour are undoubtedly advanced than its unbranched forms (*Rhizoclonium, Chaetomorpha* and *Urospora*) with diffused vegetative growth and having no division of labour (Iyengar, 1951). Thus, ancestry of Cladophorales is more convincing in the unbranched stock i.e., Ulotrichales.

Cladophorales possesses life cycle showing isomorphic alternation of generation. Invariably Ulotrichales is characterized by haplont type of life cycle and that of Siphonales by digenetic and monogenetic (Godward, 1966). Latter has predominating diploid generation without existence of haploid generation. Derivation of Cladophorales with isomorphic alternation of generation form Ulotrichales with haploint type of life cycle appears more convincing than from Siphonales with diplontic life cycle (Smith, 1938; Fritsch, 1942). Existence of isomorphic alternation of generation in *Ulva* (Foyn, 1934) and in *Enteromorpha* (Hartman, 1929; Bliding, 1936 and Ramananthan, 1939) lend further support to this view.

The following cytological features also go in favour of the derivation of Cladophorales from Ulotrichales.

Derivation of multinucleate coenocytic segments of Cladophorales seems more convincing from uninucleate segment of Ulotrichales than from coenocytic filaments of Siphonales (Blackman, 1900; Oltmann, 1904; Chapman, 1962). Infrequent existence of uninucleate condition in *Cladophora* (Brand, 1901), *Rhizoclonium* (Brand, 1898; Verma, 1969) and *Spongomorpha* (Chadefud, 1936) may be taken as the relic of ancestral habit.

The Ulotrichales are normally aneuploids and that of Cladophorales are euploids. Latter appear to have been evolved from *Microspora stagnorum* and *M. amoena* with reported polyploid numbers 16 and 32 (Sarma, 1958). Sporadic report of aneuploidy in Cladophorales (Balakrishnan, 1961; Patel, 1961; Hartmann, 1929) again recall ancestral feature.

Ulotrichales with smaller chromosomes (0.25–2 µm in length) and predominating metacentric chromosomes (Godward, 1966) may be considered primitive and further that it might has given rise to relatively bigger chromosomes (0.5–6.9 µm in length) of Cladophorales with median to subterminal centrosomere through phylogenetic increase in their length (Audulov, 1932) followed by the rearrangement of its segments.

## SYSTEMATIC POSITION

Taxonomic history of the order Cladophorales is full of conflicting ideas as much as the systematic position of the taxa included underneath (Fritsch, 1935). Without entering into the controversial taxonomic history of this order, one can adhere to any of th following three taxonomic treatments.

(i) Inclusion under Ulotrichales

(ii) Inclusion under Siphonocladales along with coenocytic forms of Siphonales

(iii) Independent existence as an order Cladophorales

Aforesaid taxonomic treatments of the multitudinous algal forms belonging to three orders in question rest mainly on the taxonomically significant informations on life history, reproduction and morphology. Setting aside all these controversies, Mattox & Stewart (1984) recognised a new class—Ulvophyceae including all the members of erstwhile Ulotrichales, Cladophorales and Siphonales at par with Chlorophyceae on the basis of sharing distinctive features of motile cells and cytokinetic characters. Basal bodies and microtubular roots in the motile cells are not exactly opposite but are slightly displaced when it is viewed from the top. This displacement (rotation) is counter clockwise in Ulvophyceae against clockwise in Chlorophyceae. Scales if present on the body of motile cells, not on flagellar plasmamembrane are in one layer only. During vegetative cell division, mitotic spindles are enclosed by a persisting nuclear membrane, spindle poles are associated with centrioles. New septa are formed by centripetal in growth of the plasma-membrane at least during initial stages. They, however, retain independent status of the order Cladophorales under Ulvophyceae but with certain modifications. It was earlier splitted in two separate orders Acrosiphonales and Cladophorales in view of the differences enumerated below in Table 1.

**Table 1. Morphological distinction of the order Acrosiphonales and Cladophorales (Mattox & Stewart, 1984)**

| | | Acrosiphonales | Cladophorales |
|---|---|---|---|
| 1. | Pyrenoids | Poly-pyramidal | Bilenticular |
| 2. | Cell wall | Consisting of randomly arranged microfibrils that lack crystalline cellulose. | Consisting of lamellae in which microfibrils are more on less parallel containing crystalline cellulose. |
| 3. | Life history | Hetromorphic, diplohaplontic | Isomorphic, diplohaplontic. |

Mattox & Stewart (1984) included *Acrosiphonia, Sponmorpha* and *Urospora* under the order Acrosiphales giving stress over several morphological and ultrastructural features of quadriflagellate zoospores studied in *Urospora* species that are unique for all green algal zoids studied.

1. Flagellar axonemes do not contain 9 peripheral microtubular doublets but 9-single microtubules.
2. Flagella have two wings, each containing 4 to 5 electron-dense plates that are projected from individual peripheral microtubules. The flagella are unable to bend.
3. The transition zone does not contain a single transverse septum but a series of lamellae. The stellate structure is altogether absent.
4. The X—and Y—roots in the cruciate microtubular root system are probably equal and consist proximlly of 4-microtubules.
5. An unusually complex system of electrodense material and stirred fibers link basal bodies to each other, to microtubular roots, and to an amorphous plate in the zoospore papilla.
6. Two rhizoplasts are far more prominent extending from the ventral sides of the basal bodies downward.

Presence of very peculiar tail in the zoospore of both *Urospora* and *Spongomorpha* is another peculiar feature which has no parallel in Cladophorales. This tail was, however, not confirmed in other member of Acrosiphonales.

The order Cladophorales under Ulvophyceae was too conceived loosely (Mattox & Stewart, 1984) by including the members of Siphonales (Fritsch, 1935).

Karyological characteristics of the members belonging to the order Ulotrichales, Cladophorales and Siphonales/Siphonocladales do not support inclusion of Cladophorales under either of the two orders in view of the differences at the points summarised in the Tables 2 & 3.

Therefore separation of Acrosiphonales from Cladophorales proposed by Mattox and Stewart (1984) on the basis of ultrastructural and morphological features of swarmers as well as life history, does not fetch support from karyological findings.

The karyological findings also do not permit separation of Acrosiphonales from Cladophorales (Mattox and Stewart, 1984) in view of the gross similarities between the members of the two orders under reference at the points mentioned below:

1. Nuclei with equal number of nucleoli and chromocentres like bodies.
2. Events of mitoses.
3. Bigger chromosomes with distinct localized centromere ranging from median to subterminal in position.
4. Chromosome complements are without abrupt size differences and karyotypes are graded in size.
5. Distinct polyploid series in multiple of basic set of 6-chromosomes.
6. Most common haploid chromosome numbers n = 12 and 18.

**Table 2. Karyological distinction of the order Ulotrichales and Cladophorales**

| Parameters | Ulotrichales | Cladophorales |
|---|---|---|
| Nuclear organisation | Chromatin reticulum in outer nucleus is prominent | It is inconspicuous |
| Cells | Uninucleate | Commonly mulitnucleate |
| Haploid chromosomal number | Commonly 10 to 16 (Aneuploid count) | Predominently 12 and 18 |
| Ploidy | Aneuploidy | Euploidy |
| Chromosome length | 0.25-2 μm-one or two chromosomes at the periphery of metaphase plate considerably long. | 0.5-6.9 μm-no wide variations in length among chromosome complements of a karyotype. |
| Centromere | invariably median | median to subterminal |
| Meiosis | sporic or gametic | sporic or zygotic |
| Sexual reproduction | Isogamy to oogamy | Isogamy to anisogamy |
| Life Cycle | Predominently haplontic | Predominantly isomorphic alternation of generation |
| Parthernogenesis | Resulting thallus is chimaeras containing both haploid and biploid tissue. | resulting thallus is diploid |

**Table 3. Karyological distinction of the order Cladophorales and Siphonales/ Siphonocladales**

| Parameters | Cladophorales | Siphonales/Siphonocladales |
|---|---|---|
| Nucleus | 2.5-7.5 μm in diameter | 2-6 μm in diameter |
| Mitosis | No reduction in size of nuclei after successive divisions | Apparently reduction in size of nuclei due to rapid succession of nuclear division (Godward, 1966) |
| Nucleoli and chromocentre like bodies | Equal in number | One nucleolus and numerous chromocentre like bodies. (Godward, 1966) |
| Chromsome size | large | small |
| Centromere | median to subterminal | median |
| Polyploidy | Euploidy | Aneuploidy |
| Haploid Chromosome number | Invariably 12 and 18 | Ranges between 8-20 |
| Thallus | Predominently haploid | Predominently diploid |
| Sexual reproduction | Isogamous to anisogamous | Isogamous to oogamous |
| life cycle | Predominently isomorphic alternation of generations | Monogenetic or digenetic |

While discussing the validity of Ulvophyceae of Mattox & Stewart (1984) Sluiman (1989) has concluded—"the number of organisms studied critically in the TEM is still very limited, particularly in the relatively large order Cladophorales (only 8 species out of 419).....". For this reason it would be premature to devise a formal ordinal taxonomy dominated by ultrastructural characters and to recommend it for general use. It is only by the concerted efforts of light microscopist and electron microscopist that the taxonomy of the Ulvophyceae may be extended beyond its "alpha"—stage (Dixon, 1970) before the end of this century.

## KARYOLOGICAL FEATURES

Prevailing informations regarding the karyological features of the order is based on the observations made mainly on the four genera only viz. *Cladophora, Pithophora, Rhizoclonium* and *Chaetomorpha*. Studies are made also in the genera *Acrosiphonia, Basicladia, Spongomorpha* and *Urospora* but of only 1 or 2 species and that too, not exhaustive. Karyological features of the order Cladophorales, emerged out of the investigations made so far may be discussed under respective heads as follows:

## NUCLEAR ORGANIZATIONS

The nuclei in the genera studied so far appear to be identical being spherical, surrounded by a distinct nuclear membrane and rich in chromatin granules. In *Pithophora*, abnormal nuclei resulting from amitotic nuclear divisions have also been reported (Sinha & Verma, 1968; Verma, 1969). These abnormal nuclei differ from the normal ones with respect to their size, shape and in having nuclear vacuoles and poor chromatin material.

Prominence of chromocentre and nucleoli in the metabolic nuclei is the common feature of the order Cladophorales. Heterochromatic regions are often made out as denser parts of chromatin material in these nuclei and are known as "chromocentre" (Heitz, 1929). They get stained more intensely than other parts of the nucleus and appear either shorter or longer threads. In some cases these chromocentres may be seen associated with the nucleolus. A distinct correlation between the number of nucleoli as well as chromocentres has been reported with the level of ploidy in general (Sinha, 1958; Verma, 1969).

### Number of nuclei/cell

The cells in Cladophorales are invariably multinucleate but uninucleate condition is also not very uncommon. Latter condition has been reported in certain species of *Rhizoclonium* (Balakrishnan, 1961; Verma, 1966; Prasad *et al.*, 1973; Sarma & Bala, 1983), *Cladophora* (Brand, 1901) and in *Spongomorpha* (Chadefaud, 1936). There exists a great diversity in the number of nuclei per cell in each of the genera studied so far. The highest number of nuclei in a cell has been counted upto 120 in the genus *Cladophora* (Sinha, 1968), 400 in *Pithophora* (Verma, 1979), 170 in *Chaetomorpha* (Verma & Verma, 1976) and 100 in *Rhizoclonium* (Verma, 1966). Aforesaid variations in the number of nuclei per cell has never been recorded genera specific and hence are of no taxonomic importance. According to Brand (1898) the number of nuclei does not depend on the genus itself but largely on the cubical content of the cell. A detailed observations on more than 50 taxa of this order by first author however, did not yield any significant correlation between the dimension of the cell and the number of nuclein this order.

**Diameter of nuclei**

It has been measured invariably between 2.5 µm and 7.5 µm in general among three genera *Cladophora, Pithophora* and *Rhizoclonium* but nuclei of 12 µm diameter has been also reported in *Rhizoclonium hieroglyphicum* by Sarma and Bala (1983). The genus *Chaetomorpha* appears to be distinct from the three genera named above in possessing relatively smaller nuclei measuring from 2.5 to 4.0 µm in diameter.

The diameter of nuclei has been reported to bear significant positive correlation with the level of ploidy in all the genera studied (Sinha, 1958, '63, '65; Noor, 1965; Verma, 1969). Thus, the diploid nuclei are bigger than haploids but smaller than polyploids.

**Micronuclei**

It has been observed in freshly collected wild forms of *Cladophora fracta* and *Rhizoclonium profundum* by Verma (1969). Occurrence of these micronuclei has been interpreted as a result of the accidental misdivisions.

## AMITOSIS

It has been studied in detail by Sinha & Verma (1968) in *Pithophora cleveana.* Nuclei dividing amitotically enlarge 4-5 times of their original diameter, nucleolus disappear gradually and vacuoles/appears. Enlarged nuclei elongate, assume dumbell shape and finally drawn apart following continued stretching resulting in the formations of two daughter nuclei by constriction without involving nuclear plate formation.

## MITOSIS

Events of nuclear division as observed under light microscope in different members of the order (Sinha, 1958; Patel, 1961; Verma, 1969) have been reported to follow typically on mitotic pattern. Exception to the above has been reported in four species of *Pithophora* in which it has been reported exclusively amitotic in *P. cleveana* (Sinha & Verma, 1968) and partially amitotic in *P. zelleri, P. polymorpha* and *P. roettleri* (Verma, 1969).

Transmission Electron Microscopic (TEM) studies of cell division have been made during vegetative growth phase (Desmoschisis) in *Acrosiphonia spinescens* (Hudson & Waaland, 1983). *Urospora neglecta* and *U. wormskioldii* (Lokhorst & Star, 1983) and during sporogenesis/gametogenesis (Elentheroschisis) in *Cladophora flexuosa* (Scott &Bullock, 1976) as well as in *Cladophora glomerata* (Mc. Donald & Pickett Heaps, 1976). The overall characteristics of cell division emerged after TEM studies may be described categorically as follows:

Invariably the nuclei regardless of their position except in *Acrosiphonia*, in which only those nuclei divide that become aligned in the plane of division. Plane of division is determined at metaphase following the appearance of a small number of MTS as a "hoop" along the periphery of cell before initiation of septum formation. During mitosis, the nuclear envelope persists all along and centrioles are situated either at or lateral to the spindle poles. The latter elongates extensively at telophase giving rise to a typical dumbell-shaped spindle in *Cladophora*. The cylinder of nucleoplasm containing the interzonal spindle seems to be abscised from the two daughter nuclei at telophase and becomes a separate membrane bound compartment. How long this organelle persists after mitosis is, however, unclear.

It is independent of mitosis. The septum formation is initiated as an annular ingrowth of plasma membrane (Mc. Donald & Pickett Heaps, 1976). At metaphase, before the septum is initiated, a small number of microtubules (MTs) appear at the leading edge of the ingrowth. The taxonomic value of the septum associated MTs is questionable. Whether these MTs are simply remnants of a pre-existing cortical cytoskeleton or are formed only during septum ontogenesis is also not known. The precised origin of the new septum membrane (plasma membrane, golgi apparatus, endoplasmic reticulum) has not been established.

## MEIOSIS

Studies on meiosis in the order have not progressed to the same degree as mitotic investigations. The reasons apparently are their restricted occasion of occurrence in undefined cells. Thus, it has been reported only in about 12 species of only three genera (Table 5) viz. *Cladophora* (8 spp.) *Rhizoclonium* (2 spp.) and *Chaetomorpha* (2 spp.). Invariably it has been reported to occur preceding zoospore formation (Sinha, 1958; '63; '65; Sinha &

Ahmad, 1971; Chaudhary, 1978; Sinha & Noor 1967; Verma, 1969, 80; Verma & Verma, 1976; Noor, 1990; Sinha, 1991) showing thereby that the Isomorphic Alternation of Generation in the order Cladophorales is of common occurrence. In *Cladophora glomerata* and in its varieties, meiosis has been reported preceding gamete formation by Schussnig (1930, '51 and '54), List (1930) and Sinha (1968). Thus, this species bear the life history of diplontic nature. In the branched forms like *Cladophora*, meiosis invariably occurs in terminal or subterminal cells of branchlets.

Events of meiosis and configuration of bivalents in all the genera studied are strikingly similar to those of higher plants. Diplotene and diakinesis ascertained especially in the genera *Cladophora* and *Rhizoclonium* look very much similar to those in *Allium*. Normally it is synchronous. Almost all division figures can be observed even in a small dividing segment of the filament a diakinetic one is more frequent, probably because of its long duration. Four strandaed nature of the paired chromosomes is not clearly resolved by light microscopy, the chiasmata are observable at diplotene. The bivalents are countable at diakinesis when it is in the state of maximum contraction and are readily scattered throughout the nucleus. 12 or 18 bivalents have been frequently reported at this stage in all the three genera. Schussing (1930), Sinha & Ahmad (1973) have, however, reported the lowest number of bivalents as 6 in *Cladophora suhriana* and *Cladophora uberrima* respectively. Number of chiasmata in these bivalents has been reported to vary from 1 to 4. No univalent has been reported so far in any of the taxa studied.

Regular presence of bivalents and complete absence of univalents in the order as a whole, which is limited only to its three genera, indicates their stability.

## CHROMOSOME

Chromosomes of as many as 96 taxa belonging to 8 genera have been studied. The appearance and structure of early prophase chromosomes have been studied critically in several species of *Cladophora* and *Rhizoclonium* by Verma (1969). In many instances they present a discontinuous staining, heavily stained regions becoming interspersed with lightly stained ones as reported earlier by Godward (1954) in *Spirogyra*, Cave and Pocock (1951) in some colonial Volvocales, Sarma (1958) in *Microspora amoena* and *Ulothrix zonata* and by Prasad (1958) in *Zygnema* and *Mougeotia.* The heavily stained regions may appear as large number of granules or blocks, their discernible number decreasing due to extreme contraction of the chromosomes at later stage and finally, disappearing at metaphase. Heterochromatic regions stand out clearly as non-stained or lesser stained regions even in metaphase or anaphase chromosomes. The close approximity of the appearance and structure of the chromosomes between the members of different Chlorophycean orders further support their inclusion under a single class Chlorophyceae.

Chromatids have been observed in late prophase or at metaphase in a large number of cases (Sinha, 1958; Verma, 1969). In most cases duality of the chromosomes is some times lost probably because of the close approximation of the chromatids throughout their length accompanied by extreme shortening of the chromosomes as a whole.

As in higher plants the chromosomes are in state of maximum contraction at metaphase and are in most cases evenly distributed on the metaphase as revealed in polar view.

## Dimension of Chromosomes

The length and breadth of these metaphase chromosomes have been reported at least in 50 taxa belonging to seven genera (Table 4).

**Table 4. Length and breadth of the chromosomes reported in different genera of Cladophorales**

| Name of algae | No. of sp. studied | Length of chromosome (μm) | Breadth of chromosome (μm) |
|---|---|---|---|
| *Aerosiphonia Ag.* | one | 0.4-1.4 | 0.28-0.35 |
| *Basicladia* | one | 1.0-3.0 | - |
| *Cladophora Kuetz* | twenty | 0.7-6.2 | 0.3-16 |
| *Chaetomorpha Kutez* | four | 0.7- 2.7 | 0.4-1.3 |
| *Pithophora Wittr.* | eight | 0.8-5.4 | 0.6-1.5 |
| *Rhizoclonium Kuetz.* | fifteen | 0.5-9.1 | 0.4-3.06 |
| *Spongomorpha Kuetz.* | one | 0.7-1.6 | 0.27-0.4 |

Apparently the chromosomes of Cladophorales in general are relatively larger in contrast to many algal groups ranging between 0.4 and 9.1 μm in length and between 0.28 and 3.06 μm in breadth (Table 4). The dimension of chromosomes is however, hardly of any taxonomic importance at generic level because of the following reasons:

1. Number of taxa investigated is not sufficient enough to generate dependable data.
2. Many of the genera are still not studied.
3. The ranges of chromosome length and breadth reported for different genera are merging in each other.

However, the chromosomes of *Rhizoclonium* are invariably bigger and it is followed by *Cladophora*, *Pithophora* and *Chaetomorpha*. Since the available data on the dimension of chromosome in three genera viz, *Acrosiphonia*, *Basicladia* and *Spongomorpha* are based on the investigation of only one species each, it appears too early to draw any conclusion for them.

The orientation and configuration of chromosomes, particularly at anaphase, give evidence of the location of centromeres. The characteristic "V" and "J" shapes, indicating a localized centromeres, either median or submedian or subterminal are common in this order.

The good size of chromosomes with distinct localized centromeres provide an ample opportunity of undertaking an investigations in comparative chromosome morphology, chiasma frequency and hybridization works in this order comparable with what is done in the higher plants. But detailed studies on this line are unfortunately only of Patel (1961 & 1971) and Verma (1966, '69, '80, '82, '85 and '86).

## Chromosome Number

The chromosomes numbers ascertained so far in 96 taxa of Cladophorales have been summarized in the Table 5.

**Table 5. Chromosome numbers reported in different taxa of the order Cladophorales**

| Name of taxa | n | 2n | Authors |
|---|---|---|---|
| *Cladophora* | | | |
| *C. alpina* | | 48 | Geitler. 1936. |
| *C. callicoma* Kuetz. | 12 | | Chaudhary, 1978. |
| | 12 | 24 | Shyam, 1980. |
| *C. clavuligera* Grun | 11 | 22 | Balakrishnan, 1961. |

| Name of taxa | n | 2n | Authors |
|---|---|---|---|
| | | 12 | Verma, 1982. |
| *C. clevigera* Gruo. | 11 | 22 | Balakrishnan, 1961. |
| *C. crispata* (Roth) Kuetz. | | 24 | Mogili, 1986. |
| | | 24 | Chandrakala, 1989. |
| | | 24 | Suresh Babu, 1994a. |
| | | 24 | Sinha, 1958. |
| | | 36 | Verma, 1985. |
| *C. crystallina* (Roth) Kuetz. | | 24 | Verma, 1985. |
| *C. dichlora* Kuetz. | | 12 | Gajarja & Patel, 1980. |
| *C. flaccida* | | 12 | D. Sato, 1959. |
| *C. flavescens* | 12 | 24 | Higgins, 1930. |
| *C. flexicaulis* | 13 | 24 | Schussnig, 1938, F. |
| | 12 | 24 | Patel, 1961. |
| *C. fracta* (Dillw.) Kuetz. | 12 | 24 | Verma, 1985. |
| *C. fracta* Raben. var. *normalis* | 12 | 24 | Geitler, 1936. |
| | | 24 | Sinha, 1958. |
| *C. fracta* var. *flavescens* | 12 | 24 | Verma, 1982. |
| *C. oligoclana* Kuetz. | | | Gajaria & Patel, 1980. |
| *C. glomerata* subf. Kuzingiana (Grunow) | | | Sinha, 1968. |
| *C. glomerata* f. callicoma (Raben) | | 48 | Sinha, 1968. |
| *C. glomerata* f. genuina (Kirchnerem. Rrand) | | 72 | Sinha, 1968. |
| *C. glomerata* var. fasciculata Raben. f. fasciculata | 12 | | Sinha & Noor, 1967. |
| *C. glomerata* (L.) Kuetz. | | + 72 | Geitler, 1936. |
| | | c. 144 | Schussnig, 1954. |
| | 48 | 96 | Schussnig, 1954. |
| | 48(?) | 96 | Sinha, 1968. |
| *C. intermedia* Fosl. | | 36,48 | Verma, 1985. |
| *C. liniformis* | | 12 | Verma, 1985. |
| *C. speciosa* Saraki | 12 | 24 | Masaki. 1978. |
| *C. suhriana* | 6 | 12 | Schussnig, 1930. |
| *C. utriculosa* | | 30 | D. Sato, 1959. |
| *C. uberrima* Lambert | 6 | 12 | Sinha and, 1973. |
| | | 12 | Gajaria & Patel, 1980. |
| | 12 | 24 | Verma, 1982. |
| C. II | | 24 | Geitler, 1936. |

| Name of taxa | n | 2n | Authors |
|---|---|---|---|
| *C.* III | | 24 | Geitler, 1936. |
| *C.* IV | | 30 | Geitler, 1936. |
| *C.* V | | 24 | Geitler, 1936. |
| *C.* (Aeg.) Linneae | 12 | 24 | Sinha, 1958. |
| **MARINE** | | | |
| *C. albida* (Huds.) Kuetz. | | 24 | Patel, 1961. |
| *C. albida* (Huds.) Kuetz. refracted form | | 24 | Patel, 1961. |
| *C. expansa* Martens | 12 | 24 | Patel, 1961. |
| *C. flexuosa* (Griff.) Harv. | 12 | 12,24 | Patel, 1961. |
| | | 30 | D. Sato, 1959. |
| *C. flexuosa* (Mull.) Kuetz. | 12 | 24 | Sinha, 1963b. |
| *C. flexuose* (Mull.) Kuetz. var. sericea (C. sericea (Huds.) Kuetz. var. sercea) | | 24 | Soderstrom, 1962. |
| *C. glaucesceps* Harvey (C. sericea (Huds) Kuetz. var. biflagellata Vanden Hock) | | 22 | Soderstrom, 1965. |
| *C. gracilis* (Griff.) Kuetz. | 12 | 24 | Patel, 1961. |
| | 13 | 24 | Foyn, 1929. |
| *C. hirta* (Kuetz.) | | 24 | Patel, 1961. |
| *C. hutchinsiae* Harv. | 12 | 24 | Patel, 1961; Sinha, 1985. |
| *C. hutchinsiae* var. distans (Ag.) Kuetz. | | 24 | Patel, 1961. |
| *C. macallana* Harv. | | 12 | Patel, 1961. |
| *C. neesiorum* Kuetz. | | 30 | D. Sato, 1959. |
| | | 18 | Patel, 1961. |
| *C. pellurida* (Huds.) Kuetz. | + 18 | | Sinha, 1959. |
| | 16(?) | 33(?) | Foyn, 1929. |
| | 36 | | Patel, 1961. |
| *C. refracts* Arescl. | | 12 | Sinha, 1965. |
| *C. repestris* (L.) Kuetz. | | 12, 24 | Patel, 1961. |
| *C. rupestris* var. distorta Harv | 12 | 24 | Patel, 1961. |
| *C. sericea* (Huds.) Kuetz. | | 12, 24 | Patel, 1961. |
| *Acrosiphonia traillii* | | 12 | Patel, 1961. |
| *Baslicladia* sp. | 6 | 12 | Prasad & Kumari, 1976. |
| *B. crassa* | | 12 | Prasad & Kumari, 1978. |

| Name of taxa | n | 2n | Authors |
|---|---|---|---|
| *Chaetomorpha aerea* (Dillw.) Kuetz. | | 12 | Kornman, 1968. |
| | | 12 | Patel, 1961. |
| | | 18 | Sinha, 1958. |
| | 10 | 20 | Hartmann, 1929. |
| *Ch. linum Farlow* (attached form) | | 18, 36 | Sinha, 1958. |
| | 36 | | Patel, 1961 & Verma, 1969. |
| | 36 | | Patel, 1961. |
| *Ch. melagonium* (Wed. & Mohr.) | 12 | 12, 24 | Patel, 1961. |
| *Ch. linoides* (Ag.) Kuetz. | | 24 | Gajaria & Patel, 1980. |
| *Spongomorpha* lamosa (Roth) Kuetz. | | 12 | Patel, 1961. |
| *S. lamosa* var. uncialis (Fl. Dan.) *Kjellman* | | 30(?) | Foyn, 1929. |
| **(LOOSE FORM)** | | | |
| *Rhizoclonium tortuosum* Kuetz. | | 22 | Patel, 1961. |
| *R. implexum* (Dillw.) Kuetz. | | 24 | Sinha, 1958; Patel, 1961. |
| *R. riparium* Harv. f. (validum) | 12 | 24 | Sinha, 1958. |
| *R. riparium* (Roth) Harvey | 18 | 36 | Sinha, 1958. |
| | | 36 | Patel, 1961; Noor, 1965. |
| *Urospora vancoweriana* | 9 | | Hanic, 1965. |
| *U. wormskjoldii* Rosen. | 12 | | Hanic, 1965. |
| **(FRESH WATER FORMS)** | | | |
| *Chaetomorpha grasillis* Kuetz. | 12 | 24 | Verma & Verma, 1976. |
| *Pithophora mooreana* | | 18 | Prasad & Kumari, 1980. |
| *P. arqualis* Wittr. | | 24 | Gajaria & Patel, 1980. |
| *P. affinis* Nordst | 18 | | Jain, 1977. |
| *P. kewensis* Tilden | | 24 | Patel, 1971; Verma, 1979; Sinha & Noor, 1967. |
| | | 24, 30 | Geitler, 1936; Jain, 1977. |
| *P. oedogonia* Wittr. | | 24 | Patel, 1961; Noor, 1968; Chandramouli, 1991; Purnachandar, 1992; Venumadhav, 1993; Varalakshami, 1996. |
| *P. polymorpha* | | 36 | Verma, 1979; Patel, 1961. |
| *P. oeettleri* (Roth) Wittr. | | 36 | Verma, 1979; Patel, 1961. |
| *P. sumatrana* Erhst | | 24 | Prasad & Kumari, 1980. |
| *P. varia* Wille | 12 | | Jain, 1977. |
| *P. zelleri* | | 24 | Verma, 1979; Patel, 1961; Jain, 1977. |

| Name of taxa | n | 2n | Authors |
|---|---|---|---|
| *Rhizoclonium arboreum* Zelle | 16 | | Prasad & Kumari, 1980. |
| *R. arbareum* Zelle | 11 | 22 | Patel, 1961. |
| *R. arenpsum* (Carm.) Kuetz. | 12 | 24 | Sinha & Ahmad, 1973. |
| *R. arenasum* var. accidentale Kuetz. | 12 | 24 | Sinha & Ahmad, 1973. |
| *R. casparyi* Harv. | 18 | 36 | Sinha & Ahmad, 1973; Prasad & Kumari, 1980. |
| *R. casparyi* Harv. | | 24 | Verma, 1986. |
| *R. crassipelitum* W. et. G.S. West | | 24 | Gajaria & Patel, 1980. |
| | 30 | | Prasad & Kumari, 1980. |
| *R. crispum* Kuetz. | | 24 | Verma, 1980. |
| *R. dimorphum* (Witt.) Stock. | 12 | 24 | Verma, 1980. |
| | 18 | | Prasad & Kumari, 1980. |
| | 11 | 22 | Patel, 1971. |
| *R. hieroghlyphicum* (Ag.) Kuetz. var. macromeres | | 24 | Verma, 1969. |
| *R. hieroglyphicum* var. Kochianum (Kuetz) Stock. | | 24 | Gajaria & Patel, 1980. |
| *R. hieroglyphicum* (Ag.) Kuetz. and Natar | | 24 | Geitler, 1936, Sin. |
| *R. hieroglyphicum* (Ag.) Kuetz. and Natar | | 30 | Peterschilka, 1923. |
| *R. hieroglyphicum* (Ag.) Kuetz. and Natar | | 24, 48 | Patel, 1961, |
| *R. hieroglyphicum* (Ag.) Kuetz. and Natar | | 24, 36 | Balakrishna, 1961. |
| *R. hookeri* Kuetz. | 22 | 46 | Sarma & Bala, 1983. |
| | 14 | | Prasad & Kumari, 1980. |
| | | 18 | Sinha, 1958. |
| *R. implexum* (Otllw.) Kuetz. | 12 | | Sinha & Noor, 1961. |
| | | 24 | Sinha, 1958; Patel, 1961. |
| | | 12 | Verma, 1986. |
| R. IV | | 24 | Geitler, 1936. |
| *R. pachydermum* Kjellman | 12 | | Prasad & Kumari, 1980. |
| | | 24 | Verma, 1966 |
| *R. profundum* Brand | | 48 | Sinha, 1958. |
| | | 36 | Verma, 1986. |
| *R. ripaium* Harv. var. validu, (Gobi) Fosile | | 36, 72 | Verma, 1969. |
| | | 24, 48 | Gajaria & Patel, 1980. |

| Name of taxa | n | 2n | Authors |
|---|---|---|---|
| *R. riparim* (Roth) Harvey | | 24 | Verma, 1986. |
| *R. stagnale* Wols | | 24 | Verma, 1980. |
| *R. sulfuratum* Brand | | 48 | Sinha, 1958. |
| *R. tortuosum* (Kuetz.) Stock. | 20, 30 | | Prasad & Kumari, 1980. |
| | | 22 | Patel, 1961. |
| | | 24 | Sinha, 1958. |
| | | 36 | Verma, 1966. |
| *R.* "M" | | 30 | Geitler, 1936. |

As shown in the table under reference the lowest chromosome number has been ascertained as 2n = 12 and the highest as 2n = 144. Between these two extremes lie almost all the counts falling in multiple of 6 showing predominance of euploidy. This is quite unlike other orders of the green algae in particular and algae in general except Charophyta which too have two distinct euploid series, one with basic chromosome number (3) and the other with (7) corresponding with two tribes Nitelleae and Chareae, respectively. The order Cladophorales is rather more coherent group of green algae in this respect, all genera being knit together through only one basic chromosome number (6) proposed originally by Geitler (1936) and subsequently confirmed by Sinha and Ahmad (1973) and by Prasad and Kumari (1976) through recording the desirable lowest count of (6) in *Cladophora uberrina* and *Basicladia chelonum*, respectively.

It is further apparent form the critical analysis of the Table 5 that the haploid numbers found are predominently 12 or 18; counted at mitosis in the haploid plant or as 12 or 18 bivalents at meiosis in the diploid plant where at mitosis 24 or 36 chromosomes are seen. Of the two haploid numbers (n = 2 and 18) reported in all the four genera *Cladophora, Pithophora, Rhizoclonium* and *Chaetomorpha*, 18 is rare being of sporadic occurrence in each genus except *Chaetomorpha* in which it is rather more frequent. The haploid number 18 has not been reported so far in the genera *Acrosiphonia, Basicladia, Spongomorpha* and *Urospora*.

## EUPLOIDY

On the basis of the data available on the chromosome numbers reported in the taxa of Cladophorales investigated so far (Table 5) which is still meager, it can be concluded that there is prevalence of euploidy in this order as a whole and hence it can be taken as one of the effective means of speciation as well as the evolution within the group. Spontaneous induction of euploidy in nature has been reported through a variety of ways and hybridization between the co-existing species of related genetic set up is one of them. Frequent associations of such Cladophoralean species in nature is quite common and frequently occurring dioecious nature of sexual reproduction among them offer a good opportunity for hybridization but no work has been done on this line.

A total of nine chromosome counts in multiple of "six" has bee reported in Cladophorales i.e. 12, 18, 24, 30, 36, 48, 72, 96 and 144 (Table 5). Of these counts, 12 and 24 seems most common being shared frequently by all the eight genera studied so far. The higher polyploid counts 72, 96 and 144, on the other hand have been reported exclusively in *Cladophora glomerata* (Sinha, 1968 and Schussnig, 1954) and 72 in *Rhizoclonium riparium* var. *validum* (Verma, 1969). Therefore, it can be concluded that only these two species have reached to the highest level of polyploidy. Critical review of the Table 5 further reveal that the higher polyploids are relatively most common in British forms than in Indian forms. An inverse relationship between the chromosome length and level of polyploids has been reported in Charophyta on the pattern of higher plants (Sarma & Ramjee, 1971). No such relationship has been confirmed in Cladophorales.

The relative frequency of the polyploidy has been calculated highest in *Cladophora* (29%) followed by *Rhizoclonium* and *Chaetomorpha* (20%), *Pithophora* (16%), *Spongomorpha* (6%) and *Acrosiphonia, Basicladia* and *Urospora* (3%). Although it seems premature to draw any conclusion on the basis of scored frequency in view of very poor number of investigated taxa at least in the last four genera named above, *Cladophora, Rhizoclonium* and *Chaetomorpha* appear more prone to the polyploidy. Their wide distributions bringing exposure to the diverse ecological niches probably help them to establish in nature.

## ANEUPLOIDY

It is not of common occurrence in the order Cladophorales being reported only in very restricted cases. Although it has been reported in four species of *Cladophora* and *Rhizoclonium* each as well as in one species of *Pithophora* and *Chaetomorpha* each (Table 5), the confirmed aneuploid counts both at haploid (n = 11) and diploid (2n = 22) levels have been determined only in *Cladophora clavuigera* (Balakrishnan, 1961), *Rhizoclonium dimorphum* (Patel, 1971) and in *P. mooreana* (Patel, 1961). Another aneuploid count n = 10 and 2n = 20 has been reported by Hartmann (1929) in *Chaetomorpha aerea.* It is interesting to note that all the cases of aneuploidy reported above are of nullistmic nature and no case of trisomics has been reported so far in the Cladophorales. In view of the above findings following two conclusions can be drawn:

Infrequent occurrence of aneuploidy in this order appears to be the result of occasional mitotic disorder taking place at the time of gamete formation. Existence of same species frequently at different ploidy levels lend support to this assumption.

Aneuploidy is of little importance in regard to the speciation and evolution within the order Cladophorales.

## KARYOTYPE

Karyotypic details of the order are mainly due to Verma (1969 ...) who has studied 26 taxa belonging to four genera viz. *Cladophora, Pithophora, Rhizoclonium* and *Chaetomorpha.* It is therefore, meagre but informative for deducing generic inter-relationship and the evolution within the group.

The Karyotypes of Cladophorales in general are characterized by the chromosomes with distinct localized centromere ranging from median to subterminal in position and are without any abrupt size difference among the chromosome complements. The gross similarities in the karyotypes of four genera under reference suggest their closeness and possibilities of one having been derived from other. However, the taxonomic distinctness of these genera is retained in the following minute structural details of the karyotype:

1. Size difference among chromosomes of the same complement.
2. Total chromatin length.
3. Length of chromosomes.
4. T.F. percent.
5. Coefficient of variations among chromosome complements.

Cladophoralean genera are divided in two broad categories—branched and unbranched forms. *Cladophora* and *Pithophora* apparently fall under the first category and Rhizoclonium as well as *Chaetomorpha* under the second category. The branched forms have been considered advanced and the unbranched forms as primitive (*vide infra*).

Karyotypic data of 26 taxa studied by Verma (1969) have been summarized in the Table 6.

**Table 6. Range of variations in the karyotypic details among the investigated species of the four genera.**

| Name of the Genus | Chromosome (mean length μm) | Coefficient of variations (%) | | T.F. (%) | |
|---|---|---|---|---|---|
| | | **Range** | **Mean** | **Range** | **Mean** |
| *Cladophora* | 1.6-4.53 | 18.9-(29.3-427) | 33.99 | 20.6-45.3 | 36.43 |
| *Pithophora* | 1.23-3.88 | 23.6-28.8 | 26.41 | 41.6-50.00 | 44.67 |
| *Rhizoclonium* | 0.96-4.66 | 19.1-29.7 | 25.72 | 39.4-50.00 | 44.63 |
| *Chaetomorpha* | 2.1-2.35 | 21-28 | 25.69 | 48.00-50.00 | 49 |

A karyotype is said to be asymmetrical and advanced when there is wide variation in the length of its chromosome complements as expressed in higher percentage of coefficient of variations. On the other hand karyotype having lower TF% is predominated by the sub-metacentric and subcentric chromosomes and that having higher TF% by the metacentric chromosomes. First condition is considered as advanced and the second one as primitive. Adhering to the above considerations of advancement and primitiveness of the karyotype (Stebbins, 1950), branched thallus of *Cladophora* with karyotype showing comparatively higher coefficient of variations and lower TF% may be considered advanced than the unbranched forms *Rhizoclonium* and *Chaetomorpha* possessing karyotype with lower coefficient of variations and higher TF% as also conceived on the basis of morphological features (*vide infra*).

*Pithophora* is the branched form and hence it has close resemblance with *Cladophora* than with *Rhizoclonium* and *Chaetomorpha*. But its karyotype with lower coefficient of variations and higher TF% falling within the range recorded for the two unbranched forms (Table 6) claim its closeness more with *Rhizoclonium* and *Chaetomorpha* than with *Cladophora*. Retainment of primitive karyotype by this genus may be considered as an outcome of the complete suppression of sexual reproduction in it. Reporting of lower photosynthetic rate by Ganesan (1967) and amitotic division as exclusive and/or partial means of karyokinesis (Sinha and Verma, 1968) in this genus further add to the primitiveness of *Pithophora*.

**ACKNOWLEDGEMENT**

Authors are very much thankful to Dr. K.K. Sahu for the pain he took in computerized typing of the manuscript and to Dr. H.N. Das for reference collection.

## REFERENCES

Audulov 1932. Karyo-systematische untersuchung der Familie Gramineen. *Bull. Appl. Bot. Suppl.* **44**: 428.

Balakrishnan, M.S. 1961. A cytotaxonomic investigation of Indian members of the Cladophoraceae. *Jour. Indian Bot. Soc.* (Prof. S.P. Agharkar Comm. Vol.) pp. 34-40.

Blackman, F.F. 1900. The primitive and the Flagellate. *Ann. Bot.* **14**: 649-689.

Bliding, C. 1936. Uber die Fortplanzungs korpereininger marinan, Cladophora Arten. *Svensk. Bot. Tidskr.* **30**: 529-436.

Boergesen, F. 1914. The marine algae of the Danish West Indies. *Danak. Bot. Arkiv.* pp. 2.

Brand, F. 1898. Culturesche mit Zwei Rhizoclonium Arten. *Bot. Centralbl.* **73**: 193-202.

Brand, F. 1901. Uber einige verhaltnisse des Banes und wachsthums von Cladophora. *Beihefte. Botan. Centralbl. Bd.* X, Heft. pp. 8.

Carter, N. 1919. The cytology of the Cladophoraceae. *Ann. of Bot.* (Lond.) **33**(132): 467-478.

Cave, M.S. and Pocock, M.A. 1951. Katyological studies in Volvocaceae. *Amer. J. Bot.* **38**(10): 800-811.

Chadefaud, M. 1936. Le cytoplasme des algues vertes at des algues brunes, etc. *Rev. Algol.* **8**: 286.

Chandrakala, G. 1989. Mutagenic studies on two species of Cladophorales. Ph.D. Thesis, Kakatiya University, Warangal.

Chapman, V.J. 1954. The Siphonocladales. *Bull. Bot. Club.* **81**(1): 76-82.

Chapman, V.J. 1962. The Algae Mc. Millan Publication, London.

Chandramouli, K. 1991. Cyto-morphological studies on certain members of Cladophorales. Ph.D. Thesis, Kakatiya University, Warangal.

Chaudhary, B.R. 1975. Cytological and cytotaxonomic investigations of green algae with particular reference to Ulotrichales. Ph.D. Thesis, Banaras Hindu University.

Chaudhary, B.R. 1978. Meiotic chromosome number in *Cladophora callicoma* Kuetzing. *Israel J. Bot.* **27**: 212.

Dixon, P.S. 1970. A critique of the taxonomy of marine algae. *Ann. N.Y. Acad. Sci.* **175**: 617-622.

Feldman, J. 1938. Sur la classification de l'ordre des Siphonocladales. *Rev. Gen. Bot.* **50**: 454-571.

Fritsch, F.E. 1927. Some aspects of the present day investigations of Protophyta. *Pres. Address, Brit. Assoc. Sect.* K, Leeds; pp. 176-190.

Fritsch, F.E. 1935. The structure and Reproduction of the Algae. **1**: 229-248.

Fritsch, F.E. 1942. Studies in the comparative morphology of the algae. II. The algal life cycle. *Ann. Bot. N.S.* **6**: 533-663.

Foyn, B. 1929. Vorlaufige mitteilund uber die sexualitat und generations wechsel von *Cladophora* und *Ulva. Ber Deutsch. Bot. Ges.* **47**: 495-506.

Foyn, B. 1934. Lebenszyklus, Cytologie und sexualitat der Chlorophyceae, *Cladophora suhriana* Kuetzing. *Arch. Protistenk.* **83**: 1-56.

Gajaria, S.C. and R.J. Patel. 1980. Taxonomical and karyological studies of fresh water Cladophorales of Gujarat. *Jour. Indian Bot. Soc.* (Suppl.) **59**: 5.

Geitler, 1936. Vergleichende untersuchungen uber den feinenen kern und chromosmen bau der Cladophoraceen. *Planta.* **35**: 530-578.

Godward, M.B.E. 1954. The diffuse centromere or polycentric chromosomes in *Spirogyra. Ann. Bot. Lond.* N.S. **18**: 143-156.

Godward, M.B.E. 1966. The Chromosomes of the Algae. Edward Arnold (Publisher) Ltd. London.

Hartmann, M. 1929. Uber die sexualitat und dern generations wechsel von *Chaetomorpha* und *Enteromorpha. Ber. Deutsch. Bot. Ges.* **47**: 485-494.

Heitz, E. 1929. Heterochromatin, Chromazentren W. chromomeren. *Ber. Dtch. Bot. Ges.* **47**: 274-284.

Hudsan, P.R. and Walland, J.R. 1974. Ultrastructure of mitosis and cytokinesis in the multinucleate green alga Acrosiphonia. *J. Cell. Biol.* **62**: 274-294.

Iyengar, M.O.P. 1951. Manual of Phycology, Chlorophyta, *Chronica Botanica.* pp. 21-67.

Jain, Vijay Kumari, 1977. Cytotaxonomic studies in Cladophorales. Ph.D. Thesis, Lucknow University, Lucknow.

Kornmann, P., 1968. Wachstum einer *Chaetomorpha*—Art von list/sylt Helgol and Wissen Meeresunter **18**: 194.

List, H. 1930. Die Entwichlungsgeschichte von *Cladophora glomerata* Kuetz. *Arch. Protistenk.* **72**: 453-481.

Korochikoff, A. 1927. On the validity of the genus *Schizomeris* Kuetz. *Arch. Russ. Protistol.* **6**: 71-82.

Masaki, K. 1978. Cytological observations on *Cladophora speciosa* Saraki. *Bull. Fac. Fish Hokkoido Univ.* **29**: 322.

Mattox, K.R. and Stewart, K.D. 1984. Classification of the green algae, a concept based in comparative cytology, In Irvine, D.E.G. & John, D.M. (Eds.) systematics of the green algae. Academic Press, London. pp. 29-72.

Mc. Donald, K.L. and Pickett-Heaps, J.D. 1976. Ultrastructure and differentiation in *Cladophora glomerata.* I Cell division. *Amer. J. Bot.* **63**: 592-601.

Mogili, T. 1986. Effect of certain pharmaceutical drugs on eukaryotic algae. Ph.D. Thesis, Kakatiya University, Warangal.

Neeraja, S. 1992. Impact of heavy metals and gamma rays on *Rhizoclonium hieroglyphicum* (Ag.) Kuetz. Ph.D. Thesis, Kakatiya University, Warangal.

Nizamuddin, Md. 1964. Phylogenetic position of Siphonocladiales. *Trans. Amer. Microscop. Soc.* **83**(3): 282-290.

Noor, M.N. 1965. Cytotaxonomic and cultural studies of common members of Chlorophyceae of Chotanagpur. Ph.D. Thesis, Ranchi University.

Noor, M.N. 1968. A new record on the cytology of *Pithophora oedogonia* Wittrock from India. *Phykos.* **7**(1 & 2): 58-61.

Noor, M.N. 1990. On the karyology of two forms of *Cladophora glomerata.* Perspectives in Phycology (Ed. V.N. Raja Rao) Today and Tomorrow, Printers and Publishers, New Delhi, pp. 59-62.

Oltmanns, F. 1904. Siphonocladiales, Morphologie und Biologie der Algen. I. 1st edit., Jena.

Oltmanns, F. 1923. Ibid., 3, 2nd edition, Jena.

Padmaja, C. 1989. Experimental and cytological studies in *Rhizoclonium hieroglyphicum* (Ag.) Kuetz. Ph.D. Thesis, Kakatiya University, Warangal.

Papenfuss, G.F. 1951. Problems in the classification of marine algae. *Sevensk. Bot. Tidsk.* **45**(1): 4-11.

Patel, R.J. 1961. A cytological and taxonomic study of British Cladophorales. Ph.D. Thesis, London, Univ., London.

Patel, R.J. 1971. Cytotaxonomic studies on *Pithophora kewensis* Wittr. *Phykos.* **10**(1 & 2): 18-23.

Peterschilka, F. 1923. Uber die kernteilung und die Vielkernigkeit und uber die Beziechungen Zwischen epiphytisums und kern—Zahl bei *Rhizoclonium hieroglyphicum* Kuetz. *Arch. Protistenk.* **47**: 325-349.

Prasad, B.N. 1958. Ph.D. Thesis, University of London, London.

Prasad, B.N., Dutta, Shipra and Jain, Vijay Kumari. 1973. Cytological studies in the genus *Rhizoclonium. Phykos.* **12**(1 & 2): 106-113.

Prasad, B.N., Dutta, Shipra and Jain, Vijay Kumari. 1976. Observation on the cytology of *Basicladia* sp. *Phykos.* **15**(1 & 2): 1-4.

Prasad, B.N., Dutta, Shipra and Jain, Vijay Kumari. 1978. Cytological studies on *Basicladia crassa* Hoffmann and Tilden. *Natl. Acad. Sci. Letters.* **1**(3): 89.

Prasad, B.N., Dutta, Shipra and Jain, Vijay Kumari. 1980a. Contributions to the cytology of the genus *Rhizoclonium. Cytologia.* **45**: 583-591.

Prasad, B.N., Dutta, Shipra and Jain, Vijay Kumari. 1980b. On the cytology of three Indian species of Pithophora. *Phykos.* **19**(2): 186-189.

Purnachander, G. 1992. Studies on the effect of pharmaceutical drugs and Irradiation on *Pithophora oedogonia* Wittrock. Ph.D. Thesis, Kakatiya University, Warangal.

Ramanathan, K.R. 1939. On the mechanism of spore liberation in *Pithophora polymorpha* Wittr. *Jour. Indian Bot. Soc.* **18**: 25-29.

Sandhya Sinha, 1991. A study on the cytology and life history of Rhizoclonium dimorphum Wittr. (A member of Cladophorales). *J. Appl. Biol.* **1**(1 & 2): 1-5.

Sarma, T.A. and Bala, Vinod. 1983. Cytological studies on the genus Rhizoclonium Kuetz. *J. Indian Bot. Soc.* **62**(2): 113-119.

Sarma, Y.S.R.K. 1958. Chromosome numbers in Ulotrichales and allied groups. *British Phycological Bull.* **6**: 22-24.

Sarma, Y.S.R.K. and Agrawal, S.B. 1980. Effect of Gamma-rays on the karyology of *Rhizoclonium hieroglyphicum* (Ag.) Kuetz. *Phykos.* **19**(2): 190-196.

Schmitz, F. 1983. Die Chromatophoren der Algen, Verh. Nt. Ver. Preuss. Rheinlandeu Westfalen, *Bonn.* **40**: 1-180.

Schussnig, B. 1930. Das chromosmen cyclus *Cladophora glomerata. Ost. Bot. Zeitschrift.* **79**(3): 273-277.

Schussnig, B. 1951. Der kern Phasenwechsel von *Cladophora glomerata. Svensk. Bot. Tids. Krift.* **45**: 597-602.

Schussnig, B. 1954a. Nuclear phase alternance and differentiation of species in the genus *Cladophora. Proc. Sec. Internat. Bot. Conf. Paris.* **17**: 69-73

Schussnig, B. 1954b. Gonidiogeneis, gametogenese und Meiose bei *Cladophora glomerata* (L.) Kuetz. *Arch. Protistnek.* **100**: 287-322.

Scott, J.L. and Bullock, K.W. 1976. Ultrastructure of cell division in *Cladophora*. Pregametangial cell division in the haploid generation of *Cladophora flexuosa*. *Can. J. Bot.* **54**: 1546-1560.

Shyam, R. 1980. On the life cycle, cytology and taxonomy of *Cladophora callicoma* from India. *Amer. J. Bot.* **67**(5): 619-624.

Sinha, J.P. 1958a. Chromosome numbers and life cycles in members of Cladophorales. *Brit. Phycological. Bull.* **6**: 24-27.

Sinha, J.P. 1958b. Cytological and cultural studies of Cladophorales and Oedogoniales. Ph.D. Thesis, Univ. of London, London.

Sinha, J.P. 1963. Cytotaxonomical studies on *Cladophora flexuosa* (Griff.) Harv. a marine spp. *Cytologia.* **28**(11): 1-11.

Sinha, J.P. 1965. Cytotaxonomical studies on *Cladophora Hutchinsiae* Harv. and *C. refracta* Aresch. two marine spp. *Cytologia.* **30**(4): 375-384.

Sinha, J.P. 1968. Cytotaxonomical studies on *Cladophora glomerata* four fresh water forms. *Cytologia.* **32**: 507-518.

Sinha, J.P. and Ahmad, S. 1971. Cytology and life history of *Cladophora uberrima* Lambert from Ranchi. Proc. 58th Indian Sci. Cong. (Bot. Abstr.).

Sinha, J.P. and Ahmad, S. 1973a. A study on the cytology and life history of *Cladophora uberrima* Lambert from Ranchi. India. *Cytologia.* **38**: 99-105.

Sinha, J.P. and Ahmad, S. 1973b. Cytological observations in three species of *Rhizoclonium* Kuetz. In: Advancing frontiers in cytogenetics (from P.N. Mehra, Comm. Vol.). pp. 254-258.

Sinha, J.P. and Akhaury, K.D.N. 1970. Cytological studies on two species of *Rhizoclonium* Kuetz. Proc. 57th Indian Sci. Congr. (Bot. Abstr.).

Sinha, J.P. and Noor, M.N. 1967. Chromosome numbers in some members of Chlorophyceae of Chotanagpur plateau, India. *Phykos.* **6**(1 & 2): 106-109.

Sinha, J.P. and Verma, B.N. 1968. On the occurrence of amitotic nuclear division in *Pithophora cleveana* Wittr. *Phykos.* **7**(1 & 2): 139-143.

Sluiman, H.J., Roberts, K.R., Stewart, K.D., Mattox, K.R., Lokhorst, G.M. 1982. The flagellar apparatus of the zoospore of *Urospora penicilliformis* (Chlorophyta). *J. Phycol.* **18**: 1-12.

Smith, G.M. 1938. The freshwater algae of the United States. New York.

Smith, G.M. 1955. Cryptogamic Botany 1. Algae and Fungi. Second Edition, New York.

Soderstrom. J. 1965. Remarks on some species of *Cladophora* in the sense of Van den Hock and of Soderstrom. *Bot. Mag.* **8**: 169.

Umamaheswar Rao, B. 1991. The effect of certain antibiotics on *Rhizocloniuum hieroglyphicum* (Ag.) Kuetz. M.Phil. Dissertation, Kakatiya University, Warangal.

Varalaxmi, B. 1996. Mutagenic studies in a filamentous alga. Ph.D. Thesis, Kakatiya Univ. Warangal.

Venkataraman, G.S. and Natarajan, K.V. 1959. Propionocarmine squash technique and chromosome spreading in algae. *Stain Technol.* **34**: 233.

Venumadhav, N. 1993. Karyomorphological studies on two green filamentous algae. Ph.D. Thesis, Kakatiya University, Warangal.

Verma, B.N. 1966. Cytological studies on four species of Rhizoclonium Kuetz. Jour. of Ranchi Univ. Vol. III. pp. 57-63.

Verma, B.N. 1969. Cytological and cultural studies on some members of the order Cladophorales. Ph.D. Thesis, Ranchi University, Ranchi.

Verma, B.N. 1979. Cytological studies in four species of *Pithophora* Wittr. *Cytologia.* **44**: 29-38.

Verma, B.N. 1980. Karyotypic analysis in three species of *Rhizoclonium* Kuetz. *Cytologia.* **45**: 433-440.

Verma, B.N. 1982. Karyotype analysis in 3 species of *Cladophora* Kuetz. *Cytologia.* **47**(1): 137-145.

Verma, B.N. 1985a. Karyomorphological studies on three species of *Cladophora* Kuetz. *Cytologia.* **50**(1): 25-30.

Verma, B.N. 1985b. The cytologyof *Cladophora liniformis* Kuetz. and *C. fracta* (Dillw.) Kuetz. *Cytologia.* **50**: 49-53.

Verma, B.N. 1986a. New counts for *Rhizoclonium riparium* complex. *Cytologia.* **51**(3): 507-512.

Verma, B.N. 1986b. Cytotaxomic studies in the genus *Rhizoclonium* Kuetz. *Cytologia.* **61**: 177-183.

Verma, B.N. and Verma, S.K. 1976. A note on the cytology and life history of *Chaetomorpha gracilis* Kuetz. *Phykos.* **15**(1 & 2): 89-94.

# Section - 3

Recent Trends in Algal Taxonomy
Vol. I. Taxonomical Issues
Editors : Vidyavati and A.K. Mahato
APC Publications Pvt. Ltd., New Delhi, 1999, pp. 129-139

# *Ultrastructure vs. Taxonomy of Desmids*

Vidyavati
Hydrobiology Laboratory, Department of Botamy
Kakatiya University, Warangal-506009

**ABSTRACT**

Taxonomical studies of desmids are being carried out from various parts of India and abroad. In the beginning most of the studies were carried out with the help of light microscope but in the last 30 years or so the electron microscope has brought a completely new dimension to our knowledge of the desmid ultra-structure. Such studies have provided us with many vital keys in understanding specially, the taxonomic features of desmids. In this paper an attempt has been made to review the work of desmid ultrastructural studies and its impact on the taxonomy.

## INTRODUCTION

In the past 30 years or so the electron microscope has added a completely new dimension to our knowledge of the algal ultrastructure which has come to have a place of its own and has provided us with many vital keys in understanding interrelationships and phylogeny of algae. Algal members have extraordinary diversity both, in morphology and cell chemistry. The fine structural data is surely as important as the biochemical data for any consideration of the classification of algae and for attempting to analyse their phylogenetic relationships. The development of freeze-etching or freeze-fracturing during mid sixties provided a technique which made it possible for the surfaces of cells, cell-walls, membranes and organelles to be examined. Thus, we have now an increasing number of papers in which attempts are being made to correlate the structure of membranes, etc. The latest major development, scanning electron microscopy (SEM) has as yet made little general impact in our knowledge of algal ultrastructure. Certainly, it has already become a vital tool for diatom taxonomists and has also been used for the study of the surface of desmids and dinoflagellates. With the development of specimen preparation techniques and the improved resolution of the microscope, SEM will no doubt become of more general use. SEM is also used for observing and photographing walls and theca. In the electron microscope the narrow beam of electrons is adjusted to strike the outside of the cell and then get reflected from it. The SEM also has a great depth in the focus which is about 300 times that of the light microscope. The image shows a three dimensional appearance as the resolving power is quite good. SEM shows a very good range of magnifications and the specimen mounted on a stub in a specimen chamber can be tilted and rotated.

In 1950 some details of the ultrastructure began to be revealed. However, a great improvement took place in the early sixties with automatic ultramicrotomes and diamond knives the great period for fine structure began. In 1973, Dodge published a book on "the fine structure of algal cells". In this book he has aimed to provide a comprehensive review of the structure of eucaryotic algal cells. He has provided enough description of organelles etc. Dodge (1985) has also published an atlas of Dinoflagellates. SEM has proved to be of considerable value in the taxonomy of Dinoflagellates as it clearly shows the characters visible by light microscopy & also adds finer details of the cell wall structures. In his course of research over the past six years Dodge has collected 25,000 scanning electron micrographs of Dinoflagellates. Pickett-Heaps (1975) in his book "Green algae" has mentioned various structures of eukaryotic algal cells as they pass through different stages in their life-cycles at the light and electron microscopical levels.

Many ultrastructural studies have been carried out on members of Chlorophyceae. There have been well over 300 papers dealing with the fine structure of green algae. Recent developments in all aspects of SEM have made this technique reliable and useful for the study of a wide variety of cells. It has been used primarily to study surface characteristics of cells and has bcome an invaluable tool for studying the taxonomy and morphology of many algal members.

Spectacular advances in algal taxonomy have not necessarily been limited to contributions made by TEM and SEM. Electron microscopy showing the ultrastructural differences among the plastids has confirmed the major distinction among brown, red and green algae. It is only quite recent that electron microscopy has been introduced in algal cytology and it has opened new vistas in understanding the taxonomy, phylogeny and evolution of different algal groups. However, a combination of both, LM and EM studies is urgently required for better application of the cytological criteria to algal taxonomy and to draw firm cytotaxonomical conclusions.

While ultrastructural and cytological studies have demonstrated unequivocally their significance in solving many conflicting problems of algal taxonomy, the physiological and biochemical approaches might further offer substantial promise to establish the taxonomy of algae on firmer grounds.

Brown (1945) published the first algal electron micrographs. He employed the shadow-casting technique to examine the flagella of various algal members Prof. Irene Manton published in 1950 the male gamete of *Fucus* and since then she has produced over 65 informative and illustrated papers on algal fine ultrastructure. Further, Manton and her colleagues investigated the structure of flagella in numerous algae and lower plants. During this period descriptive work on algal cell-walls was made specially, on diatom cells. One advantage with the diatom frustules for electron microscopic studies is that it requires no special preparation other than thorough cleaning. Whereas, in other cases of living biological material there are specimen preparation techniques involving fixation, dehydration and other processes. Electron microscopical autoradiography and other cyto-analytical techniques have not yet been much used with algae.

The procaryotic organisms (blue-green algae) are quite distinct from the true algae and they are, in any case, the subject reviewed by Lang (1969).

A useful technique in electron microscopy is the generation of three dimensional images of fixed specimens. Such images can be obtained from sectioned material using transmission electron microscopy, particularly, with thick sections and a high-voltage microscope, scanning microscopy provides easily and conveniently obtainable stereoviews of whole microscopic objects.

## DESMID ULTRASTRUCTURAL STUDIES

Desmid ultrastructural studies have not only revealed new characters and intricate details of the cell structure which have proved invaluable aids in taxonomy, but also led to a clearer understanding and sometimes a reassessment of the higher taxonomic hierarchies. Extensive ultrastructural studies will permit better founded taxonomic decisions of these groups. Scanning electron miscroscopy has permitted more precise information with regard to the surface structure and cell coverings, revealing, thereby, new perspectives on generic concepts and taxonomic affinities of many groups of algae especially, Desmids and Diatoms. In recent years Vidyavati (1995) has reviewed the ultrastructural studies among Conjugales.

In saccoderm desmids (Mesotaeniaceae) the cells mainly exhibit smooth walls without any ornamentation and hence not a suitable material for their study under SEM. These are unicellular organisms with simple form and internal structure compared to other members of the Conjugales. The cell-walls lack the elaborate ornamentation evident in placoderm desmids. Hardly any work has been done on the ultrastructure of these organisms. Pickett-Heaps (1975) studied with difficulty the ultrastructure of *Netrium* and *Cylindrocystis*.

Gerrath (1969) has done a taxonomic correction based on cell-wall ultrastructure in *Penium spinulosum*. He has also studied the connecting strands in *Cosmocladium saxonicum* and has discussed its taxonomy (Gerrath, 1970). Gerrath (1975) has studied cell wall and zygote wall of a saccoderm desmid, *Cylindrocystis brebissonii* and *Bambusina brebissonii* and Chardard (1964) on the ultrastructure of *Mesotaenium cladoriorium* and *Penium margaritaceum*. Mix (1972) has presented fine structure and cell wall of Mesotaeniaceae and Gonatozygaceae members under scanning electron microscope. Previously, ultrastructure of certain Conjugales was studied with special reference to desmids (Chardard, 1965 & 1966; Chardard and Rouiller, 1957; Drawert, 1966; Drawert and Metzner-Kuster, 1961; Drawert and Mix, 1961a-g, 1962a-c, 1963; Lyon, 1969; Pickett-Heaps, 1973, 1974; Mix, 1973, 1975; Stephen *et al.*, 1976; Coesel, 1984; Vidyavati, 1984, 1986; Noguchi and Ueda, 1985, 1988; Vidyavati, 1985, 1995; Vidyavati and Sathaiah, 1985; Coute and Tell, 1981; Coute and Thegezien, 1987 and Federica and Vecchi, 1989).

The surface ornamentation of the cells is of taxonomic importance. Such studies might possibly point to the fact that much of the currently confused state of desmid taxonomy can only be resolved by similar detailed studies of populations. Various aspects of wall structure on the genus *Penium* have been reported by Mix (1967), Mix *et al.* (1983) and Gerrath (1969). Taxonomic revision of the Conjugatophycean family Peniaceae on the basis of cell wall ultrastructure, has been studied by Kouwets and Coesel (1984). Coesel and Delfos (1986) observed and compared the characters between *Actinotaenium didymocarpum* and *A. diplosporum* Dutch material under SEM.

In Desmidiaceae, the cell wall consists of two layers a thin inner layer composed mainly of cellulose and an outer firmer and thicker layer composed of cellulose impregnated with pectic substances which often include iron compounds. External to the outer layer is a narrow or broad diffluent mucilage layer. The mucilage is secreted through a large number of pores which traverses both the layers. Some planktonic desmids have a wide mucilage envelope. Pores were not found in the walls of any of the Mesotaeniaceae, although there is a copious production of these desmids. This mucilage is often very difficult to remove completely and appears as small surface deposits under the SEM. SEM studies of some placoderm species showed excessive mucilage interference (Vidyavati, 1982a-f; 1983a-e; Vidyavati *et al.*, 1983). Hence, it was thought desirable to try and establish certain methods to avoid mucilage interference (Vidyavati, 1983a, Vidyavati & Dodge, 1984). After trying various methods for removal of the mucilage, it was found that pretreatment with the glusulase preparation was usually effective in cleaning the cells and enhancing their appearance under the SEM. However, it was found that freshly subcultured cells were better fixed for such studies, compared to old cultures which accumulates lot of mucilage and thus observes most of the surface ornamentation. These observations may be further useful for the taxonomic identification of the desmids.

Some of the workers have studied ultrastructural observations on *Closterium* sp. under SEM (Brandham and Godward, 1965; Hogetsou and Shibook, 1978a & b; Mix, 1969; Dubios-Tylski, 1973, 1975 & 1978; Fairbanks & Tod, 1973; Vidyavati, 1984 and Noguchi, 1988).

Many ultrastructural studies have been made on different species of the genus *Cosmarium* (Mix, 1966; Lott *et al.*, 1972; Chardard, 1970 and 1974; Gerrath, 1970; Pickett-Heaps, 1972; Martha *et al.*, 1980; Vidyavati, 1982a-d, 1983a-c, 1985 & 1986; Vidyavati and Sathaiah, 1983, 1984a & b; Vidyavati and Sulek, 1988, 1989 and Vidyavati *et al.*, 1983).

Vidyavati (1983d & e) has reported morphology and cell division studies under SEM of *Euastrum verrucosum*. There have been well over 14 papers dealing with fine structure of the genus *Micrasterias* (Drawert and Mix, 1961a & b; Drawert and Kalden, 1967; Staechelin and Kiermayer, 1970; Kiermayer, 1970; Kiermayer and Doedel, 1976; Kies, 1970; a & b; 1973; Ueda, 1972a

& b; Noguchi, 1978 and Noguchi and Ueda, 1988). Neuhaus and Kiermayer (1981 & 1982) studied the formation and distribution of cell wall pores in desmids. Numerous bundles of micro-tubules have been reported in *Micrasterias*, immediately after cell division (Kiermayer, 1968). Pickett-Heaps (1974) has presented scanning electron micrographs of five species of *Staurastrum* and one species of *Cosmarium*, five species of *Micrasterias* and three fairly stable cultured variants of *M. thomasiana* (Uni-, tri-radiate and diploid forms). Vidyavati (1982e & f) has studied morphological and cell division aspects of *Staurastrum gracile* under SEM. Cell wall structure and filament formation in the placoderm desmid *Hyalotheca* sp. was worked out by Krupp (1980).

As in other fields, SEM may prove of value in the taxonomy of desmids. Many aspects of semicell ornamentation become expressed during expansion of the primary wall but pores were only established in the secondary wall. Scanning electron microscopy confirms that the primary wall in contrast to the pattern of pores that distinguishes the secondary wall. As the taxonomic identification of desmid species is based not only on the measurements but also on the ornamentation of the cells SEM may prove of value in the taxonomy of desmids. The symmetry of the cell and its degree of ornamentation are controlled phenomena, as previously been clearly demonstrated by Waris and Kallio (1964).

Recently, Vidyavati (1985 & 1995) has reviewed the work on the scanning electron microscopy of certain desmids. As the taxonomic identification of desmid species is based not only on the measurements but also on the ornamentation of the cells, SEM may prove of value in the taxonomy of desmids. SEM reveals clearly the plasticity of desmid morphology, of obvious importance to taxonomists. Such variability obviously needs also to be taken into account in any experimental studies of semicell morphogenesis.

## REFERENCES

Brandham, P.E. and Godward, M.B.E. 1965. Ultrastructure of the nucleus of the green algae, *Closterium ehrenbergii*. J. Roy. *Microse. Soc.* **84**: 499-507.

Brown, H.P. 1945. Ohio *J. Sci.* **45**: 247-301.

Chardard, R. 1964. Etude de l'ultrastructure de deux algues Zygophycees, *Mesotaenium cladariorium* et. *Penium margaritaceum. Rev. Cytol. Biol. Veget.* **27**: 77-93.

Chardard, R. 1965. Nouvelles observations sur linfrastructure de deux Algues Desmidiales *Cosmarium lundellii* et. *Closterium acerosum. Rev. Cytol. Biol. Veget.* **28**: 15-30.

Chardard, R. 1966. Aspects infrastructure aux de lautvantagonisme Chez deux Desmidiales. *Rev. Cyt. Biol. Veget.* **29**: 336-356.

Chardard, R. 1970. Action des solutions hypertonique sur l'ultrastructure dun algue verte, *Cosmarium lundelli. Bull. Soc. Bot.* France. Mem. Colloque de cytologie expt. **10**: 59-81.

Chardard, R. 1974. La paroi de *Cosmarium lundellii* ultrastructure et essai de localization cytochemique des constituents. *C.R. Acad. Sci.* (Paris). **278**: 609-612.

Chardard, R. and Rouiller, C. 1957. L'ultrastructure de trois algues Desmidicees. Etude au microscope electronique. *Rev. Cytol. Biol. Veget.* **18**: 153-178.

Coesel, P.F.M. 1984. Taxonomic implications of SEM revealed cell wall sculpturing in some small sized desmid species (Chlorophyta, Conjugatophyceae). *Acta. Bot. Neerl.* **33**: 385-398.

Coesel, P.F.M. and Delfos, A. 1986. New and interesting cases of conjugating desmids from Lapland. *Nord. J. Bot.* **6**: 363-371.

Conte, Alain and Yves—Thegezien 1987. SEM description of some desmids (Algae, zygophyceae) from French Guiane (South America). *Rev. Hydrobiol. Trop.* **19**(1): 31-44.

Dodge, J.D. 1973. The fine structure of algal cells. Academic press London and New York.

Dodge, J.D. 1985. Atlas of Dinoflagellates. Ferrand Press, London, Blackwell Scientific Publications, Palo Alto.

Drawert, H. 1966. Der mikrospektralphotometrische nachweis von chlorophyll-b in *Closterien* und einige Mesotaeniceen. Ber Deutsch *Bot. Ges.* **19**: 403-407.

Drawert, H. and Kalden, G. 1967. Licht und elektronmikroskopische untersuchungen an Desmidiacean. XIII. Mitteilung; Der Gestal tswandes des Nucleolus in Interphasekern von *Micrasterias rotata—Mitt. Staatsinst. Allg. Bot. Hamburg.* **12**: 21-27.

Drawert, H. und Metzner-Kuster I. 1961. Licht und elektronmikroskopische untersuchungen an Desmidiaceen. I. Zellwand und Gallerstruckturen bei einigen *Arten. Planta.* **56**: 213-228.

Drawert, H. und Mix, M. 1961a. Licht und electronmikroskopische untersuchungen an Desmidiaceen. II. Hullgallerte und Schleimbildung bei *Micrasterias, Pleurotaenium* und *Hyalotheca. Planta.* **56**: 237-261.

Drawert, H. und Mix, M. 1961b. Licht und electronmikroskopische untersuchungen an Desmidiaceen. III. Der Nucleolus im Interphase kern von *Micrasterias rotata* Flora. **150**: 185-190.

Drawert, H. und Mix, M. 1961c. Licht und electronenmikroskopische untersuchungen an Desmidiaceen IV. Mitt. Beitrage zur elektronenmikroskopischen struktur des Interphasekernes von *Micrasterias rotata.* Z. Naturforschg. **16b**: 546-551.

Drawert, H. und Mix, M. 1961d. Licht und elektronenmikroskopische untersuchungen an Desmidiaceen V. Mitt. Uber die variabilitat der chloroplastenstruktur bei *Micrasterias rotata.* Planta. **56**: 648-665.

Drawert, H. und Mix, M. 1961e. Licht und elektronenmikroskopische untersuchungen an Desmidiaceen VI. Mitt. Der Einfluss von Antibiotica auf die Chloroplasten struktur bei *Micrasterias rotata* Planta. **57**: 51-70.

Drawert, H. und Mix, M. 1961f. Licht und elektronenmikroskopische untersuchungen an Desmidiaceen VII. Mitt. Der Golgi-Apparat von *Micrasterias rotata* nach Fixierung mit kaliumpermanganat und Osmiumtetroxyd. *Mikroskopie.* **16**: 207-212.

Drawert, H. und Mix, M. 1961g. Licht und elektronenmikroskopische untersuchungen an Desmidiaceen VIII. Mitt. Der Chondriosomen von *Micrasterias rotata.* Flora. **151**: 487-508.

Drawert, H. und Mix, M. 1962a. Licht und elektronenmikroskopische Untersuchungen an Desmidiaceen IX. Mitt. Die struktur der Pyrenoide von *Micrasterias rotata.* Planta. **58**: 50-74.

Drawert, H. und Mix, M. 1962b. Die sphaerosomen in elektronenmikroskopischen. Bild Ber Dtsch *Bot. Ges.* **75**: 128-134.

Drawert, H. und Mix, M. 1962c. Licht und elektronmikroskopische, untersuchungen an Desmidiaceen X. Mitt. Beitrage zur kenntnis der Hautung von Desmidiaceen. *Arch. Mikrobiol.* **42**: 96-100.

Drawert, H. und Mix, M. 1963. Licht und elektronenmikroskopische Untersuchungen an Desmidiaceen XI. Mitt. Die struktur von Nucleolus und Golgi-Apparat bei *Micrasterias denticulata Breb. Portugal. Acta. Biol. Ser. A.* **7**: 17-28.

Dubois-Tylski, T. 1973. Aspects ultrastructurer de linduction sexuelle chez *Closterium moniliferum* (Bory) Ehr. Beih Nova *Hedwigia.* **42**: 91-101.

Dubois-Tylski, T. 1975. Aspects ultrastructuraux de l'induction sexvelle chez *Closterium moniliferum* (Bory) Ehrbg. (Desmidiaceae) Mollenhauer, D. Erstes internationales symposium fur desmidiaceenforschung.

Dubois-Tylski, T. 1978. Evolution ultrastructurale de la zygospore du *Closterium moniliferum* (Bory) Ehr. au cours de sa germination. *Rev. Algol. N.S.* **13**: 211-224.

Fairbanks, Tod. R. 1973. The ultrastructure of *Closterium acerosum* University of Vermont, Burlington, VT.

Federica Cheli and Luisa de Vecchi, 1989. An ultrastructural and cytochemical study on conjugatophycean cell wall. *Caryologia.* **42(**2): 127-137.

Gerrath, J.F. 1969. *Penium spinulosum* (Wolle) Comb. nov. (Desmidiaceae) a taxonomic correction based on cell wall ultrastructure. *Phycologia.* **8**(2): 109-118.

Gerrath, J.F. 1970. Ultrastructure of the connecting strands in *Cosmocladium saxonicum* de Bary (Desmidiaceae) and a discussion of the taxonomy of the genus. *Phycologia.* **9**(3 & 4): 209-.

Gerrath, J.F. 1975. Notes on Desmid ultrastructure I cell wall and zygote wall of *Cylindrocystis brebissonii* Meneghinii II. The replicate division septum of *Bambusina brebissonii* Kutzing. *Beih. Nova. Hedwigia.* **42:** 103-113.

Hogetsou and Shibook, H. 1978a. The change of pattern in microfibril arrangement in the inner surface of the cell wall of *Closterium acerosum* during cell growth. Planta. **140**: 7-14.

Hogetsou and Shibook, H. 1978b. Effect of colchicine on cell shape and on microfibril arrangement in the cell wall of *Closterium acerosum.* Planta. **140**: 15-18.

Kiermayer, O. 1968. The distribution of microtubules in differentiating cells of *Micrasterias denticulata* Breb. Planta. **83**: 223-236.

Kiermayer, O. 1970. Elektronmikroskopische untersuchungen zum problem der cyctomorphogenese von *Micrasterias denticulata* Breb. *Protoplasma.* **69**: 97-132.

Kiermayer, O. and Doedel, B. 1970. Elektronmikroskopische untersuchungen zum problem der cytomorphogenese von *M. denticulata* II. Einfluss von Vitalzentrifugierung auf form bildung und Feinstruktur. *Protoplasma.* **87**: 179-190.

Kies, L. 1970a. Elektronenmikroskopische untersuchungen uber Bildung und struktur der zygotenwand bei *Micrasterias papillifera* I. Das Exospor. *Protoplasma.* **70**: 21-47.

Kies, L. 1970b. Elektronenmikroskopische untersuchungen uber Bildung und struktur der zygotenwand bei *Micrasterias papillifera* II. Die Struktur von Mesospor und Endospor. *Protoplasma.* **71**: 139-146.

Kies, L. 1973. Elektronenmikroskopische untersuchungen uber die konjugation bei *Micrasterias papillifera* (Desmidiaceae) Bein. *Nova Hedwigia.* **42**: 139-154.

Kouwets, F.A.C. and Coesel, P.F.M. 1984. Taxonomic revision of the conjugatophycean family Peniaceae on the basis of cell wall ultrastructure. *J. Phycol.* **20**: 555-562.

Krupp, J.M. 1980. Cell wall structure and filament formation in the placoderm desmid, *Hyalotheca* sp. *Jour. of Phycol.* pp. 16.

Lang, N.J. 1969. In: The fine structure of algal cells. Edt. J. Dodge, 1973. Academic Press, London and New York.

Lott, J.N.A., Harris, G.P. and Turner, C.D. 1972. The cell wall of *Cosmarium botrytis*. *J. Phykol.* **8**: 232-236.

Lyon, T.L. 1969. Scanning electron microscopy: a new approach to the Desmidiaceae. *J. Phycol.* **5**: 380-382.

Manton, I. 1950. In: The fine structure of algal cells. Edt. J. Dodge 1973. Academic Press, London and New York.

Martha D. Berliner and Elizabeth Guth. 1980. A transmission and scanning electron microscopy study of division and protoplast formation in *Cosmarium botrytis*. *Jour. of Phycol.* pp. 16.

Mix, M. 1966. Licht und elektronenmikroskopische Untersuchungen an Desmidiaceen XII. Zur Feinstruktur der zellwande und Mikrofibrillen einiger Desmidiaceen von *Cosmarium* type. *Arch. Mikrobiol.* **55**: 116-133.

Mix, M. 1967. Zur feinstruktur der cellwande in der gattung *Penium* (Desmidiaceae). *Ber. Dtsch. Bot. Ges.* **8**: 715-721.

Mix, M. 1969. Zur feinstruktur der Zellwande in der Gattung *Closterium* Desmidiaceae unter besondezer Berucksichtgung der Porensystem. *Arch. Microbiol.* **68**: 306-325.

Mix, M. 1972. Zur feinstruktur der zellwande bei Mesotaeniaceae and Gonatozygaceae mit einen fergleichenden Betrachtung der Verschiedenen Wandtypen der Conjugatophyceae und uber deren systematischen Went. *Arch. Mikrobiol.* **8**: 197-220.

Mix, M. 1973. Die feinstruktur der zellwande der conjugaten und ihre systematische Bedeutung. *Bein. Nova Hedwigia.* **42**: 179-194.

Mix, M. 1975. Die feinstruktur der zellwande bei *Genicularis spirotaenia* (De Bary) De Bary Mitt. *Inst. Alla. Bot. Hamburg.* **17**: 27-38.

Mix, M., Dietrich Lorich and Elspeth Man Shart, 1983. Fine structure and chemical composition of the outer layer of the cell wall of *Penium spirostiolatum, Gonatozygon aculeatum* and *Closterium ehrenbergi. Mitt. Inst. Allg. Bot. Hamburg.* **18**(6): 5-22.

Neuhaus, G. and Kiermayer, O. 1981. Formation and distribution of cell wall pores in Desmids. In: Cytomorphogenesis in plants. Kiermayer ed. Springer-Verlag, Vienna, New York.

Newhaus, G. and Kiermayer, O. 1982. Roster elektronenmikroskopische untersuchungen an Desmidiaceen die poren und ihre verteilungs muster. *Nova Hedwigia.* **36**: 499-568.

Noguchi, T. 1978. Transformation of the golgi apparatus in the cell cycle especially at the resting and earliest developmental stages of a green alga, *Micrasterias americana. Protoplasma.* **95**(1 & 2): 73-88.

Noguchi, T. 1988. Numerical and structural changes dictyosomes during zygospore germination of *Closterium ehrenbergii. Protoplasma.* **147**(2 & 3): 135-142.

Noguchi, T. and Ueda, K. 1985. Cell wall plasmamembranes and dictyosomes during zygote maturation. *Protoplasma.* **128**: 64-71.

Noguchi, T. and Ueda, K. 1988. Cortical microtubules and cortical microfilaments in the green alga, *Micrasterias pinnatifida. Protoplasma.* **143**(2 & 3): 188-192.

Pickett-Heaps, J.D. 1972. Cell division in *Cosmarium. J. Phycol.* **8**: 343-360.

Pickett-Heaps, J.D. 1973. Stereo scanning electron microscopy of desmids. *J. Micros.* **99**: 109-116.

Pickett-Heaps, J.D. 1974. Scanning electron microscopy of some cultured desmids. *Trans. Amer. Microsc. Soc.* **93**(1): 1-23.

Pickett-Heaps, J.D. 1975. Green algae. Structure, Reproduction and evolution in selected Genera. Sinaver Associates. Inc. Publisher, Sunderland, Massachusetts.

Staechelin, L.A.S. and Kiermayer, O. 1970. Membrane differentiation in the golgi complex of *Micrasterias denticulata* Breb. *J. Cell Sci.* **7**: 787-792.

Stephen B. Gough, Garwin, T.W. and Wodkerling, J. 1976. On processing fields and culture samples of desmids (Desmidiales) Chlorophyta for scanning electron microscopy. *Br. Phycol. J.* **11**: 245-250.

Ueda, K. 1972a. Electron microscopical observations of the Golgi apparatus in the cell cycle of a green alga, *Micrasterias americana Protoplasma.* **87**: 145-162.

Ueda, K. 1972b. Electron microscopical observations on nuclear division in *M. americana. Bot. Mag.* (Tokyo). **85**: 263-271.

Vidyavati. 1982a. Cell ornamentation in *Cosmarium bioculatum* Breb. under SEM. *Curr. Sci.* **51**(11): 569-570.

Vidyavati. 1982b. *Cosmarium subtumidum* under scanning electron microscope. *Phytomorphology.* **32**(2 & 3): 191-193.

Vidyavati. 1982c. Cell ornamentation of *C. formosulum* Hoff. under the scanning electron microscope. *Proc. Indian Natn. Sci. Acad. B.* **48**(5): 632-634.

Vidyavati. 1982d. Division in *C. formosulum* Hoff. under the scanning electron microscope. *Life Sci. Adv.* **1**(2): 151-155.

Vidyavati. 1982e. *Staurastrum gracile* Ralfs. under SEM. *J. Indian Bot. Soc.* **61**: 444-446.

Vidyavati. 1982f. Cell division in *S. gracile* Ralfs. under the scanning electron microscope. *Proc. Indian Acad. Sci.* (Plant Sci). **91**(5): 443-447.

Vidyavati. 1983a. Mucilage interference in *C. cucumis* (Corda) Ralfs. under scanning electron microscope. *Phykos.* **22**(1 & 2): 61-63.

Vidyavati, 1983b. Cell division in *C. contractum* Kirchn. under scanning electron microscope. *Geobios New Rept.* **2**(1): 39-41.

Vidyavati. 1983c. Surface ornamentation in *C. praemorsum* Breb. *Indian J. Bot.* **6**(1): 95-97.

Vidyavati. 1983d. *Euastrum verrucosum* Ehrenb. under scanning electron microscope. *J. Indian Bot. Soc.* **62**: 326-329.

Vidyavati. 1983e. *Euastrum verrucosum* Ehrenb. Division under scanning electron microscope. Curr. Sci. **52**(10): 492-493.

Vidyavati. 1984. Scanning electron microscopy of two species of Closterium. *J. Indian Bot. Soc.* **63**: 445-447.

Vidyavati. 1985. Scanning electron microscopy of certain Desmids. A Review. Adv. in Appl. Phycology eds. A.C. Shukla and S.N. Pandey. Internatl. Society for plant and Environ, Kanpur. pp. 23-38.

Vidyavati. 1986. *Cosmarium cucumis* (Corda) Ralfs. under scanning electron microscope. *J. Indian Bot. Soc.* **65**(4): 502-504.

Vidyavati. 1995. Ultrastructural studies among Conjugales: A Review. In: *Biology of Conjugales*, Edit. Vidyavati and M.N. Noor. Printwell Publishers, Jaipur. pp. 168-187.

Vidyavati and Dodge, J.D. 1984. Mucilage interference in desmids under SEM. *Proc. Indian Acad. Sci.* (Plant Sci.). **93**(5): 561-565.

Vidyavati and Sathaiah, G. 1983. Division in *Cosmarium botrytis* menegh under SEM. *Indian J. Bot.* **6**(2): 198-201.

Vidyavati and Sathaiah, G. 1984a. *Cosmarium botrytis* Menegh under the light and scanning electron microscope. *Proc. Indian Acad. Sci.* (Pl. Sci.) **93**(4): 459-465.

Vidyavati and Sathaiah, G. 1984b. *Cosmarium praemorsum* Breb. under light and scanning electron microscope. *J. Pl. & Nature.* **1**(2): 87-91.

Vidyavati and Sathaiah, G. 1985. Cell division in desmids under scanning electron microscope. *Arch. F. Hydrobiol.* **105**(2): 239-249.

Vidyavati, Sathaiah, G., Digamber Rao, B. and Reddy, Y.R. 1983. *Cosmarium botrytis* Menegh. under SEM. *Plant and Nature.* **1**(1): 62-64.

Vidyavati and Sulek, J. 1988. *Cosmarium wembaerensis* Schmidle under SEM. *Curr. Sci.* **57**(15): 861-862.

Vidyavati and Sulek, J. 1989. Surface ornamentation of two *Cosmarium* species under SEM. *Phykos.* **27**: 161-166.

Waris, H. and Kallio, P. 1964. Morphogenesis in *Micrasterias*. *Adv. Morph.* **4**: 45-82.

Recent Trends in Algal Taxonomy
Vol. I. Taxonomical Issues
Editors : Vidyavati and A.K. Mahato
APC Publications Pvt. Ltd., New Delhi, 1999, pp. 141-154

# *Taxonomy of Zygnemataceae : Problems and Prospects*

A.N. Kargupta
University Department of Botany, L.N. Mithila University
Darbhanga, Bihar

## INTRODUCTION

The Zygnemataceae are common green algae found in nearly every kind of freshwater habitats on every continent except Antarctica (Hoshaw, 1968; Hoshaw and Mc Court, 1988). They have drawn the attention of scientists for almost 200 years because of their beauty, abundance and unique mode of reproduction. They are filamentous, unbranched and characterized by distinctive chloroplasts. Flagellated cells are altogether absent in their life cycle. Non-flagellated gametes without being released to surrounding water are brought together through a typical process of conjugation (scalariform, lateral, direct lateral or geniculation). The zygote, a product of fertilization of gametes, develops into a thick walled zygospore which undergoes meiosis. Three nuclei of the meiotic products disintegrate and the surviving one leads to formation of new filament. The life cycle pattern is therefore haplobiontic type (Bold and Wynne, 1985).

## TAXONOMIC STATUS

Earlier taxonomists used pigmentation, nature of food reserve and growth-habit as the primary basis for grouping of algae into major divisions. Further classifications were however, based on the reproductive features, presence or absence of flagella in gametes and their mode of conjugations. Since the time of Bohlin (1901) different proposals were put forward on the status of these conjugating forms. Bohlin (1901) put members of Zygnemataceae in the order Conjugatae of Chlorophyceae. Oltmann (1904, 1922), Pascher (1914, 1931) and Printz (1927) followed Wille (1909-11) who elevated the order Conjugatae to a class equivalent to Chlorophyceae. West (1904) supported Blackman (1900) and

Blackman and Tansley (1902) who included members of Zygnemataceae under a class 'Akontae' of green algae. Fritsch (1935) renamd the order as Conjugales within the class Chlorophyceae. Smith (1950, 1955) divided the order Zygnematales (= Conjugales) into three families: Zygnemataceae, Mesotaeniaceae and Desmidiaceae, Transeau (1951) followed Smith (1950) and Randhawa (1959) simply renamed Zygnematales as Zygnemales and Zygnemataceae as Zygnemaceae. The order Conjugales was elevated to a class Conjugatophyceae by Fott (1959) and Round (1963) which almost corresponds with the Conjugatae propounded by Wille (1909) and Zygophycees by Chadefaud and Emberger (1906). Round's classification was however, justified on cytological grounds by Sarma (1964).

Recent work on mitosis, cytokinesis ad flagellar apparatus (= kinetid) in the green algae has resulted in proposals for a major restructuring of systematics of the green algae at the class level and reinterpretation of their deemed relationship to ancestors of higher plants. Of the five classes Mattox and Stewart (1984) proposed, the Charophyceae was considered most closely related to higher plants. They placed Zygnemataceae and other conjugating algae in the order Zygnematales within the class Charophyceae. In this classification the distinction of classes is based on mode of cell division, the occurrence of phragmoplast or phycoplast at telophase, a collapsing or persistent interzonal spindle and position of centrioles at metaphase. The phragmoplast has been reported in *Spirogyra* (Fowke and Pickett-Heaps, 1969; Pickett-Heaps, 1975) and *Mougeotia* (Pickett-Heaps and Wetherbee, 1987). The investigated species of *Mougeotia* (Bech-Hansen and Fowke, 1971), *Spirogyra* (Fowke and Pickett-Heaps, 1969a) and *Sirogonium* (Hoshaw and Wells, 1982) possess an intact nuclear envelope in the early stages of cell division. This is in contrast to charophycean characters showing dispersal of nuclear envelope (Mattox and Stewart, 1984) during division. The study on distribution of enzymes (Syrett and Al-Houty, 1984) also supports this classification. But Hoshaw *et al.* (1988) did not favour the placing of Zygnemataceae in Charophyceae until more is known about the noncharophycean characters of conjugating algae. From the scheme of Mattox and Stewart (1984) one can infer that (1) conjugating algae had flagellate ancestors, (2) these flagella were of charophycean types and (3) flagella were lost entirely in course of evolution.

## CONVENTIONAL TAXONOMY

Traditional taxonomy relies primarily on characters of field collected materials from nature. Morphological key characters of the group are (1) dimension of vegetative cells, (2) nature of septa (plane, replicate, semi-replicate, colligate, unduliseptate), (3) number of chloroplast, (4) shape and orientation of the chloroplast within the cell (ribbon shaped, axile plate, stellate type), (5) features of conjugation and (6) zygospore wall structure, dimension, shape and colour. Identification of individuals upto species/varieties requires mature zygospores in collected filaments. New species or taxa are described only when the set of characters displayed by a filament population does not fit the combination of characters assigned to any known species. Morphological variations in field collected materials made proliferation of new species descriptions.

Zygnemataceae comprises twelve genera (Randhawa, 1959; Kadlubowska, 1984) consisting of nearly 850 species (Kadlubowska, 1984; Kargupta and Sarma, 1992; Kargupta and Ahmad, 1995; Kargupta and Jha, 1995). Three new genera have recently been described by Ahmad and Goldstein (1971) and, Mahato and Mahato (1994) but their taxonomic recognitions still remain questionable. *Lloydina* Ahmad *et* Goldstein was not included in Kadlubowska (1984) possibly because it was a result of cultural observations of the soil from Kanke, Ranchi. The generic status of *Sangirellum* Mahato et Mahato and *Trigonum* Mahato et Mahato remains doubtful because they are very much like *Spirogyra* except in their chloroplast. The nature of the chloroplast can however, be ascertained in a newly emerged filament from zygospore after germination, besides the fact that it can be deformed (when attacked by Chytrids, the fungi), reoriented or modified under different environmental conditions. The author has even observed a net like appearance of the chloroplast in *Spirogyra* when attacked by fungi and an annular shape when the chloroplast is very short and single in number. *Hallasia*, *Pleurodiscus* and *Entransia* were considered genera of doubtful status (Yamagishi, 1963; Bourrelly, 1966). *Hallasia* was placed under *Zygnemopsis* (Bourrelly, 1966) and *Pleurodiscus* was treated synonymous with *Zygogonium*. The vegetative filament of *Sirogonium* can easily be confused with *Spirogyra* as straight chloroplasts are not uncommon in the latter. Similarly some species of *Mougeotia* resemble very much with *Mougeotiella* except in characters of zygosporangia. *Temnogametum* cannot be distinguished from *Mougeotia* simply on the basis of chloroplast. On the other hand if cytoplasmic residues were present in the gametangia of *Debarya*, it would have been very difficult to distinguish it from *Mougeotia* as they

**Table 1. Distinctive characters of genera under *Zygnemataceae***

| Genera | Vegetative | | Reproductive Characters | | | | | |
|---|---|---|---|---|---|---|---|---|
| | Chloroplast | Pyrenoid | Conjugation canal | Cytoplasmic residue | Pectic subs. in gametangia | Progametangial cell | Gelatinous envelope around zygospore | Zygospor-angium |
| Mougeotiopsis | Quadrate, margin thick | - | + | - | - | - | - | - |
| Debarya | Single, ribbon like | + | + | - | + | - | - | - |
| Mougeotia | 1-2, plate like | + | + | + | - | - | ± | ± |
| Temnogametum | Single, plate like | + | - | - | - | + | + | - |
| Lloydiana* | 2, plate like | + | - | - | - | - | - | - |
| Zygnemopsis | 2, stellate type | + | + | - | + | - | - | - |
| Zygnema | 2, stellate type | + | + | - | - | - | - | - |
| Zygogonium | 1-2, pillow shaped | + | + | + | - | ± | ± | + |
| Pleurodiscus* | 2, disk shaped | + | + | - | - | - | + | + |
| Trigonum* | Single, triangular | - | + | - | - | - | - | - |
| Entransia* | 1-2, laminate | + | N O T K N O W N | | | | | |
| Sangirellum* | 1, band shaped/ girdle shaped | + | + | - | - | - | - | - |
| Spirogyra | 1-16, ribbon like spiral | + | + | - | - | ± | - | - |
| Sirogonium | 2 to many, ribbon like, straight | + | - | - | - | + | - | - |
| Sirocladium* | 2, flat, plate like | + | - | - | - | + | - | - |

* Genera of doubtful status.

tally well in all other characters. The revival of the genus *Temnogyra* by Yamagishi (1963) does not get support from Printz (1927), Smith (1950), Transeau (1951), Randhawa (1959) and Bourrelly (1966). *Temnogyra* (= species of *Spirogyra* in which there occurs an unequal division and the short cells so formed with a greater part of the chloroplast behave as a fertile cell in conjugation) most likely serves as a connecting link with *Sirogonium*. It is interesting to note that in *Sirogonium* (*S. vandalurensis* Iyengar var. *macrospora* Kargupta *et* Jha—unpublished), the conjugation tube is prominently formed by both the gametangia and no sterile cells are cut off during gametangial differentiation. These features can be thought as probable archaic features of the genus tallying well with *Spirogyra*.

The generic status of *Sirocladium* Randhawa was questioned by Singh (1945) and Sinha and Noor (1969-70). It was even interpreted as terrestrial variety of *Sirogonium* (Sinha and Noor, 1969-70). Hoshaw (1968) also expressed doubt about the generic status of *Sirocladium*. According to him three species of *Sirocladium* possess features of *Mougeotia* and *Sirogonium*.

From a review on the taxonomy of the group (Transeau, 1951; Randhawa, 1959; Gauth.-Lievre, 1965; Kadlubowska, 1971, 1984; Kargupta and Sarma, 1992; Kargupta and Jha, 1995) it appears that there has been a gradual increase in the description of new species (Fig. 1). This may be accounted for an extensive study in different geographical areas and phenotypic variations resulted spontaneously in nature. Because of unabated species proliferations experimental phycologists, by necessity, isolate individual strains into culture and based on their variations seen in culture they re-describe or augment description of various species. However it becomes a difficult situation for a taxonomist when materials collected from nature fit the description for one species but later develop a range of morphological characters in culture that tallies with several conventionally defined species. Such discoveries and reports (Mc Court and Hoshaw, 1990) have necessitated the understanding of the bio-logical properties of a species in the face of divergent views of what a species is.

## PLASTICITY RELATED TO TAXONOMY AND SPECIES CONCEPT

Phenotypic plasticity is difficult to distinguish from genetic variability which often reuslts in taxonomic confusion. Phenotypic plasticity can be defined as the amount by which expressions of individual characteristics are altered due to changes in environmental conditions.

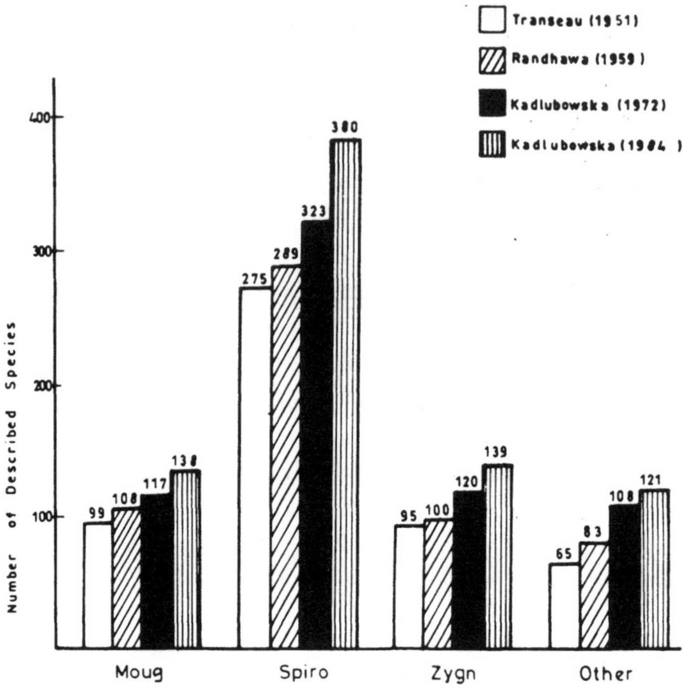


Fig. 1.Histogram showing the increase in species numbers in genera of Zygnemataccae as cataloged in monographs from 1951 to 1984. Code to genera: Moug. = *Mougeotia.* Spiro. = *Spirogyra,* Zygn. = *Zygnema.* "Other" includes nine additional genera recognized in this family (see monographs cited in the figure).

These variations are no less significant in evolution. In conventional monographs (Transeau, 1951; Randhawa, 1959; Kadlubowska, 1984) taxonomic descriptions are highly conservative. Taxonomists often look for characters which are discontinuous, stable and easily handled in keys and descriptions. These characters are considered "good" and held much importance in evolutionary studies.

In Zygnemataceae, there are no discontinuities among allied species particularly, in respect of width of vegetative cells, chloroplast number, contribution of gametangia in the formation of conjugation tube, length and width of conjugation tube, inflation or swelling of female gametangia. The diversity and plasticity of characters is more pronounced in species of *Spirogyra* than other most common genera (*Zygnema, Mougeotia*) of the group. This may possibly be accounted for its being the largest and most common genus of the family. From a perusal on the distribution of taxa in India and abroad (Gauth.-Lievre, 1965; Kadlubowska, 1984; Kargupta and Sarma, 1992) it appears clearly that species with variable characters i.e. showing more plasticity are widely distributed than those having relatively less variabilities. It is observed that species of *Spirogyra* with single chloroplast usually show less variations in other key characters compared to species with variable number of chloroplasts. The wider distribution of species of *Spirogyra* with intermediate breadth is possibly due to its variation in chloroplast number. This variation in the number of chloroplast might reflect its adaptations to a wider variety of microhabitats and environmental conditions.

Greater than expected number of chloroplast has been observed in some species of *Mougeotia* (*M. prona, M. victoriensis*). The morphology of the chloroplast in *Mougeotia robusta* exhibits an unusual character of being deeply cleft. An increase in the number of chloroplast (*Z. circumcarinatum*), length and width of the chloroplast (*S. pratensis*) and even change in the direction of spiraling of the plastid has been observed in culture (Butterfass, 1957; Hoshaw, 1968).

The characters of conjugation tube, their shapes and dimensions are considered important characters in delimitation of taxa. But in many cases these characters vary considerably in different gametangia of the same conjugating pair. Examples are seen in *Spirogyra polymorpha* Kirchner, *S. tubifera* Kargupta *et* Jha (in press) having variable length of conjugation tubes (Jha, 1995).

In conjugating filaments of some species (*S. circumvarians* Kargupta *et* Jha) there is a wide range of variation towards contribution by either of the gametangia in the formation of conjugation tube. Somewhere it is entirely contributed by male gametangium or female gametangia alone while at others it is formed by equal contributions of both the gametangia. This raises a doubt as to the strength of this character in taxonomy.

The inflation or swelling of female gametangia is an another character to distinguish taxa. In this respect our observations (Kargupta and Jha—unpublished) in *S. polymorpha* var. *bolpurensis* Kargupta *et* Sarma, *S. paradoxa* Rao and *S. circumvarians* Kargupta *et* Jha (in press) clearly reveal inflation of varying degrees all along the length of female gametangial filaments (Kargupta and Sarma, 1992; Jha, 1995). There are instances where receptive gametangia in the same conjugating pair vary so much so that they fit well to descriptions of more than one species. Normally conjugation is observed between filaments of similar width and the same number of chloroplast but one may find filaments of markedly different characteristics undergoing conjugation (*S. crassa, S. communis*) (Andrews, 1911). Such an occurrence might throw light on heterothallism or interspecific conjugation.

(*S. crassa, S. communis*) (Andrews, 1911). Such an occurrence might throw light on heterothallism or interspecific conjugation. It deserves mention here that conventional monograph never describes morphology of male and female gametangial filaments undergoing conjugation.

The scalariform conjugation is the usual mode of reproduction in the genus *Spirogyra* but due to certain environmental condition like drying out of water and overcrowding of population in the habitat even species conjugating normally by scalariform means may employ other modes of reproduction including lateral and direct lateral conjugation through pores in cross walls. Hence delimitation of species purely on the basis of mode of conjugation seems confusing. It is supported by our observations that in species like *S. condensata, S..parvula, S. biformis* besides lateral conjugation, scalariform conjugation can occur and for some others like *S. corrugata, S. hymerae, S. pseudospreeiana* besides scalariform conjugation, the lateral conjugation also (Jha, 1995). Similar observation was made by Simons (1995).

Among some allied species the discontinuities are not found on characters of filament width and dimensions of zygospores (*S. laxa* Kuetz., *S. semiornata* Jao, *S. weberi* Kuetz., *S. inflata* (Vaucher) Kuetz., *S. juergensii, S. singularis* Nordstd., *S. silvicola*) (Kargupta and Sarma, 1992; Jha, 1995). This poses a serious problem in morphotaxonomy as the breadth character within a range has been given priority in identification.

Some new taxa have been erected based on characters of rhizoids or rhizoid like protuberances but their appearance is mostly circumstantial and a response of contact stimulus. Our study with Zygnemataceae (Kargupta and Sarma, 1992; Jha, 1995) reports the formation of rhizoids in some species of *Spirogyra* and *Zygnema* (*S. submaxima* Trans., *S. jaoi* Ley, *S. rhizopus* Jao, *S. rhizobrachialis, S. dubia* Kuetz., *Z. pawhuskae* Taft) in most of which there are no previous records of such outgrowths.

The wide distribution of the three common genera due to similar habitat preferences results a high frequency of co-occurrence. *Spirogyra* occurs alone in most collections than with *Zygnema* and *Mougeotia*. Co-occurring filaments of *Spirogyra* collected from the field often are described as different species if they differ in quantitative characters. Simons (1987) used the term "species complex" to refer such group of co-occurring species purely on ecological standing. Allen (1958) observed a 'species complex' for

*Spirogyra* pratensis Trans. which consisted filaments of varying ploidal levels (haploid, diploid and tetraploid forms) derived from the clonal culture. Hoshaw *et al.* (1985) referred to related filament types in a species complex as "morphotypes". They observed changes in width of filaments in relation to ploidal changes in Spirogyra spp. From an original clone three separate species (*S. singularis, S. communis, S. fragilis*) were identified, the width range of each species fitting well with the range described in Transeau (1951). From cultural studies with clonal materials (Hoshaw *et al.*, 1985) it is suspected that "species complexes" at any given site may be generated year after year or even more often. If this is so, gene flow between different filament width population of different ploidal levels may be quite substantial. If such ploidal variations occur in nature, it may cause significant effect on phenotypic plasticity, thereby, allowing the population the ability to adjust to stress and strain of the environment of the habitat.

Similar co-occurring species of *Mougeotia* (*M. floridana, M. poinciana, M. transeaui, M. longiarticulata*) were reported by Transeau (1951), Islam (1972) and Kargupta (1989) from Florida, Bangladesh and West Bengal, respectively. Their resemblance and co-existence might indicate their relationship to a common ancestor while their differences in quantitative characters might have been due to possible gene-flow among them. The genus *Spirogyra* collected from nature sometimes appears to consist of some population that are sexual while others are not, and sometimes these two types of populations may have very similar morphology. It is also difficult to ascertain whether *Spirogyra* filaments are sexually compatible with similar looking filaments.

Polyploidy is supposed to be widespread in *Spirogyra* (Hoshaw and Mc Court, 1988) and it may be responsible for much of morphological plasticity of filaments. Species complexes have yet to be found in genera other than *Spirogyra*. However, in the light of cultural observations the definition of species may require to be redefined and importance should attach to features other than the quantitative ones.

These cultural observations never lessens the importance of conventional taxonomy either. Because it is essential to know the extent of phenotypic variations which enable plants to adapt to disruptive selection pressure of the environment and thus confers the evolutionary process.

## PROSPECTS

The documentation of morphological variations would be a

worthwhile pursuit so long as the sources of such variations are investigated. The researches on morphological plasticity, sexual compatibility and phylogenetic relationship are required if any progress is to be made in the systematics of *Spirogyra* and other genera of Zygnemataceae.

It is a fact that polymorphism in Zygnemataceae has been responsible for the description of many ecotypes (ecomorphs), development stages and aberrant forms as unjustified "taxa". The clarification of taxonomy of the group and eventual reduction of deluge of described taxa to a much smaller number of well defined species is not only a challenge to taxonomists but also of great importance for possibility of their future utilization in research and biotechnology.

To attain this objective one is required to find out relationship among taxa at morphological, cytological and biochemical levels. The recent observation on linear correlation of characters like nuclear volumes, number of chloroplast and filament width in *Spirogyra* spp. (Mc Court *et al.*, 1986; Jha, 1995) might be a step forward.

The application of DNA-hybridization technique can reveal genetic identity or close relationship among different morphotypes of the same species and to a limited extent interspecies relationship. Also by using ribosomal RNA sequencing, the phylogenetic relationship among species and genera may be assessed as has been done for several Chlorococcalean algae.

The ultrastructural studies and the analysis of chloroplast-DNA and allozymes can reveal a good deal about evolution of the group.

But from the data on significant variations in the nature and number of chromosomes in different species (Godward, 1966) it can be presumed that the mechanism of transference of characters through chromosome has not been perfected in the group. This is why the plasticity in their characters both at morphological and cytological levels are quite evident. For retentivity and continuity of morphological characters the cytological stability becomes a necessity but in genera under reference (*Spirogyra, Mougeota, Zygnema*) there are evidences which are indicative of the fact that cytological stability has not been achieved to a degree where it acts as a controlling factor for directive evolution. It may also be said that different parameters responsible for evolution are still active at different degrees in the lower plants and they can be treated as mini-laboratory of the nature.

The reliance on any single character how important it be on taxonomy of the group creates more confusion. The zygospore wall structure even on light microscopic observation can help much in solving conflicts in taxonomy of at least some species. Numerical approach on the variables can reduce further proliferation of species and produce a better picture of species structure.

## ACKNOWLEDGEMENTS

The author is thankful to Professor Late R.W. Hoshaw, R.M. Mc Court, both from USA, J. Simons, The Netherland, and J.Z. Kadlubowska, Poland for inspiration and sending valuable reprints on the group. Thanks are due to Mr. Ram Naresh Jha, Lecturer, Sarisabpahi College, Madhubani who worked out Zygnemataceae of North Bihar in my laboratory and to UGC for providing financial assistance in the form of Career Award.

## REFERENCES

Ahmad, A. and Goldstein, M.E. 1971. *Lloydiella*, a new genus of the Zygnemaceae. *Phycologia*. **10**(1): 1-6.

Ahmad, A. and Goldstein, M.E. 1972. *Lloydina*, a new generic name for the later homonym *Lloydiella* Ahmad and Goldstein (Zygnemaceae). *Phycologia*. **11**(1): 217.

Allen, M.A. 1958. "The biology of a species complex" in *Spirogyra*. Ph.D. Thesis, Indiana Univ., pp. 240.

Andrews, F.M. 1911. Conjugation in two different species of *Spirogyra*. *Bull. Torrey. Bot. Club*. **38**: 299.

Bech-Hansen, C.W. and Fowke, L.C. 1972. Mitosis in *Mougeotia* spp. *Can. J. Bot.* **50**: 1811-1816.

Blackman, F.F. 1900. The primitive algae and the flagellatae: An account of modern work bearing on the evolution of the algae. *Ann. Bot.* **14**: 647-689.

Blackman, F.F. and Tansley, A.G. 1902. A revision of the classification of the green algae. *New Phytol.* **1**: 17-24, 47-48, 67-72, 89-96, 114-120, 133-144, 163-168, 189-192, 213-220, 238-244.

Bohlin, K. 1901. Utkast till de grona Algernas och Arkegoniaternas fylogeni. Lund. pp. 1-43.

Bold, H.C. and Wynne, M.J. 1985. Introduction to the Algae. 2nd edn., Prentice-Hall, Inc., Englewood Cliffs, New Jersey, pp. 720.

Bourrelly, P. 1966. Les Algues O'Eau Douce. Vol. 1. N. Boubee & Cie.pp. 511.

Butterfass, T. 1957. Uber das Vorkommen umgekehrt gewundener chromatophoren bei *Spirogyra*. *Protoplasma*. **48**: 419-421.

Chadefaud, M. and Emberger, L. 1960. Trate de Botanique Systematique. Tome 1. Chadefaud, M. Les Vigelaux non Vascularis, Paris.

Folt, B. 1959. Algenkunde, Gustav Fischer Verlag, Jena. pp. 581.

Fowke, L.C. and Pickett-Heaps, J.D. 1969a. Cell division in *Spirogyra* I. *Mitosis*. *J. Phycol.* **5**: 240-259.

Fowke, L.C. and Pickett-Heaps, J.D. 1969b. Cell division in *Spirogyra* II. *Cytokinesis*. *J. Phycol.* **5**: 273-281.

Fritsch, F.E. 1935. The structure and reproduction of the Algae. Vol. 1, Cambridge University Press.

Gauth.-Lievre, L. 1965. Zygnemataceae Africanes. *Beih. Zur. Nova Hedwigia.* **20**: 1-120.

Godward, M.B.E. 1966. The chromosomes of the algae. Edward Arnold (Pub.) Ltd., London.

Hoshaw, R.W. 1968. Biology of the filamentous conjugating algae. In: Algae, Man and the Environment (Ed. O.F. Jackson) Syracuse University Press, Syracuse, New York, pp. 135-184.

Hoshaw, R.W. and Mc Court, R.M. 1988. The Zygnemataceae (Chlorophyta): a twenty year update of research. *Phycologia.* **27**(4): 511-548.

Hoshaw, R.W. and Wells, C.V. 1982. Systematics of the Zygnemataceae (Chlorophyceae) IV. Nuclear cytology of *Sirogonium melanosporum*, a species with large chromosomes. *Trans. Amer. Microsc. Soc.* **101**: 276-286.

Hoshaw, R.W., Mc Court, R.M. and Wang, J. 1988. Phylum Conjugatophyta. In: Handbook of Protoctista (Edt. L. Margulis, J.O. Corliss, M. Melkonian & D.J. Chapman). Jones & Bentlett Publ. Inc., Beston, Massachusetts. pp. 163-179.

Hoshaw, R.W., Wang, J.C., Mc Court, R.M. and Hull, H.M. 1985. Ploidal changes in clonal cultures of *Spirogyra communis* and implications for species definitions. *Amer. J. Bot.* **72**: 1005-1011.

Jha, R.N. 1995. Ecological and cytomorphological observations in some Zygnemataceae of North Bihar. Ph.D. Thesis submitted in B.R.A. Bihar University, Muzaffarpur.

Kadlubowska, J.Z. 1972. Zygnemaceae, Tom 12A. Flora Slodkowodna Polski. Polska Akademia Nauk Instytut Botaniki, Krakow. pp. 431.

Kadlubowska, J.Z. 1984. Susswasserflora von Mitteleuropa Band 16, Chlorophyta VIII—Conjugatophyceae 1 (Zygnemales) Gustav Fischer - Verlag, Stuttgart, pp. 532.

Kargupta, A.N. 1987. Taxonomic investigations of freshwater filamentous green algae of Birbhum and Murshidabad disricts of West Bengal. Ph.D. Thesis, Burdwan Univ., West Bengal.

Kargupta, A.N. 1989. Indian Zygnemataceae—1. The genus *Mougeotia* from West Bengal. *Biojournal.* **1**(1): 97-105.

Kargupta, A.N. and Ahmad, M.R. 1995. The genus *Spirogyra* (Chlorophyceae: Zygnematales) from Mithilanchal, Bihar. In: Biology of Conjugales (Edt. Vidyavati & Noor, M.N.). Printwell Publ., Jaipur. pp. 195-218.

Kargupta, A.N. and Jha, R.N. 1995. Taxonomy of Indian Zygnemataceae: A Review. In: Biology of Conjugales (Edt. Vidyavati & Noor, M.N.). Printwell Publ., Jaipur. pp. 104-129.

Kargupta, A.N. and Sarma, P. 1992. The genus *Spirogyra* (Chlorophyceae: Zygnemataceae) from West Bengal, India. Bibliotheca *Phycologica.* (with 229 figs.). **91**: 1-75.

Mahato, A.K. and Mahato, P. 1994. *Sangirellum taeniforme* and *Trigonum indicum*, two new members of Zygnemaceae from India. *Phykos.* **33**(1-2): 113-124.

Mattox, K.R. and Stewart, K.D. 1984. Classification of the green algae: A concept based on comparative cytology. In: Systematics of the Green Algae (Ed. D.E.G. Irvine & D.M. John). Academic Press, Inc. Orlando, Florida. pp. 29-72.

McCourt, R.M. and Hoshaw, R.W. 1990. Noncorrespondence of breeding groups, morphology and monophyletic groups in *Spirogyra* (Zygnemataceae: Chlorophyta) and the application of species concepts. *Systematic Botany.* **15**(1): 69-78.

Miller, R.D. and Hoshaw, R.W. 1974. Cell width as a taxonomic character with special reference to *Zygnema circumcarinatum* Czurda. *Br. Phycol. J.* **9**: 145-148.

Oltmarins, F. 1904. Morphologie und Biologie der Algen, Jena.

Oltmanns, F. 1922. Morphologie und Biologie der Algen, 2nd edn., Jena.

Pascher, A. 1914. Uber Flagellaten und Algen. *Ber. Dtsch. Bot. Ges.* **32**: 136-160.

Pascher, A. 1931. Systematische Ubersicht uber die mit Flagellaten in Zussamenhang stebenden Algenreihen und Versuch einer Einreihungdieser Algenstamme in die Stammae des Pflanzenreiches. *Beih. Bot. Centralbl.* **48(Abt. B)**: 317-332.

Pickett-Heaps, J.D. 1975. Green Algae: Structure, Reproduction and Evolution in selected genera. Sinauer Associates Inc. Sunderland, Massachusetts, pp. 606.

Pickett-Heaps, J.D. and Wetherbee, R. 1987. Spindle function in the green alga *Mougeotia*: absence of anaphase. A correlate with postmitotic nuclear migration. *Cell Motility Cytoskeleton.* **7**: 68-77.

Printz, H. 1927. Chlorophyceae nebst Conjugatae, Heterokontae and Charophyta in Engler and Prantl's Die Naturlichen Pflanzenfamilien. *Zweite Auft. Leipzig.* **3**: 463.

Randhawa, M.S. 1959. Zygnemaceae, ICAR, New Delhi, pp. 478.

Round, F.E. 1963. The taxonomy of the Chlorophyta. *British Phycol. Bull.* **2**: 224-235.

Sarma, Y.S.R.K. 1964. Cytology in relation to systematics of algae with particular reference to Chlorophyceae. *The Nucleus.* **7**: 127-136.

Simons, J. 1987. *Spirogyra* species and accompanying algae from dune waters in The Netherlands. *Acta Bot. Neerl.* **36**: 13-31.

Simons, J. 1995. *Spirogyra* and accompanying Zygnematacean algae: Some morphological aspects, and distribution and ecology in The Netherlands. In: *Biology of Conjugales* (Edt. Vidyavati & Noor, M.N.), Printwell Publ., Jaipur. pp. 39-58.

Singh, R.N. 1945. On Randhawa's "A critical review of some recently created new species of Indian Zygnemales." *Proc. Ind. Acad. Sci. B.* **22**: 383-386.

Sinha, J.P. and Noor, M.N. 1969-70. A critical review on *Sirocladium kumaoense* Randhawa (*Sirocladium kumaoense* Randh., *Sirogonium sticticum* (Engl. Bot.) Kuetzing var. *terrestres* var. nov.) *J. Ranchi Univ.* **6-7**: 198-204.

Smith, G.M. 1950. The freshwater algae of The United States. 2nd ed., McGraw-Hill, New York.

Smith, G.M. 1955. Cryptogamic Botany. Vol. 1. Algae and Fungi. Mc Graw-Hill, New York.

Syrett, P.J. and Al-Houty, F.A.A. 1984. The phylogenetic significance of the occurrence of urease/urea amidolyase the glycollate oxidase/glycollate dehydrogenase in green algae. *Br. Phycol. J.* **19**: 11-21.

Transeau, E.N. 1951. The Zygnemataceae. Ohio State Univ. Press, Columbus, pp. 327.

West, G.S. 1904. A treatise on the British Freshwater Algae, Cambridge, pp. 327.

Wille, N. 1909-11. Conjugate und Chlorophyceae in Engler and Prantl's Die Naturlichen Pflanzenfamilien. Nachtrage Zum Teil, I, Abt. 2, Leipzig, pp. 136.

Yamagishi, T. 1963. Classification of the Zygnemataceae. *Sci. Rep.* Tokyo Kyoika, Daigaka. **11**: 191-210.

Recent Trends in Algal Taxonomy
Vol. I. Taxonomical Issues
Editors : Vidyavati and A.K. Mahato
APC Publications Pvt. Ltd., New Delhi, 1999, pp. 155-175

# *Taxonomic Proposals for Systematics of the Zygnemacean Algae*

A.K. Mahato* and Pushpa Mahato
Department of Botany, P.T.P.S. College, Patratu
Bihar-829 119


## ABSTRACT

Since the establishment of the most interesting green algal family, Zygnemaceae (Meneghini, 1838), primarily 19 genera have been described till now. Of these, the distinguishing characters of *Hallasia* Rosenvinge, *Mougeotiella* Yamagishi, *Neozygnema* Yamagishi and *Temnogyra* Lewis are not sufficient enough to warrant their taxonomic recognition. Remaining 15 genera are either isogamous or anisogamous. Anisogamous genera lack all possible types of specialized reproductive characters and hence, they are separable by only their vegetative characteristics, especially by the chloroplast structure. On the other hand, isogamous genera can be delineated by their specialized reproductive characteristics only. Among the anisogamous genera, *Spirogyra* Link and *Sirogonium* Kützing need emended definitions to accommodate some species of unusual characters. Considering the present need, definitions of these two genera have been emended. We think, traditionally used taxonomic characters are not sufficient to explain the gross morphology of a taxon, hence on the basis of an extensive study of Zygnemacean algae, some additional taxonomic characters have been proposed here for speciation. These characters are: length and width of the chloroplast, margin of the chloroplast, shape of the chloroplast, nature of turning of the chloroplast, pyrenoids in the chloroplast, shape and size of the conjugation canal, shape of the island, orientation of zygospores in the gametangia and space occupied by a zygospore in the receptive gametangium.


## NUMBER OF GENERA IN ZYGNEMACEAE

Members of the family Zygnemaceae are characterized by simple or often branched filaments, symmetrical cells, elaborate chloroplasts and amoeboid gametes. Among these the most important character which keeps the family apart from all other green algae is the conjugation of amoeboid gametes. Since the establishment of the family primarily 19 genera are described till now. These are: *Mougeotiopsis* Palla, 1893; *Debarya* (Wittrock 1872) emend. Transeau, 1934; *Mougeotia* Agardh, 1824; *Mougeotiella* Yamagishi, 1963; *Temnogametum* W. and G.S. West,

* Present address : Department of Botany J.R.S. College, Jamalpur, Munger, Bihar.

1897; *Lloidiella* Ahmad and Goldstein, 1971, 1972; *Hallasia* Rosenvinge, 1924; *Zygnemopsis* (Skuja) Transeau, 1934; *Zygnema* Agardh, 1824; *Neozygnema* Yamagishi, 1963; *Zygogonium* Kützing, 1843; *Pleurodiscus* Lagerheim, 1895; *Entransia* Hughes, 1943; *Sangirellum* A.K. and P. Mahato, 1994; *Trigonum* A.K. and P. Mahato, 1994; *Spirogyra* Link, 1820; *Temnogyra* Lewis, 1925; *Sirogonium* Kützing, 1843 and *Sirocladium* Randhawa, 1941.

Among these genera, the genus *Hallasia* is very much similar to the genus *Zygnemopsis*. It differs from its closely related genera *Zygnemopsis* and *Zygnema* only by the production of 1-3 sporelings on germination of aplanospores and frequent occurrence of upto 7 chloroplasts. Recently, A.K. Mahato (1993) has recorded production of two sporelings from a germinating zygospore in *Spirogyra condesata* (Vaucher) Kützing and he stated that production of 1-3 sporelings is a part of abnormal reproduction. A.K. and Pushpa Mahato (1993) further stated that increase in the number of chloroplasts in the cells of *Zygnemopsis* and *Zygnema* is a common phenomenon. Each chloroplast divides into two or often more than two during the cell division resulting in the increase in number of chloroplasts. In view of the above facts they suggested inclusion of the monotypic genus *Hallasia* in its closely related genus *Zygnemopsis*. Randhawa (1959) in his monograph though has described *Hallasia* as a separate genus, advised to include it in *Zygnemopsis*. Yamagishi (1963) and Bourrelly (1966) also placed this genus in synonymy with *Zygnemopsis*.

Lewis (1925) described a peculiar species of *Spirogyra* under the generic name *Temnogyra* Lewis in which gametes are formed in special gametangial cells, but Printz (1927) disregarded taxonomic recognition of *Temnogyra* and renamed the species *T. collinsii* Lewis as *Spirogyra collinsii* (Lewis) Printz. Randhawa (1959) followed Printz (1927) and regarded *T. collinsii* as a synonym of *S. collinsii*. Later workers also did not recognize *Temnogyra* as a new genus (Czurda, 1932; Kolkwitz and Krieger, 1941; Transeau, 1951; Bourrelly, 1966; Usha Devi and Panikkar, 1991; A.K. and Pushpa Mahato, 1993). Species of such type of unusual character are not uncommon in *Spirogyra* and at least three have been described alone from the Indian subcontinent (Patel and Ashok Kumar, 1969; Ashok Kumar, 1972; Usha Devi and Panikkar, 1991).

During the study we have observed three peculiar types of species in *Spirogyra* apart from the normal type (Figs. 1). These are: (i) species with male progametangial cells (ii) species with female progametangial cells and (iii) species with both male and female progametangial cells. Species of third category are included in the genus *Temnogyra* (Lewis, 1925; Patel, 1980), but all these species, viz., *S. collinsii* (Lewis) Printz; S. tuwensis Patel and Ashok Kumar; *S. madhuensis* Ashok Kumar and *S. patelensis* Usha Devi and Panikkar possess typically *Spirogyra*-type of chloroplasts. Hence, they are unquestionably species of *Spirogyra*. Thus, the genus *Temnogyra* does not merit a separate generic name.

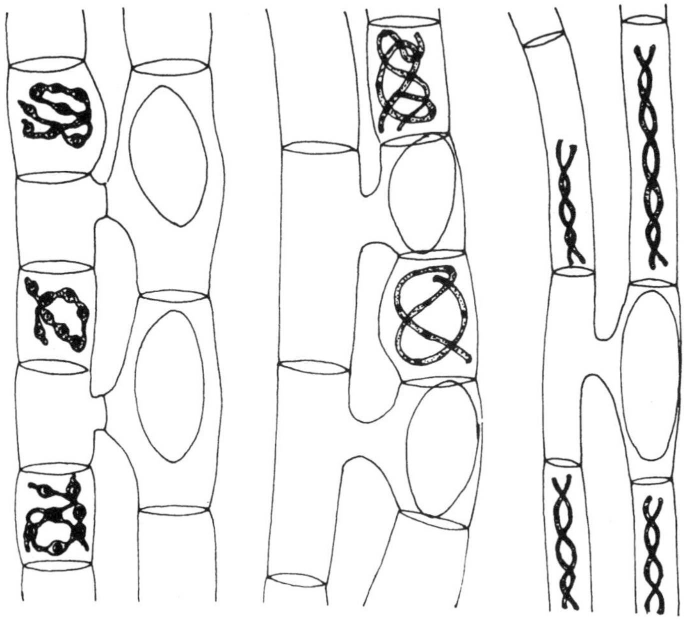

(i) (ii) (iii)

Fig. 1. Unusual types of *Spirogyra* spp. (i) species with male progametangial cells (ii) species with female progametangial cells (iii) species with both male and female progametangial cells.

Likewise, *Mougeotiella* and *Neozygnema* were also separated from their closely related genera *Mougeotia* and *Zygnema* respectively, by single specialized reproductive character (Yamagishi, 1963).

From the above discussion it is evident that distinguishing characters of *Hallasia*, *Mougeotiella*, *Neozygnema* and *Temnogyra* are not sufficient enough to warrant their taxonomic recognition.

## DELINEATION OF GENERA

Genera of Zygnemaceae are separated from one another by: structure of the chloroplasts, pyrenoids in the chloroplasts, progametangial cells, cytoplasmic residue in the gametangia, pectic cellulose colloid in the gametangia, gelatinous substance surrounding zygospores and a zygosporangium (Transeau, 1951; Randhawa, 1959; Yamagishi, 1963; Gauthier-Livre, 1965). However, Czurda (1932) disregarded all these characters except the chloroplast structure.

All the 15 genera of the family Zygnemaceae are divisible into two groups, viz., isogamous and anisogamous on the basis of their sexuality. *Lloydina, Trigonum, Sangirellum, Spirogyra, Sirogonium, Sirocladium* and probably *Entransia* are anisogamous. All the specialized reproductive characters, viz., cytoplasmic residue in the gametangia, pectic substance in the gametangia, gelatinous substance surrounding zygospores and zygosporangia are absent from these genera. Hence, these can be separated from one another by only their vegetative characteristics, like: structure of the chloroplasts and presence or absence of pyrenoids and oil globules in the chloroplasts. Distinguishing vegetative characters of these anisogamous genera are summarized in Table 1.

**Table 1. Summary of the distinguishing characters of anisogamous genera of Zygnemaceae**

| Genera | Chloroplasts | Pyrenoids | Conjugation-type |
|---|---|---|---|
| *Lloydina* | Two, axile, ribbon-like | many, scattered | *Spirogyra*-type |
| *Entransia* | 1-2, laminate, often with finger-like processes | many, scattered | unknown |
| *Sangirellum* | Single, girdle-shaped or band-shaped | many, scattered | *Spirogyra*-type |
| *Trigonum* | Single, axile, triangular, with rounded ends or polsterform-triangular | absent | *Spirogyra*-type |
| *Spirogyra* | One to many, ribbon-like, spiral, rarely straight | many | Phsiological anisogamy, conjugation canals present, progametangial cells generally absent |
| *Sirogonium* | Two to many, narrow, delicate, straight or nearly straight, rarely spiral | many | Morphological anisogamy, conjugation canals generally absent, progametangial cells present. |
| *Sirocladium* | Two, flat, broad, plate-shaped or ribbon-like, margin smooth or lacicinate | many, scattered | *Sirogonium*-type |

Of the anisogamous genera, *Sirogonium* sometimes possesses spiral chloroplasts and primitive type of conjugation canals. On the other hand a few species of *Spirogyra* reveal *Sirogonium*-type of characters, like: straight chloroplasts and progametangial cells. In these cases, other associated vegetative and reproductive characters should be considered to delineate these two genera. Species of *Spirogyra* having progametangial cells always possess spiral chloroplasts and distinct conjugation canals. On the other hand, species of *Sirogonium* mostly show geniculation. Moreover, shape of the gametangia is also an useful criterion for delineation of *Sirogonium* from *Spirogyra*.

Genera like *Mougeotiopsis, Debarya, Mougeotia, Temnogametum, Zygnemopsis, Zygnema, Zygogonium* and *Pleurodiscus* are isogamous. However, *Mougeotia, Zygnema* and *Zygogonium*, especially *Zygnema* show a clear tendency towards anisogamy. Of these genera, only *Mougeotiopsis* and *Pleurodiscus* can be distinguished by their chloroplast structure. *Mougeotiopsis* possesses a single, quadrate, flat or plate-like axile chloroplast with slightly thickened and granulated margin (Randhawa, 1959) whereas, in *Pleurodiscus* chloroplasts are distinctly disc-shaped (Tiffany, 1936; A.K. and Pushpa Mahato, 1993). Remaining isogamous genera possess basically two types of chloroplasts, viz., axile plate-like and axile stellate to ovoid-type. *Debarya, Mougeotia* and *Temnogametum* are characterized by former type whereas, *Zygnemopsis, Zygnema* and *Zygogonium* possess latter type of chloroplast.

*Debarya* is separated from *Mougeotia* by the presence of pectic cellulose colloid in the gametangia. Moreover, in *Mougeotia* only a part of the cytoplasm of a cell is used in the production of a gamete and hence, cytoplasmic residue is left in the gametangia. The genus *Temnogametum* is distinguished from *Mougeotia* and *Debarya* by the presence of progametangial cells and a mucilaginous layer around the zygosporangia. Another important character which separates *Temnogametum* from *Mougeotia* is whole of the protoplast of a gametangium forms a gamete and no cytoplasmic residue is left.

*Zygnemopsis, Zygnema* and *Zygogonium* are not easily distinguishable in their vegetative phase as they possess similar type of chloroplasts. Ofcourse, Fritsch (1916) and G.S. West and Starkey (1915) stated tht chloroplasts of *Zygogonium* are of special type, where the cells contain a single axile chloroplast which is deeply constricted and sometimes twisted at the middle, with one pyrenoid in each of the irregular lobes. The chloroplast divides into two just prior to cell division. The cell sap is often rich in tannins. The statement is true only in case of a terrestrial species. Freshwater species reveal two stellate, semistellate, ovoid or even sometimes an axile, plate-like chloroplast in each cell (A.K. and Pushpa Mahato, 1993, 1995). So, *Zygnemopsis*, *Zygnema* and *Zygogonium* are separable only by their specialized reproductive characteristics. Like *Mougeotia*, in *Zygogonium* only a part of the protoplasm forms a gamete and cytoplasmic residue is left in the gametangia. In addition to this, zygospores are formed in the zygosporangia. *Zygnema* lacks all possible types of specialized reproductive characters. Like *Debarya*, in *Zygnemopsis* pectic substance is deposited in the gametangia and in most of the cases zygospores are extended into both the gametangia. Distinguishing reproductive characters of isogamous genera are summerized in Table-2.

**Table 2. Summary of the distinguishing characters of isogamous genera of Zygnemaceae**

| Chloroplasts | Genera | Cytoplasmic residue left in the gametangia | Pectic substance in the gametangia | Progametangial cells | Zygosporangia | Gelatinous substance surrounding zygospores |
|---|---|---|---|---|---|---|
| Single, axile, | *Mougeotiopsis* | — | — | — | — | — |
| flat plate- | *Debarya* | — | + | — | — | — |
| shaped | *Mougeotia* | + | — | — | ± | ± |
| Two, axile, | *Temnogametum* | — | — | + | — | + |
| stellate, | *Zygnemopsis* | — | + | — | — | — |
| ovoid or | *Zygnema* | — | — | — | — | — |
| disc-shaped | *Zygogonium* | + | — | ± | ± | ± |
| | *Pleurodiscus* | — | — | — | — | + |

We (A.K. and Pushpa Mahato, 1993) have clearly defined all the 15 genera of the family Zygnemaceae, but preceding discussion evidences that *Spirogyra* and *Sirogonium* need some emendation to accommodate some species of unusual characters. Hence considering the present need, definations of these two genera are emended as follows:

**Spirogyra (Link, 1820) emend. nov.**

Chloroplasts one to many, ribbon-like, spiral, *rarely straight*, pyrenoids many in each chloroplast; conjugation scalariform as well as lateral, physiologically anisogamous, conjugation brought about by conjugation canals, *progametangial cells rarely present*, whole of the protoplasm forms a gamete in each gametangium, other specialized reproductive characters absent; *monoecism rare.*

**Sirogonium (Kützing, 1843) emend. nov.**

Chloroplasts 2—many, narrow, delicate, straight or nearly straight, *rarely spiral*, pyrenoids many in each chloroplast, cell wall lacks an external pectose layer; conjugation strictly scalariform, morphologically anisogamous, conjugation brought about by geniculation of gametangia, *conjugation canals rarely formed*, whole of the protoplasm forms a gamete in each gametangium, progametangial cells present, other specialized reproductive characters absent; *monoecism frequent.*

## SPECIATION IN ZYGNEMACEAE

Members of Zygnemaceae are cosmopolitan in distribution and have been reported from every continent except Antarctica. Of the 15 genera, *Spirogyra*, *Zygnema* and *Mougeotia* are ubiquitous in distribution comprising more than 700 species. The monographs compiled so far recorded ever increasing species of these three genera as given below.

| Genera | species in the monographs | | |
|---|---|---|---|
| | Transeau, 1951 | Randhawa, 1959 | Kodlubouska, 1984 |
| *Mougeotia* | 99 | 108 | 138 |
| *Zygnema* | 95 | 100 | 139 |
| *Spirogyra* | 275 | 289 | 386 |

Now the question arises whether these genera include such a large number of species or some of them are being created on account of wrong interpretation? Both may be possible, but the latter part of the question arises other pertinent questions that are the traditionally used taxonomic characters sufficient enough to interpret the gross morphology of a taxon or additional characters are needed? Hoshaw (1980) suggested that additional studies should be conducted to establish the validity of presently described species. We also think, characters used for speciation are not sufficient enough to explain the gross morphology of a taxon and ultimately new species are being created on account of minor variation in much used taxonomic characters like cell width, size of the zygospores, ornamentation of the spore wall, etc. so intensive studies should be conducted to gather additional taxonomic characters and statistical data for speciation.

The traditionally used taxonomic characters for speciation in Zygnemacean algae are: (i) length and width of vegetative cells (ii) shape of zygospores (iii) length and width of zygospores (iv) colour of zygospores and (v) ornamentation of the spore wall. Some additional characters like number of chloroplasts and their turns, nature of formation of conjugation canals and shape of the receptive gametangia are also considered in case of *Spirogyra* and *Sirogonium.*

Among these the much used taxonomic characters, viz., length and width of vegetative cells and length, width and colour of zygospores are variable. Uspenskii (1934) observed that cell width in *Spirogyra* can be made to fluctuate appreciably above and below normal values by increasing or decreasing the growth rate of filaments. Allen (1958) found the occurrence of a species complex of three filament width groups in a clonal culture of *Spirogyra pratensis* Czurda. She confirmed these width groups as being haploids, diploids and tetraploids. She called these series of filaments a "species complex". It is different from "polyploid complex" of higher plants (Grant, 1981). Wang *et al.* (1986) have used th term "species complex" to refer to a series of strains, derived from a single clone that differ in ploidy and morphology. Simons (1987) used the term "species complex" to refer to a group of co-occurring species whose taxonomic descriptions differed only in quantitative characters. Mc Court *et al.* (1986) and Hoshaw *et al.* (1987) stated that polyploid species complex in *Spirogyra* is common and wide spread in nature. Harada and Yamagishi (1984a, b) studied mitosis and meiosis in some species of *Spirogyra.* They observed pairs of homologous chromosomes in vegetative cells of four species and concluded that each was diploid in vegetative condition. Their studies support the hypothesis that polyploidy is common in *Spirogyra.* Allen (1958), Hoshaw *et al.* (1985) Wang *et al.* (1986) demonstrated that ploidal change in *Spirogyra* resulted in morphological change. Lewis (1980) observed that cell volume usually increases with ploidal increase. Price (1976) stated that DNA content is generally correlated with the cell volume in higher plants. Shuter *et al.* (1983) observed the same result in unicellular eukaryotes. Miller and Hoshaw (1974) concluded from their experimental results that cell width is not a valid character for *Zygnema circumcarinatum* Czurda. Hoshaw (1980) suggested that variation in cell width due to polygenic factors and change in ploidy demonstrate the uncertainties of speciation in Zygnemacean algae. He further suggested for additional studies to establish the validity of presently described species.

The zygospore wall of *Spirogyra* and most of the Zygnemacean algae is composed of four or five layers. Of which outer and innermost layers are of a thin and hyaline nature and composed of cellulose and/or pectic substances (Ogawa, 1982; Simons *et al.*, 1982). The mesospore consists of two or three layers. In three layered mesospore, the layers are termed outer, median and inner mesospore. The ornamentation of median mesospore is an important criterion for speciation. Hoshaw (1980) studied zygospore wall structure in *Sirogonium* and concluded that because of its conservative nature, ornamentation of zygospore wall should be treated as a major criterion for speciation. Hull *et al.* (1985) studied zygospore wall structure in *Spirogyra* and *Sirogonium* and considered its validity for identification of species.

Types of ornamentation of the spore wall are very less in number in comparison to the number of species, so a good number of species of a genus is characterized by the same type of spore wall ornamentation. In this case, the character, zygospore wall structure would not be of much help.

Ecological, cytological, genetical and physiological factors either alone or together alter the phenotype of a particular species. It is not possible to study all these factors at a time by a systematist, but one can try to consider all visible characters in detail, i.e. the gross morphology. After handling a large number of samples of Zygnemacean algae (1982-1996), we are able to say that some of the characters like shape, margin and nature of turning of chloroplasts; shape of conjugation canals; shape of islands surrounded by gametangia and conjugation canals; shape of receptive gametangia; shape of zygospores and orientation of zygospores in the gametangia show their conservative nature. Moreover, amount of the photosynthetic apparatus in vegetative cells and the area occupied by the zygospores in the receptive gametangia are also constant for a particular species.

As we have stated above, traditionally used taxonomic characters are not sufficient enough to explain the gross morphology of a taxon, hence we suggest to consider following vegetative and reproductive characters in addition to preexisting ones.

### Length and width of the chloroplast

When chloroplasts are flat, plate-like or ribbon-shaped, their length and width vary from species to species within a genus. The chloroplast width gives a distinct morphology not only to the chloroplast itself, but also to the respective vegetative cell. On the basis of width, chloroplasts of *Spirogyra* are divisible into four types, viz., flat ribbon-shaped, wide tubular, narrow tubular and thread-like (Fig. 2). Percentage of the chloroplast material in a vegetative cell is also an important factor. It should be calculated by using the formula: $CHL_1 = \frac{l \times w \times n \times 100}{L \times W}$

(Where $CHL_1$ = percentage of the chloroplast material, l = mean length of chloroplasts of a cell, w = mean width and n = number of chloroplasts. L and W stand for length and width of the vegetative cell respectively).

$$\overline{CHL}\ \text{(Arithmetic mean)} = \frac{CHL_1 + CHL_2 ------ CHLn}{n}$$

(Here n stands for number of observations).

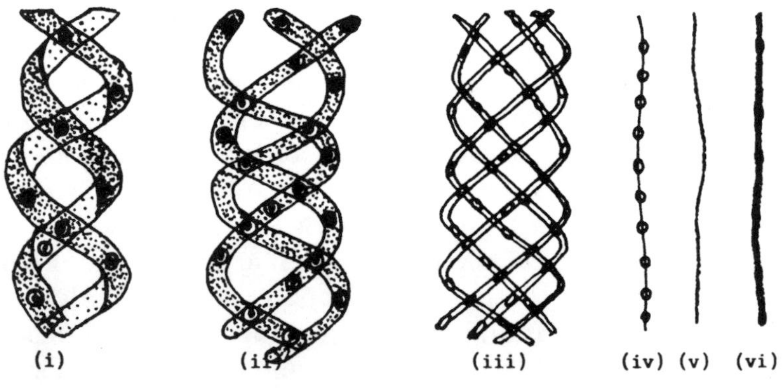


Fig. 2. Chloroplast width in *Spirogyra*. (i) flat, ribbon-shaped (ii) wide-tubular type (iii) narrow-tubular type (iv) - (v) thread-types.

### Margin of the chloroplast

Like width, chloroplast margin is also variable (Fig. 3). Several types of margins are seen in the species of *Spirogyra*. These are: smooth, undulate, dentate, crenate, lobed, spiny and incised-type. The margin of the chloroplast also play an important role in chloroplast morphology. So it should be compared with the described species during the identification of a taxon.

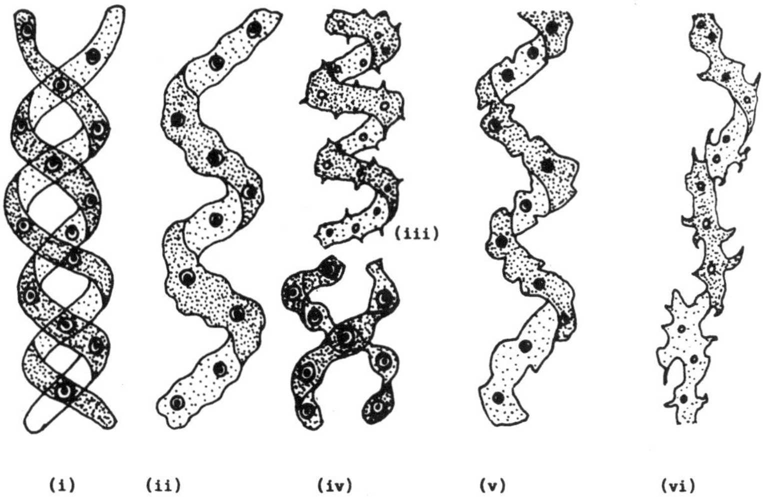


Fig. 3. Chloroplast margin in *Spirogyra*. (i) smooth (ii) undulate (iii) dentate (iv) lobed (v) incised (vi) spiny.

## Nature of turning of the chloroplast

The genus *Spirogyra* stands apart from all other Zygnemacean algae by the nature of turning of its chloroplasts. Basically, two types of turnings are met with in the species of *Spirogyra* (Fig. 4), viz., spiral or twisted-type and angular-type. Rarely chloroplasts are irregular in their distribution. In case of spiral-type, the average length of turns of chloroplasts in a cell is correlated with the cell length in a particular species. A relation can be established by measuring these two factors. It is seen that the ratio obtained from these two data is constant for a particular species. It evidences that the length of turn is directly proportional to the cell length in a species. In angular-type the magnitude of the angle making at the point of turning seems to be a more useful criterion than the number of turns. Mostly it varies from $30^0$ - $60^0$. In some exceptional cases the angle may be larger than $75^0$ where the chloroplasts are nearly straight or smaller than $30^0$, in close angular-type.

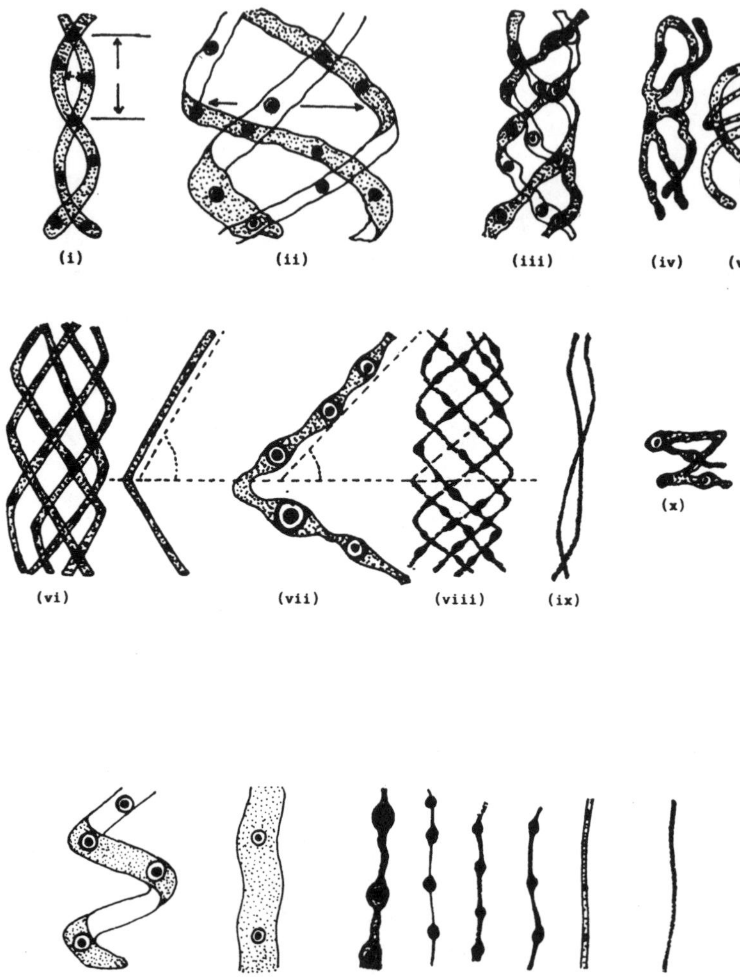


Fig. 4. Nature of turning of chloroplast in *Spirogyra*. (i) - (iii) spiral types (iv) - (v) irregular types (vi) - (x) angular types.

## Pyrenoids in the chloroplast

Pyrenoids are the special organs of the cells which persist even under conditions of starvation (Randhawa, 1959). Structurally, the pyrenoid consists of a central proteinaceous core sheathed by minute starch plates. According to Smith (1950) they play important role in starch formation. Czurda (1928) has shown a definite relation between the amount of cytoplasm and the total pyrenoid substance in a cell.

In other words, there is a definite relation between the cell volume and the total pyrenoid substance. The percentage of pyrenoid substance present in a cell can be calculated by using the following formula:

$$p_1 = \frac{\Pi r^2 n \times 100}{L \times W}$$

(Where r = mean radius of pyrenoids, n = total number of pyrenoids in a cell and L and W stand for length and width of the cell, respectively).

$$\bar{p} \text{ (arithmetic mean)} = \frac{p_1 + p_2 + - - - - p_n}{n}$$

(Here n stands for number of observations)

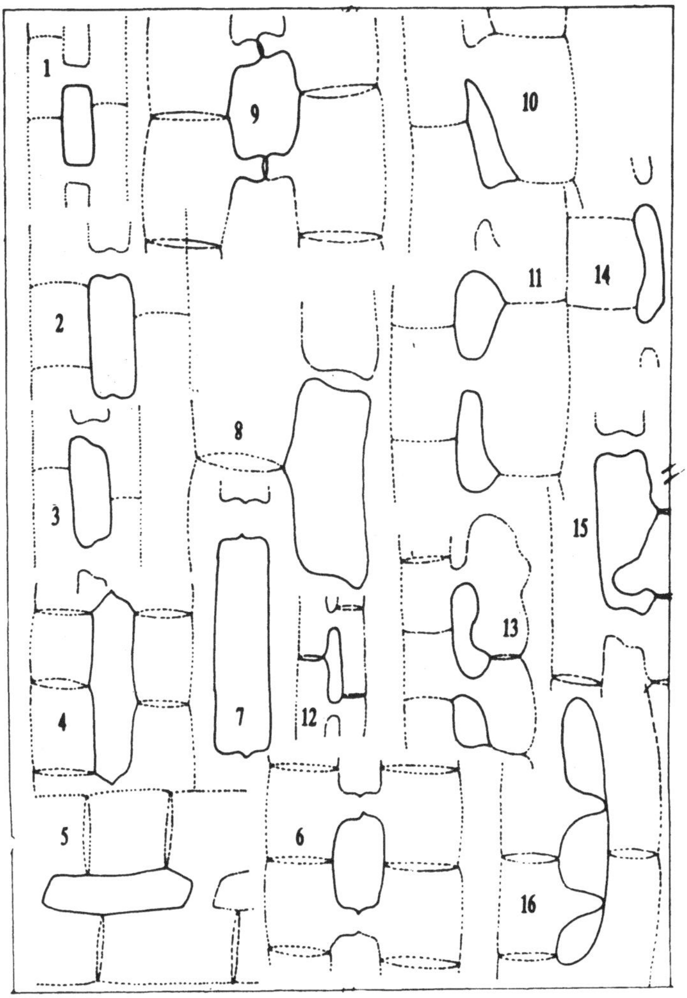


Fig. 5. Shape of the chloroplast in *Spirogyra.* (i) - (ii) normal types (iii) - (vi) beaded types (vii) pitted type (viii) threaded type.

On the other hand, the size of the pyrenoids in relation to chloroplast width is equally important as it imparts a distinct morphology to the chloroplast of a species. On the basis of the size of pyrenoids chloroplasts of *Spirogyra* are of four types (Fig. 5):

(i) *Normal type*: where chloroplasts are ribbon-shaped or wide tubular-type and pyrenoids are either equal or slightly smaller than the chloroplast width.

(ii) *Beaded type*: chloroplasts are ribbon-shaped to thin thread-like, but pyrenoids are larger than the chloroplast width, imparting a beaded appearance to the chloroplast.

(iii) *Pitted*: chloroplasts are narrow, tubular-type and pyrenoids are caryoid-like in appearance, i.e., proteinaceous core is either viscous mass or absent.

(iv) *Threaded*: chloroplasts are thin thread-like and delicate with ill-defined pyrenoids.

## CONJUGATION CANAL

In the monographs (Transeau, 1951; Randhawa, 1959; Kadlubouska, 1972, 1984) the only criterion which is taken into account for speciation is the nature of formation of conju-gation canals, i.e., whether canals are formed by only the male gametangia or by both the gametangia. Other important characteristics like shape and size of the canals have not been considered. Each species is characterized by a distinct type of conjugation canal. Shape and size of the canal depends upon shape and size of the papillae produced by male and female gametangia. Mostly, papillae are dome-shaped, in some cases they are either conical or button-shaped or even tubular. Generally, the female papillae are shorter and wider than the male, though it is not always true. Part shared by each papilla in the formation of a conjugation canal is an important factor as it gives a definite shape to the canal together with shape of the papillae. In Addition to this, it also indicates the degree of sexuality of each of the mating filament. Broadly, canals are of four types in *Spirogyra*. These are: tubular, cup-shaped, socket-shaped and wide. (Fig. 6). Tubular-type canals are again divisible into: short-tubular, long-tubular, straight-tubular, oblique-tubular, constricted-tubular and inflated-tubular. In tubular type and wide type, canals are formed by both the game-tangia whereas, cup-shaped canals are formed mostly by only the male gametangia. In socket-type the male papillae are conical in shape.

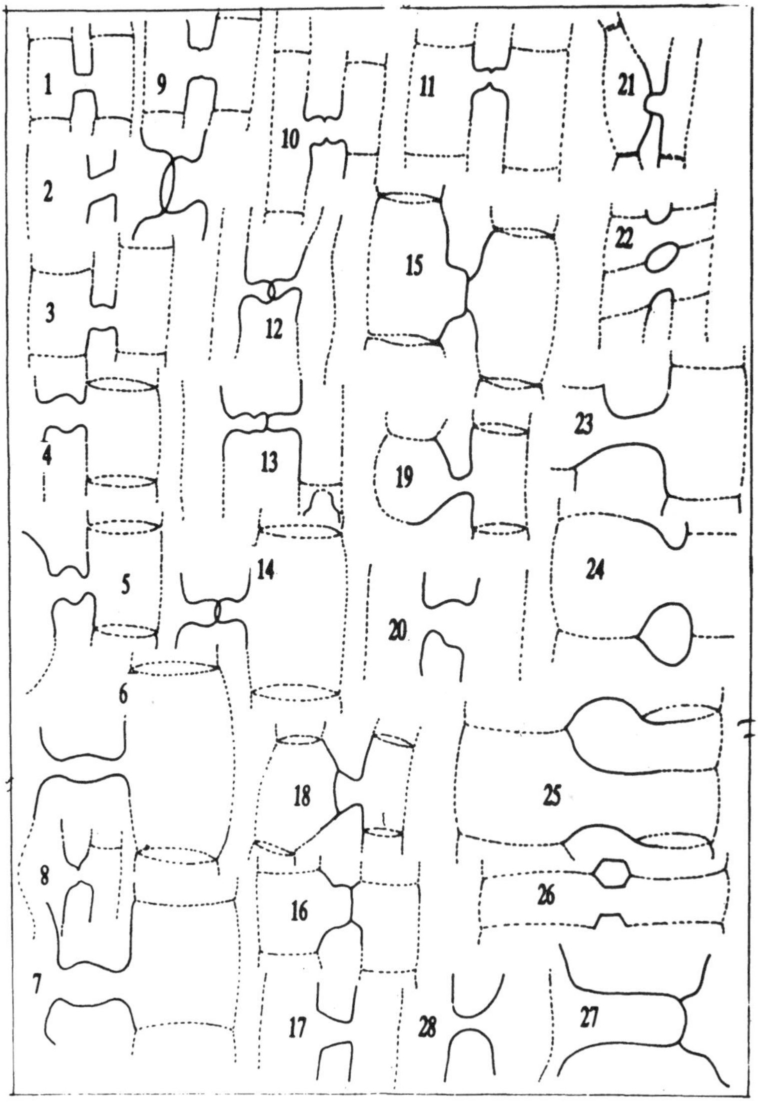

Fig. 6. Types of conjugation canals in *Spirogyra*. 1-24: tubular types, 15-16, 21: socket shaped, 17-20: cup shaped, 22-26: wide types, 27: long-tubular type.

## SHAPE OF ISLANDS

The shape of the space surrounded by conjugation canals and gametangia of mating filaments which may be termed "island", varies from species to species (Fig. 7). Undoubtedly, it is an important criterion for speciation in Zygnemacean algae. Except in a few cases, only one type of island is formed in a species. In rare cases, mixed-type islands are formed. Types of islands found in the members of Zygnemaceae are: rectangular, constricted-rectangular, compressed-rectangular, quadrate, compressed-quadrate, pentagonal, hexagonal, sub hexagonal, triangular with or without rounded ends, trapezoid, ovate, compressed-ovate, oblong-ovate, obovoid, obpyriform, oblong, cordiform, rhomboid, lanceolate, spathulate and fusiform.

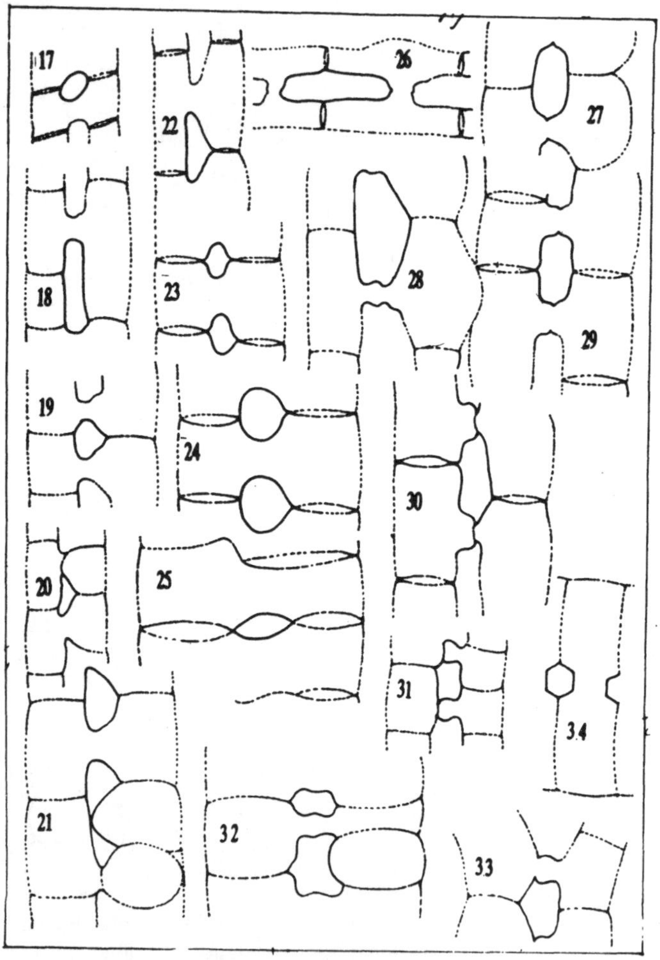


Fig. 7. Types of islands in *Spirogyra*.

## ORIENTATION OF ZYGOSPORES

Zygospores are formed either in the conjugation canals or in one of the gametangia. In anisogamous species, where zygospores are formed in one of the gametangia, they are oriented in various ways. In some cases they lie either parallel or make right angle with the long axes, but in other cases zygospores make some angles with the long axes which varies from $0^0$ to $90^0$ (Fig. 8). Average magnitude of this angle is definitely an important criterion for speciation, especially in *Spirogyra* and *Sirogonium*.

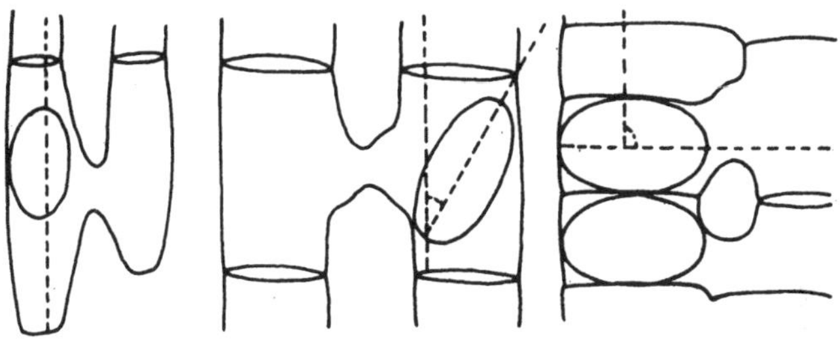


Fig. 8. Orientation of zygospores in the gametangia. (i) parallel to the long axis (ii) zygospore making an angle of $30^0$ with the long axis (iii) zygospore making right angle with the long axis.

## SPACE OCCUPIED BY ZYGOSPORES IN THE RECEPTIVE GAMETANGIA

When zygospores are similar in shape and size in two or more different species,the percentage of area occupied by the zygospores in the receptive gametangia seems to be an useful criterion for speciation. It indicates the size of the zygospores in comparison to the size of the receptive gametangia. It should be calculated by using the formula:

$$Zg_1 = \frac{\text{Area of the zygospore} \times 100}{\text{Area of the gametangia}}$$

$$\overline{Zg} \text{ (Arithmetic mean)} = \frac{Zg_1 + Zg_2 + \text{-------} Zg_n}{n}$$

The area of the receptive gametangia and of the zygospores should be calculated by using the appropriate formulae. When an appropriate formula is not available, it is suggested to plot these sketches on the graph paper. By counting the divisions areas of these two can be calculated very easily.

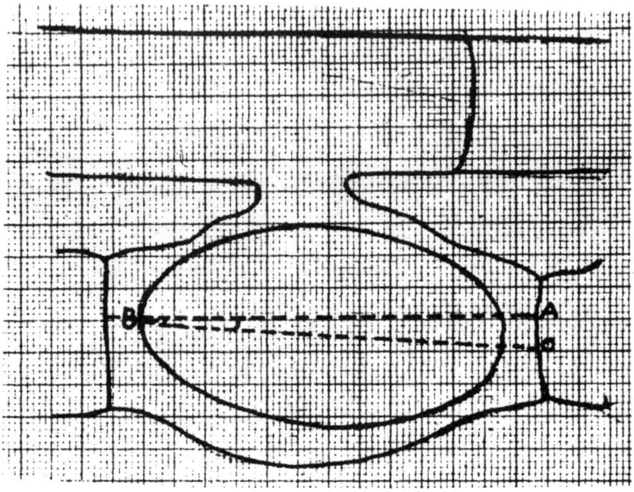

Fig. 9. Space occupied by a zygospore in the receptive gametangia. The zygospore also showing an angle of $5^0$ with the long axis of the receptive gametangia.

## SPECIATION IN ISOGAMOUS GENERA

In isogamous genera, especially in *Mougeotia, Temnogametum, Zygnema* and *Zygogonium* the pectic substance plays an important role in the evolution of various genera as well as in the evolution of different species within a genus (A.K. and Pushpa Mahato, 1995). Progressive elimination, i.e. selection pattern of pectic substance is very interesting in the species of these genera and this criterion may be used safely for speciation. In *Zygnema spiralc* Fritsch, *Mougeotia capucina* (Bory) Agardh, *M. prone* Transeau, *M. irregularis* W. and G.S. West, *M. tambaramensis* Iyengar, *M. trapeziformis* Iyengar and *M. tropica* (W. and G.S. West) Transeau intervening space of the gametangia is filled with the pectic colloid. In *Zygnema pseudomougeotia* Gauthier-Livre, *Z. algeniense* Gauthier-Livre, *Mougeotia victoriensis* G.S. West, *M. indica* Randhawa and *M. kunjukuttii* Patel and Ashok Kumar each sporangium is surrounded by an extensive layer of pectin which extends even beyond the outer sides of the gametangium. The amount of pectic substance is remarkably decreased and each spore is surrounded by only a thin layer of pectin in *Zygnema carinatum* Taft, *Zygogonium pectosum* Taft and in many species of *Mougeotia*, viz., *M. pectosa* Transeau, *M. pawhuskae* Taft, *M. maltae* Wittrock, *M. caimani* transeau, *M. cherokeana* Taft and *M. seminoleana* Tiffany. The outer sporangial wall of *Zygogonium pectosum* is pectic in nature whereas, in *Z. stephensiae* Transeau, *Mugoeotia disjuncta* Transeau, *M. chlamydata* Prescott and *M. oedogonioides* Czurda median spore wall is pectic. In *Zygnema circumcarinatum* Czurda and *Mougeotia atubulosa* Kolkwitz and Krieger at the time of spore maturation the tube wall is changed to pectic compounds. The pectic substance is very little in amount and confined only to the gametangial wall in *Mougeotia africana* (G.S. West) Transeau and *M. islamii* Patel and Ashok Kumar. It is almost eliminated from reproductive phase and shifted to vegetative phase in *Zygnema gangeticum* Rao where it forms a sheath upon the vegetative cell, a case more or less similar to higher members of this family.

## DISCUSSION AND CONCLUSION

From the preceding discussion it is evident that polyploidy is common in the members of Zygnemaceae, especially in *Spirogyra* (Mc Court *et al.*, 1986; Hoshaw *et al.* 1987; Harada and Yamagishi, 1984a, b). In addition to this, a genotype produces different morphotypes in different ecological conditions. Now the question arises itself, whether the characters like chloroplast morphology of a vegetative cell, amount of pyrenoid substance in a cell, thickness of the cell wall, shape and size of the conjugation canals, shape of the islands, space occupied by the zygospores in the receptive gametangia, etc. change with the cell width in different morphotypes of a species complex or not? If the change in quantitative characters come in the same proportion and statistical ratios display the same constant numbers, but characters like chloroplast morphology, shape of conjugation canals, shape of islands, shape of gametangia, shape of zygospores and ornamentation of the spore wall remain unchanged, in that case morphotypes occurring in a "species complex" are undoubtedly of the same species. In either case they may be of different species. In other cases also speciation should be done by comparing all above characters.

In conclusion, we like to say that not a single character is negligible or worthless. Each character plays a significant role to interpret a taxon, so each visible character present in a taxon needs a close observation. In addition to quantitative data and ratios, the description of a species should include sufficient number of sketches and photomicrographs, so that they may be able to explain all the morphological characters, i.e., gross morphology of that taxon.

We hope, if a revision of the taxa, described in the present monographs is made with these criteria, certainly a dramatic change will come in the number of species in each member of Zygnemaceae.

## REFERENCES


Ahmad, A. and Goldstein, M.E. 1971. *Lloidiella*, a new genus of the Zygnemaceae. *Phycologia*. **10**(1): 1-6.

Ahmad, A. and Goldstein, M.E. 1972. *Lloidiella*, a new generic name for the later homonym *Lloidiella* Ahmad and Goldstein (Zygnemaceae). *Phycologia*. **11**(2): 271.

Allen, M.A. 1958. The biology of a species complex in *Spirogyra*. *Ph.D. Thesis*, Indian a University, Bioomington, pp. 240.

Ashok Kumar, C.K. 1972. Study of Zygnematales of Gujarat. *Ph.D. Thesis*, Sardar Patal University, Vallabh Vidyanagar, Gujarat, India.

Bourrelly, P. 1966. *Les Algues D' Eau Douce Vol. I* Pub. N. Boubee & Cie, Paris, pp. 511.

Czurda, V. 1932. *Zygnemales* In: A. Pascher's süsswasserflora von Mitteleuropa. **9**: 1-232.

Fritsch, F.E. 1916. The morphology and Ecology of an Extreme Terrestrial form of *Zygnema* (*Zygogonium*) *ericetorum* (Kütz.) Hansg. *Ann. Bot.* **30**: 135-149.

Gauthier-Lievre, L. 1965. Zygnemacees Africaines. *Beih. Nova Hedwigia*. **20**: 1-210.

Grant, V. 1981. *Plant speciation 2nd edn.* Pub. Columbia University press, New York, pp. 563.

Harada, A. and Yamagishi, T. 1984a. Mitosis in *Spirogyra* (Chlorophyceae). *Japan. J. Phycol.* **32**: 1-9.

Harada, A. and Yamagishi, T. 1984b. Meiosis in *Spirogyra* (Chlorophyceae). *Japan. J. Phycol.* **32**: 10-18.

Hoshaw, R.W. 1980. Systematics of the Zygnemataceae (Chlorophyceae). II. Zygospore wall structure in *Sirogonium* and a taxonomic proposal. *J. Phycol.* **16**: 242-250.

Hoshaw, R.W. and Mc Court, R.M. 1988. The Zygnemataceae (Chlorophyta): a twenty year update of research. *Phycologia*. **27**(4): 511-548.

Hoshaw, R.W., Wang, J.C., Mc Court, R.M. and Hull, H.M. 1985. Ploidal changes in clonal cultures of *Spirogyra communis* and implications for species definition. *Am. J. Bot.* **72**: 1005-1011.

Hoshaw, R.W., Wells, C.V. and Mc Court, R.M. 1987. A polyploid species complex in *Spirogyra maxima* (Chlorophyta, Zygnemaceae), a species with large chromosomes. *J. Phycol.* **23**: 267-273.

Hull, R.M., Hoshaw, R.W. and Wang, J.C. 1985. Interpretation of zygospore wall structure and taxonomy of *Spirogyra* and *Sirogonium* (Zygnemataceae, Chlorophyta). *Phycologia*. **24**(2): 231-239.

Kadlubouska, J.Z. 1984. Chlorophyta VIII. Conjugatophyceae I (Zygnematales) In: *Süsswasserflora von Mitteleuropa, Band 16* Pub. Gustav Fischer Verlag, Stuttgart, pp. 532.

Kolkwitz, R. and Krieger, H. 1941. *Zygnemales*. In: Rabnhorst's Kryptogamen-Flora von Deutchland, Osterreich und der Schweiz, 13, Abt. 2, Lief. 1-2, Leipzig, pp. 499.

Lewis, I.F. 1925. A new conjugate from woods Hole. *Am. J. Bot.* **12**: 351-357.

Lewis, W.H. 1980. Polyploidy in species populations. In: *Polyploidy, Biological Relevance* Ed: Lewis, W.H., Plenum Press, New York, pp. 103-144.

Mahato, A.K. 1993. The germination of zygospore *in situ* in *Spirogyra condensata* (Vaucher) Kützing. *Acta Botanica Indica*. **21**: 305-307.

Mahato, A.K. 1995. Two new species of the genus *Spirogyra* (Zygnemaceae, Chlorophyceae) from South Bihar, India. *J. Econ. Tax. Bot.* **19**(2): 464-468.

Mahato, A.K. and Mahato, Pushpa. 1993. Affinities, Evolution and Taxonomy of Zygnemaceae (Chlorophyceae): A Revaluation. *Bionature*. **13**(2): 135-156.

Mahato, A.K. and Mahato, Pushpa. 1994. *Sangirellum taeniforme* and *Trigonum indicum*, two new members of Zygnemaceae from India. *Phykos.* **33**(1 & 2): 113-124.

Mahato, A.K. and Mahato, Pushpa. 1995. Role of pectic substances in the Evolution of members of Zygnemaceae. In: *Biology of Conjugales* Ed. Vidyavati and M.N. Noor, Printwell, Jaipur, India. pp. 188-194.

Mc Court, R.M., Hoshaw, R.W. and Wang, J.C. 1986. Distribution, morphological diversity and evidence for polyploidy in North American Zygnemataceae (Chlorophyta). *J. Phycol.* **22**: 307-313.

Miller, R.D. and Hoshaw, R.W. 1974. Cell width as a taxonomic character with reference to *Zygnema circumcarinatum* Czurda *Br. Phycol. J.* **9**: 145-148.

Ogawa, S. 1982. Disintegration of chloroplasts during zygote formation in *Spirogyra verruculosa. Bot. Mag. Tokyo.* **95**: 249-260.

Patel, R.J. 1980. Genus *Temnogyra* Lewis. A reconsideration of its validity. *Taxonomy of Algae*, Univ. of Madras, pp. 492-498.

Patel, R.J. and Ashok Kumar, C.K. 1969. *Spirogyra tuwensis* sp. nov. with a note on its reproduction. *Phykos.* **8**(1 & 2): 105-107.

Price, H.J. 1976. Evolution of DNA content in higher plants. *Bot. Rev.* **42**: 27-52.

Printz, H. 1927. Chlorophyceae nebst Conjugatae, heterokontae und Charophyta. In: Engler and Prantl's Die naturlichen pflanzenfamilien. Zweite Auft. 3, Leipzig, pp. 463.

Randhawa, M.S. 1959. *Zygnemaceae* Pub. ICAR, New Delhi, pp. 478.

Shuter, B.J., Thomas, J.E., Taylor, W.D. and Zimmerman, A.M. 1983. Phenotypic correlates of genomic DNA content in unicellular eukaryotes and other cells. *Am. Nat.* **122**: 26-44.

Simons, J. 1987. *Spirogyra* species and accompanying algae from dune waters in the Netherlands. *Acta. Bot. Neerl.* **36**: 13-31.

Simons, J., Van Beem, A.P. and De Vries, P.J.R. 1982. Structure and chemical composition of the spore wall in *Spirogyra* (Zygnemataceae, Chlorophyceae) *Acta Bot. Neerl.* **31**: 359-370.

Transeau, E.N. 1951. *The Zygnemataceae* Pub. The Ohio State University Press, Columbas, pp. 327.

Usha Devi, K. and Panikkar, M.V.N. 1991. A new species of *Spirogyra* Link from Kerala, South India. *J. Econ. Tax. Bot.* **15**(3): 743-745.

Uspenskii, E.E. 1934. Transformation of certain hereditary forms into other ones and the arising of new forms in the genus *Spirogyra* (Abstr.) *Biol. Abstr.* **11**: 724.

Wang, J.C., Hoshaw, R.W. and Mc Court, R.M. 1986. A polyploid species complex of *Spirogyra communis* (Chlorophyta) occurring in nature. *J. Phycol.* **22**: 102-107.

West, G.S. and Starkey, C.B. 1915. A contribution to the cytology and life history of *Zygnema ericetorum* (Kützing) Hansg. etc. *New Phytol.* **14**: 194-205.

Yamagishi, T. 1963. Classification of the Zygnemaceae. *Sci. Rep. Tokyo Kyoiku Daigaku.* **11**: 191-210.

# Section - 4

Recent Trends in Algal Taxonomy
Vol. I. Taxonomical Issues
Editors : Vidyavati and A.K. Mahato
APC Publications Pvt. Ltd., New Delhi, 1999, pp. 179-220

# *Taxonomy of Charophyta—Trends and Perspective*

Sandhya Sinha and Ichha Purak
Department of Botany, Ranchi Women's College
Ranchi-834 001, Bihar

## INTRODUCTION

Among the algae, the Charophytes, commonly known as stoneworts, are marked out for their isolation, having no clear affinity to any other group. This isolation is due to their multicellular female organ, the complex antheridium, strong apical growth displayed by the vegetative shoot and its general degree of specialization rarely observed in green algae. The existence of sterile jacket cells in the oogonium is a remarkable innovation for the algae, although this jacket is not homologous to the sterile jacket making up the archegonium in the embryophyta. Further, unlike those in Bryophyta and Trachaeophyta, the antheridia of the Charophytes are much more complex and highly specialised. In apical dominance, too, the Charophytes have achieved a degree of polarity seldom found in Thallophyta. The growth of the plant has superficial resemblance to that of much higher Equisetales, but the absence of vascular system in the Charophyta denies it a position among advanced groups such as Equisetales.

The fossil evidences reported so far indicate that Charales were thriving in the rocks before the triassic. This led Dittmer (1964) to suggest that Charophytes branched off in the distant past from some algal ancestor such as *Draparnaldiopsis* and *Stigeoclonium* and reached the present stage through specialization and differentiation not shared by other algae. But their inability in developing the vascular system and sporangial organ necessary for any progenitor of higher plants, leads a school of thought to believe that Charophytes may be treated as flourishing plants in Chlorophyta, but at the end of their line. The Charophytes, besides lacking a clear affinity towards others in the algal groups, possess characteristic morphological features as outlined above which are often conflicting in nature from those of the rest. This situation and the varying interpretations of the evidences available at present, have together, resulted in a controversial position for the Charophytes. Additional evidences and information need to be collected to ascertain their position beyond doubt. The present emphasis on investigations based on cytological, karyomorphological and biochemical grounds is a move in the direction.

## ORIGIN AND PHYLOGENY

The origin and evolutionary history of Charophytes are yet to be clearly understood, although various hypotheses have been proposed from time to time by different phycologists. The two presently prevailing views are (i) Charophyta is on par with Chlorophyta and (ii) Charophyta is a derivative of Chlorophyta; though Bold and Wynne (1978) have put a question mark on Charophyta belonging to algae at all. Our present understanding of origin and evolution of Charophyta are bound to be affected by the evidences and information emerging out of future research in the field. Chapman and Chapman (1973), for similar reasons perhaps, have stated, "Any evolutionary scheme, however, can only be regarded as tentative".

Blackman (1900) suggests that green algae have evolved from ancestral unicellular motile forms (such as *Chlamydomonas*) along the following different lines:

(i) The Volvocine line: flagellated vegetative cell division, and organisation of motile colony.

(ii) The Tetrasporine line: loss of motility except in reproductive stages, and vegetative division leading to non-motile cells.

(iii) The Chlorococcine line: absence of vegetative division together with loss of motility except in reproductive stages.

(iv) The Rhizopodal line: evolution towards a naked amoeboid type of development.

Numerous studies on evolutionary trends among Chlorophyceae along these lines have been made by various phycologists, important among them being Blackman (1900), West (1915), Fritsch (1935), Womersley and Ophel (1947), Smith (1950, 1955), Iyengar (1951), Papenfuss (1951, 1955), Desikachary and Sundaralingam (1962), Silva (1962), Pal *et al.* (1962), Round (1963), Vaidya (1963), Dittmer (1964), Wood and Imahori (1965), Klein & Cronquist (1967), Prescott (1969), Bold and Wynne (1978) and others. According to Kumar and Singh (1982), these workers have based their findings of evolutionary trends within green algae on three important characters, namely thallus organisation, sexual reproduction and life cycle.

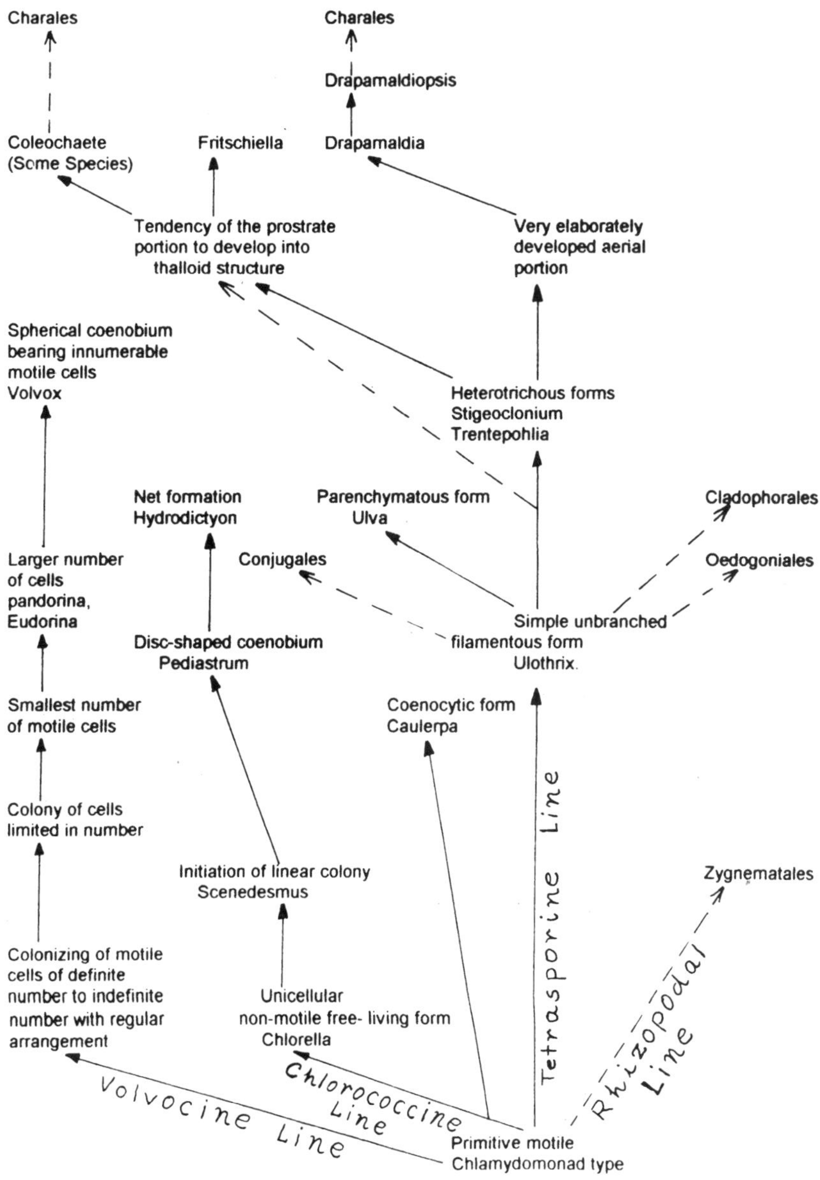


Fig. 1. A probable phylogenetic arrangement of Chlorophyta (based on Gangulee and Kar, 1982). Evolutionary trends under dispute are represented by dotted lines.

The aspects of phylogeny, interrelationship and evolution have also been discussed on cytological ground by a host of workers such as Kasaki (1964), Sarma (1964), Hotchkiss (1966), Guerlesquin (1967), Ramjee (1969), Khan and Ramjee (1970), Sawa (1965, 1974) and Bhatnagar (1988, 1989); the number of chromosomes and its morphology and karyotypes are being considered significant in this regard. However, electron microscopic studies of cell structure and comparative biochemistry reveal that cellular and biochemical level in green algae mainly determines its phylogeny and evolutionary trends (Pickett-Heaps and Merchant, 1972; Frederick *et al.* 1973; Fott, 1974; Stewart and Mattox, 1975; Pickett-Heaps, 1975).

Based on the morphological findings of various workers Gangulee and Kar (1982) and Mitra, Mitra and Choudhary (1991) have outlined the scheme of evolutionary trends among the Chlorophyta. A probable phylogenetic arrangement of Chlorophyta modified in light of recent findings is presented in Fig. 1.

As regards the Charales there are many different views on its position and origin. As shown in Fig. 1, it is generally believed that Charales have originated from *Chlamydomonas* through tetrasporine line of evolution via Ulotrichales. However, there are different opinions on the evolutionary trends thereafter. According to Desikachary and Sundaralingam (1962) Charophytes have Chaetophoralean origin as the Chaetophoralean genus *Draparnaldiopsis* resembles *Chara* in the differentiation of main filaments, formation of cortical filaments, limited growth of laterals, morphology of sex organs and its restriction to the laterals. On the other hand, the findings of recent electron microscopic and biochemical studies suggest that the Charales might have originated from certain Chaetophoralean members such as *Coleochaete*. In course of its evolution the Charales, after branching off from its green algal ancestor, developed, through specialization and differentiation, certain features not shared by rest of the algae. So most of the algologists place them in the Chlorophyta at the end of the line, as is evident from the phylogenetic arrangement suggested for Chlorophyta by Prescott (1984), which is shown in Fig. 2. But opponents of this view hold that the apical growth and complexity of sex organs go against the possible origin of stoneworts from Chaetophorales.

Similarity in morphological structure and development of sex organs between Bryophyta and Charophyceae led Papenfuss (1957) and Bold (1967) to suggest an affinity between the

Bryophyta and the Charophyceae, though the haplontic type of life cycle of Charophyceae points otherwise. Grambast (1974) is of the view that charophyta occupies an isolated position between green algae and Bryophytes. After taking note of greater elaboration of thallus structure, protection of sex organs by sterile cells and striking similarity between the antherozoids of Charales and those of higher plants, Church (1919) suggested, "*Chara* is a remnant of many probable evolutionary tendencies that failed to attain land habits". According to him, despite this failure *Chara* continued to retain some of the characters of higher plants while other members of Chlorophyta failed to retain these characters and therefore he was of the view that *Chara* represented just a "mere specialised side line" of the Chlorophyta. However, Bell and Woodcock (1968) considered that the stoneworts are highly specialised group of aquatic plants long removed from the main trends of algal evolution.

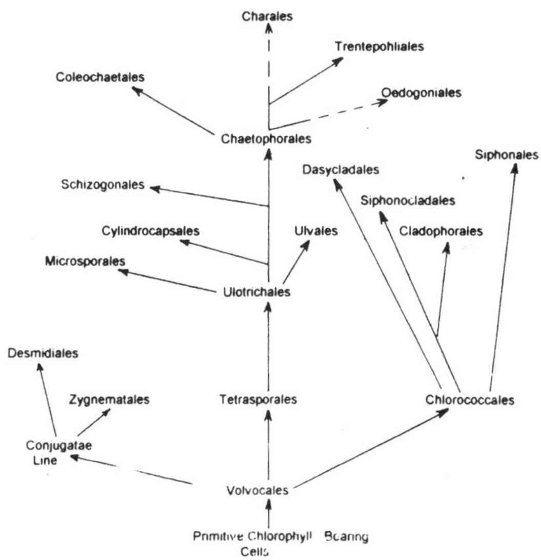


Fig. 2. The phylogenetic arrangement of Chlorophyceae (After Prescott, 1984)

## TAXONOMY

The taxonomic position of stoneworts continue to be controversial; suggestions over a wide spectrum ranging from its exclusion from algal group itself to reducing it to the rank of an order under the class Chlorophyceae have been put forward. Initially the classification of the algae were based entirely on their morphological aspects, but over the years, diverse characters, at times even combination of characters, have come into play while deciding the status of a taxon.

Early workers relied solely on their form. Later, cytology along with their life history were also used. Gradually, other characters such as the chemical nature of reserve products, the pigment complements, the submicroscopic wall and flagellar structures, the number of chromosomes and its karyotypes, nuclear and cell division, organisation of zoospores or gametes and their comparative biochemistry, have come to be employed as taxonomic tools. A view is emerging that there is a need to take a whole view of the issue wherein combination of characters should be taken into consideration. While doing this, at the moment, there appears no obvious advantage in attaching overriding importance to any one type although morphological framework is generally taken as the reference. Many different types of information and characters may emerge in future through employment of new and advanced techniques, which may require modification of the existing concept of taxonomy. The taxonomy itself has kept on acquiring new dimensions with passage of time; starting with what is known as "Classical (or orthodox or alpha) taxonomy" it now encompasses "Experimental (or Omega) taxonomy", "Chemotaxonomy" and "Numerical taxonomy".

The earlier phycologists examining the issue on morphological grounds placed stoneworts under *Equisetum*. Adanson (1763) and others placed them under monocotyledonous phanerogams (Naiadeae), while some others referred to their affinity to dicotyledonous phanerogams (Ceratophyllaceae).

Agardh (1824) grouped them along with Confervaceae: Brongniart placed them near the Filicales and the Marsiliaceae while Berkely recognised them as separate class Charales, intermediate between Muscineae and Thallophyta. Linnaeus (1778) placed them in Class Monoecia, Section Monandria. However, Smith (1800) removed them to class Monandria section Monogynia. Richard (1815) gave them the rank of an order Characeae while Bennett (1878) regarded them as abnormal form of Muscineae. Sachs (1875) raised them to the rank of a division by the side of Thallophytes and Muscineae as he regarded them distinct from the Thallophytes.

The twentieth century phycologists, recognising the algal nature of the stoneworts—as is evident from the fact that they grow in water, have little internal differentiation and are haploid almost throughout the whole life cycle—are almost unanimous in considering them among algae, but there are a few exceptions. According to Grambast (1974) stoneworts occupy an isolated position between green algae and Bryophytes. Bold and Wynne (1978), too, have raised a question mark about these plants being considered among algae.

Considering the characters of the stoneworts and noting their similarities and dissimilarities with those of various other groups, the following three hypotheses have been put forth:

1. Stoneworts should be given the status of a separate division Charophyta intermediate between the Thallophyta and the Bryophyta. Some of their characters such as equisetoid organisation of the thallus, nodal arrangement of branchlet whorls, apical growth, discoid chloroplast, complex sex-organs with sterile jacket, spirally coiled biflagellate antherozoids and the protonemal germination of oospore, which are quite distinct from those of green algae are mainly responsible for taking this group out of Chlorophyta. Further, despite similarity of complexity of sex organs stoneworts have not been placed in Bryophyta as they have haplontic life cycle whereas, the Bryophyta has haplodiplontic one. Though retaining some of the characters of higher plants, the stoneworts, in absence of vascular tissue and sporangial organ essential for progenitor of terrestrial plants, can not be placed among the higher plants.
2. Stoneworts be given the rank of a class Charophyceae along with the Chlorophyceae under the division Chlorophyta. The adherents of this view hold that the difference between the stoneworts and the main body of the Chlorophyceae are not so great as to treat them as a separate division. Infact, according to Goebel and Hofmeister, the sex organs though apparently complex, are in actual practice simple and single celled. Inspite of these differences, in many of their biological and biochemical characteristics the stoneworts resemble the green algae.
3. The third view is to treat the stoneworts as an order charales of the class Chlorophyceae. Supporters of this line base their argument on the resemblance of stoneworts to the typical green algae in their algal simplicity, cellulose cell wall, starch as food reserve, presence of both chlorophyll a & b as photo-

**Table 1. Taxonomic consideration of charophytes**

| Author(s) | Division Phyta* [Subdivision** Phytina*] | Class phyceae [Subclass Phycidae] | Order ales [Suborder ineae] | Family aceae [Subfamily oideae] | Tribe eae [Subtribe inae] | Genus |
|---|---|---|---|---|---|---|
| Wildeman, 1900 | | Charophycees | Characees | Characees | | Chara & Nitella. |
| Oltmanns, 1904 | | Chlorophyceae | Charales | | | Chara, Nitella, Lychnothamus, Tolypella & Lamprothanium. |
| Fritsch, 1935 | | Chlorophyceae | Charales | Characeae [Nitelleae & Chareae] | | Nitella & Tolypella Chara, Lamprothamnium, Lychnothamnus & Nitellopsis. |
| Tilden, 1937 | | Chlorophyceae | Charales | Characeae | | Nitella, Tolypella, Tolypellopsis, Lamprothamnus, Lychnothamnus & Chara. |
| Smith, 1938, 1951, 1958 | Chlorophyta | Charophyceae | Charales | Characeae | | Chara, Nitella, Tolypella, Lychnothamnus & two others. |
| Sachs, Groves, Allen and others, 1875, 1924, 1927 | Charophyta | | Charales | Characeae [Nitelleae chareae] | | Nitella & Tolypella Nitellopsis, Lychnothamnus, Lamprothamnium, Chara & Protochara. |

| | | | | | | |
|---|---|---|---|---|---|---|
| Pal, Kundu, Sundaralingam & Venkataraman, 1962. | Charophyta | | Charales | Characeae [Nitelleae & Chareae] | | Nitella & Tolypella Nitellopsis, Lychnothamnus, Lamprothamnium, Chara & Protochara. |
| Round, 1963 | Chlorophyta [Charophytina] | Charophyceae | Charales | | | Chara, Lamprothamnium, Nitella, Tolypella & Fossil Genera Clavator & Palaeochara. |
| Prescott, 1969 | Chlorophyta | Charophyceae | Charales | Characeae | Chareae | Chara, Nitellopsis, Lamprothamnus & Lychnothamnus. |
| | | | | | Nitelleae | Nitella & Tolypella. |
| Wood and Imahori, 1965 | Chlorophyta | Charophyceae | | Characeae | Chareae [Charinae Nitellopsinae] | Chara, Lamprothamnium & Lychnothamnus. |
| Bold and Wynne, 1978 | Charophyta | Charophyceae | Charales | Characeae | Nitelleae | Nitellopsis Nitella & Tolypella. |

* The suffixes are those prescribed in the International Code of Botanical Nomenclature.

** The names within the brackets [ ] represent the subgroup concerned within the given rank.

synthetic pigments, haploid genetic constitution of the plant, flagellar morphology and haplontic life cycle.

The above three hypotheses are based on the classifications proposed by various workers on morphological ground as shown in Table 1.

The first hypothesis derives support from Sachs (1875), Migula (1890), Groves and Bullock-Webster (1920, 1924), Pascher (1925), Groves and Allen (1927), Zaneveld (1940), Papenfuss (1955), Silva (1962), Desikachary and Sundaralingam (1962), Vaidya (1963), Pal *et al.* (1962), Round (1973), Bold and Wynne (1978) and Vashista (1995).

The second line is based on the findings of Wildeman (1900), Strasburger (1912). Smith (1938, 1957 & 1958), Wood and Imahori (1965) and Prescott (1969).

The third hypothesis is supported by Richard (1815), Oltmanns (1904), Fritsch (1935), Tilden (1937) and Iyengar (1951, 1958).

Examining the issue cytologically Sarma *et al.* (1970) support the first hypothesis with the reservation that Charophyta be not excluded from algae. Chaetophoralean origin of stoneworts as suggested by Church (1919) and Desikachary and Sundaralingam (1962) has been ruled out on cytological grounds by Sarma (1964), Sarma *et al.* (1970) and Sarma (1973) because of appreciable differences in the morphology of chromosomes between the two groups, Charophytes are generally characterised by much longer chromosomes usually with median and submedian centromeres and common occurrence of polyploidy, whereas, Chaetophorales have small chromosomes with sub-terminal and terminal centromere with aneuploidy being common.

A new system of classification of the green algae on the basis of nuclear and cell division, organisation of zoospore or gametes and their comparative biochemistry has been proposed wherein erection of a separate independent class Charophyceae similar to that of Chlorophyceae under the division Chlorophyta has been suggested by Pickett-Heaps and Merchant (1972), Frederick *et al.* (1973), Stewart *et al.* (1973), Fott (1974), Stewart and Mattox (1975), Pickett-Heaps (1975-1976), Mattox and Stewart (1984), Round (1984), Gram (1982), Graham and Bride (1975), Bhatnagar and Saxena (1990) and Gupta and Bhatnagar (1992). These electron microscopic and biochemical studies reveal that Charophyceae is characterised by features such as motile cells with flat band of microtubules referred as multilayered structure (MLS) at the flagellar bases, flagellar insertion slightly lateral, and presence of Glycolate oxidase. Also, mitotic spindle is open and like higher plants the cytokinesis is effected by a phragmoplast that proliferates from an interzonal spindle which persists at telophase. These findings favour certain Chaetophoralean members such as *Coleochaete* as ancestor of Charales.

## TAXONOMY AND PHYLOGENY WITHIN CHAROPHYTA

The evolutionary trend within the status assigned to stoneworts by different phycologists down to the level of genus is evident from Table 1. By and large the classification starting at division level from Chlorophyta/Charophyta follows the route Chlorophyceae/Charophyceae, Charales, Characeae to Nitelleae and Chareae at the subfamily/tribe level. The hypothetical evolutionary sequence of the Characeae as reported by Wood and Imahori (1965) is shown in Fig. 3.

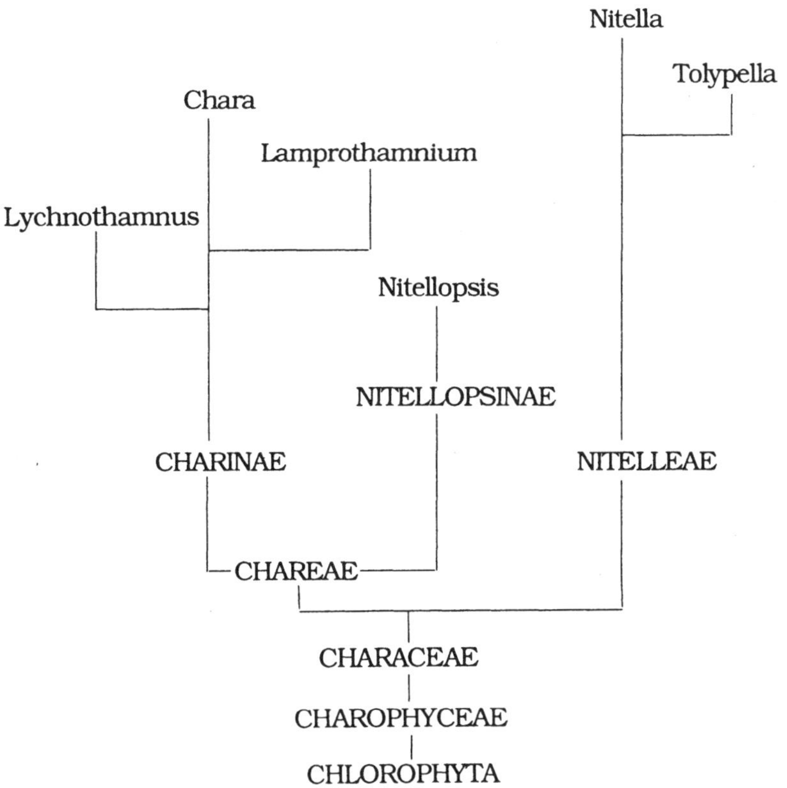


Fig. 3. Hypothetical evolutionary sequence of the Characeae. (Adopted from Wood and Imahori, 1965)

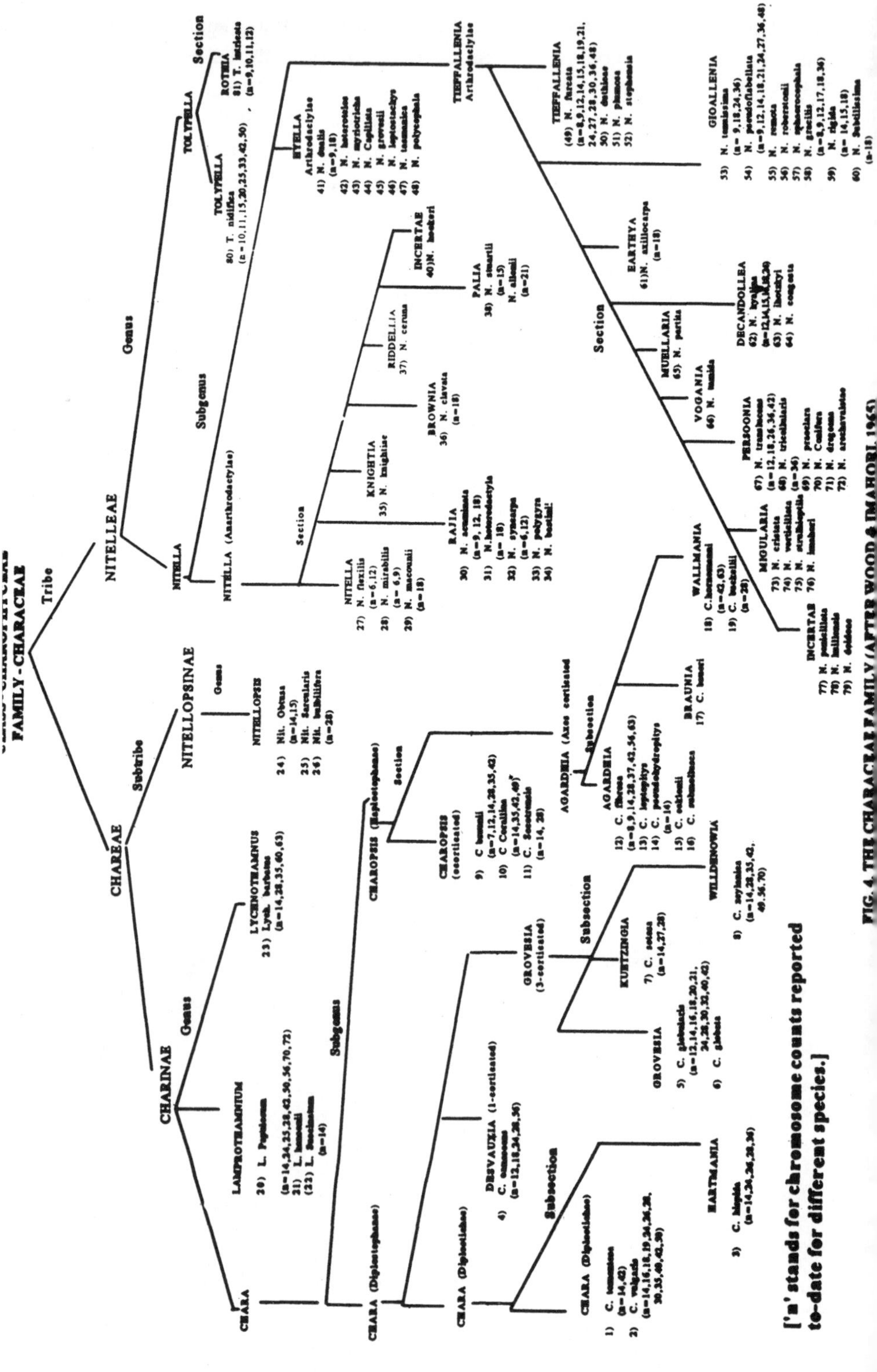


['n' stands for chromosome counts reported to-date for different species.]

FIG. 4. THE CHARACEAE FAMILY (AFTER WOOD & IMAHORI, 1965)

In early records stoneworts appear as *Equisetum*. Vaillant (1721) assigned them to new genus *Chara*. Agardh (1824) divided the genus, assigning the ecorticated species to a new genus *Nitella*. Emphasizing the coronular structure as a distinctive feature rather than Agardh's criterion based on the presence or absence of cortication, Braun (1849) recognised a third group as sub-genus *Tolypella* which was subsequently raised to genus *Tolypella* by Braun (1857). Ruprecht (1845) first separated a new sub genus which was ultimately raised to genus *Lychnothamnus* by Leonhardi (1863). Braun (1876) reported a new sub-genus *Lamprothamnus* which was later renamed as genus *Lamprothamnium* by Groves (1916). The genus *Nitellopsis* was first described by Hy (1889). Womersley and Ophel (1947) reported *Protochara* which according to Wood and Imahori (1965) appears on both experimental and herbarium analysis to represent an extreme form of *Chara* and not a distinct genus. The further evolutionary sequence down to the level of species which is based on the finding reported by Wood and Imahori (1965) is presented in Fig. 4 which also shows chromosome numbers recorded to-date in different species.

The charologists hold different views on the issue of interrelationship and evolution within the family Characeae. But, it is generally accepted that the family divides into two tribes Chareae and Nitelleae. This division is based on the number of tiers of coronary cells; one tier of five cells in the tribe Chareae and two tiers of five cells each in the tribe Nitelleae.

The concept of parallel evolution of the Chareae and the Nitelleae from a common ancestor suggested by Womersley and Ophel (1947) has received support from Desikachary and Sundaralingam (1962), Wood and Imahori (1965), Guerlesquin (1967), Ramjee (1969), Sarma *et al.* (1970) and Bhatnagar (1981, 1988). In absence of any genus which may link the two tribes, this view seems to be acceptable. It would be appropriate here to review the further evolution and interrelationship among members on morphological and cytological grounds, separately.

## MORPHOLOGICAL SCENE

The different views among the charologists on this issue may be linked to the question whether evolution is from simple to complex form or otherwise. In terms of major morphological features the simple to complex hypotheses implies that

(i) ecorticated forms of *Chara* are primitive and corticated forms are derived.

(ii) in Nitelleae the anarthrodactylous forms (= Nitella) are primitive than arthrodactylous forms (= Hyella and Tieffallenia) and

(iii) Chareae are more evolved than Nitelleae.

The findings of Womersley and Ophel (1947) lend support to this view. On the other hand, the findings of Desikachary and Sundaralingam (1962) and Wood and Imahori (1965) support the view that (i) ecorticated forms of *Chara* are more evolved than corticated forms, (ii) arthrodactylous forms in Nitelleae (Tieffallenia) are more primitive than anarthrodactylous forms (Nitella) and (iii) Nitelleae are more evolved than Chareae.

Wood and Imahori (1965) suggest that segregation of *Lamprothamnium* and *Lychnothamnus* from *Chara*, and *Tolypella* from *Nitella* (Fig. 3) have taken place recently, whereas, Daily (1967) feels that *Lamprothamnium* is an ancient genus. Guerlesquin (1967) considers *Tolypella* to be an ancient genus. According to Womersley and Ophel (1947) evolution in Chareae started from *Protochara*.

## CYTOLOGICAL SCENE

The cytologists on the otherhand, are generally in agreement over the premise by which the issue of interrelationship and evolution within Chareae and Nitelleae should be examined. Accordingly, taxon is considered relatively more advanced if it exhibits (i) higher number of chromosomes (ii) shorter size of chromosomes and (iii) higher degree and extent of polyploidy. Later, Khan and Sarma (1967 a, b, c) introduced karyological features based on length of chromosomes, breadth of chromosomes and position of centromere as additional criteria for cytological investigation of Charophytes. Accordingly, taxa with low chromosome numbers have longer chromosomes generally characterised by median and sub-median centromere and those with higher number have smaller chromosome and sub-terminal and terminal centromere. Subramaniam and Ganesan (1983) examined the issue on the basis of preponderance of chromosome counts. More recently, Sinha and Purak (1995) have considered karyotypic asymmetry, represented through the ratio of lengths of longest and shortest chromosomes and proportion of chromososmes with arm ratio less than 1.7:1, as additional basis for examining evolutionary

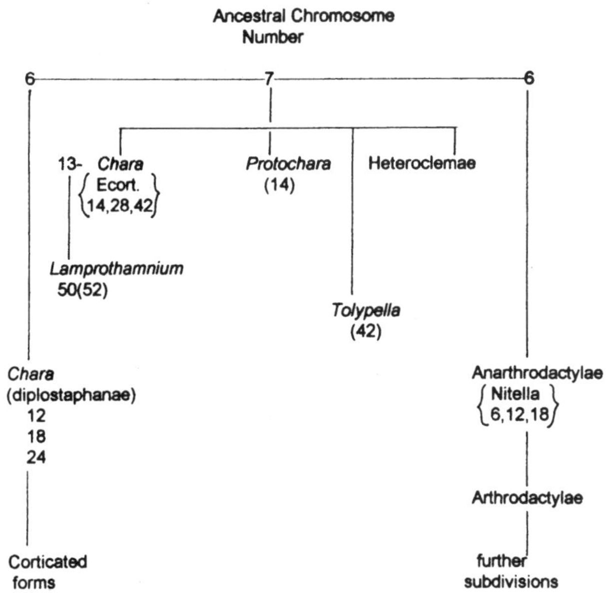


Fig. 5. Simplified scheme of evolution among Charophytes. (Adopted from Kasaki, 1964)

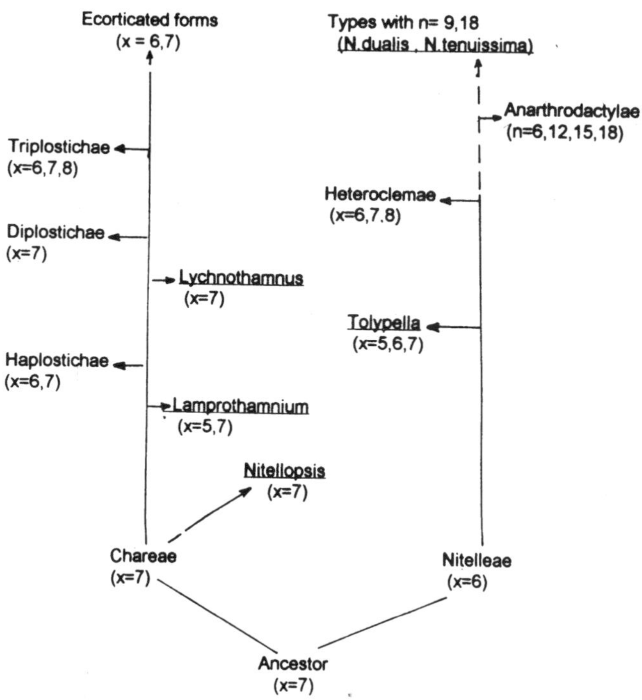


Fig. 6. Simplified scheme of evolutionary sequence in Charophytes. (Adopted from Guerlesquin, 1970)

**Table 2. Cytology of charophyta—list of contributors**

| * Mitotic chromosome count | Process of mitosis and chromosome count | Cytotaxonomy and Karyomorphology | | |
|---|---|---|---|---|
| ** 1892-1944 | 1945-1964 | 1965-1979 | 1980-1990 | 1990-1995 |
| **Abroad** | **Abroad** | **Abroad** | **Abroad** | **Abroad** |
| Scottlander, 1892; Debski, 1897; Gotz, 1899; Strasburger, 1908; Oehlekar, 1916; Earnst, 1918; Riker, 1921; Lindenbein, 1927; Schmuker, 1927; Karling, 1928; Telezynski, 1929; Walther, 1929; Muhldorf, 1930; Pekarek, 1932; Steil, 1941 and Gonclaves, 1941-42. | Rodrigues, 1945; Mendes, 1946; Muhldorf, 1946; Geitler, 1948; Delay, 1949, 1963; Corillion, 1955, 1960; Moutshen, Dahmen & Gillet, 1956; Mac Donald & Hotchkiss, 1956; Hotchkiss, 1958, 1963, 1964; Sato, 1959; Gillet, 1959; Sasaki, 1961; Imahori and Kato, 1961; Guerlesquin, 1961, 1963, 1964; Tindall and Sawa, 1964, 1965; Griffin and Proctor, 1964; Corillion, Gillet and Guerlesquin, 1959. | Griffin, 1965; Hotchkiss, 1965, 1966; Tindall, 1967; Guerlesquin, 1967, 1972, 1977; Kanahori, 1967, 1974; Mirsaidov, 1971; Proctor, 1971; Corillion and Guerlesquin, 1972; Williams and Tindall, 1973, 1975; Sawa, 1971, 1974; Peshkow *et al.*, 1973, 1974; Wood and Manson, 1977 and Bhattacharyya, 1972. | Guerlesquin, 1980, 1984.<br><br>**India**<br><br>Mukherjee & Noor, 1980; Khan & Sarma, 1981-82, 1984; Subramanian, 1983; Subramanian & Ganesan, 1983; Bharati & Chennaveeraiah, 1983; Das & Majumdar, 1984; Sarma & Bala, 1984; Labh & Verma, 1984; 1985, 1986; Noor & Purak, 1985, 1989; Purak, 1986; Jwale & Patel, 1985; Bhatnagar, 1981, 1983, 1988, 1989, 1990; Bhatnagar & Johri, 1986; Sinha & Sinha, 1989; Bharati, 1990; Ray and Chatterjee, 1986, 1987, 1988; Bandyopadhyay & Chatterjee, 1989; Bhatnagar & Gangwar, 1985. | Zanusz, 1991.<br><br>**India**<br><br>Subramanyam, Chowdary & Sarma (1991), Bhatnagar & Johri (1991), Subramanyam and Chowdary (1992) Sinha & Sinha (1992) Pundhir & Vidyavati (1993, 1994), Sinha and Purak (1995), Purak (1994), Pundhir (1994), Pundhir and Gautam (1993, 1994), Pundhir, Gautam & Vidyavati (1993). |

| | | |
|---|---|---|
| | **India** | **India** |
| | Sundaralingam, 1946; Sarma, 1964 and Sarma and Khan, 1964. | Sarma & Khan, 1964, 1965; Sarma, 1968; Sinha & Noor, 1967, 1968,1969,1971; Noor, 1968, 1969; Khan & Sarma, 1967; Verma, 1967, 1970; Sinha & Sinha, 1969; Verma, 1969; Sinha & Verma, 1969; Sinha, 1969-70; Ramjee & Sarma, 1971; Sarma, Khan and Ramjee, 1971; Chatterjee, 1971, 1975, 1976, 1979; Ahmed and Sinha, 1973; Sinha, 1974; Chennaveeraiah and Bharati, 1974; Noor and Mukherjee, 1974, 1975, 1976, 1977; Mukherjee, 1978; Ramjee and Bhatnagar, 1978; Sundaralingam and Bharatan, 1978; and Patel and Jwale, 1978, 1979. |

The rows marked by * and ** represent the main areas of work and the corresponding periods, respectively.

trends in Charophyta. The criteria being, higher the asymmetry of karyotype more advanced a taxon is. A good number of Charologists have employed cytological yardsticks for determining the evolutionary sequence; Table 2 shows the broad areas of study and the list of contributors during different periods of nearly a century of cytological investigations.

Fig. 5. represents a simplified scheme of evolution among Charophytes as proposed by Kasaki (1964). He believes the ancstral chromosome number to be (x = 7) and suggests that reduction in basic chromosome number has played a considerable role in evolution of the group.

Guerlesquin (1970) proposes her scheme of evolutionary sequence with (x = 7), which is shown in Fig. 6. She suggests *Tolypella* to be an ancient genus, in conformity with the views of Hotchkiss (1967).

Sarma *et al.* (1970) have come out with their scheme of evolutionary sequence among Charophyta (Fig. 7) and have concluded that

(i) tribe Nitelleae, characterised by comparatively longer and thicker chromosomes and basic chromosome number (n = 6) is primitive than tribe chareae in which n = 7.

(ii) because of higher degree and extent of polyploidy the corticated forms of *Chara* are more evolved than ecorticated taxa, and arthrodactylous taxa of *Nitella* are more advanced than anarthrodactylous ones.

(iii) *Lamprothamnium* and *Lychnothamnus* are primitive genera, and

(iv) genera *Nitellopsis* and *Tolypella* segregated very early from the main lines of evolution.

Sawa (1974) proposes a hypothetical scheme of Charophyte evolution from an ancestor (x = 3) and suggests that (i) the tribe Chareae, genus *Tolypella* and the genus *Nitella* have originated by aneuploidisation from a common ancestor (x = 3) (ii) phylogenetic distance between *Tolypella* and Chareae is much closer than that between *Tolypella* and *Nitella* and (iii) genus *Tolypella* has three base numbers (x = 5, 8 and 11).

Bhatnagar (1988) also considers a hypothetical ancestral chromosome number (x = 3). His scheme of evolution is shown in Fig. 8. In addition to the conclusions easily apparent from the scheme some of his other conclusions are:

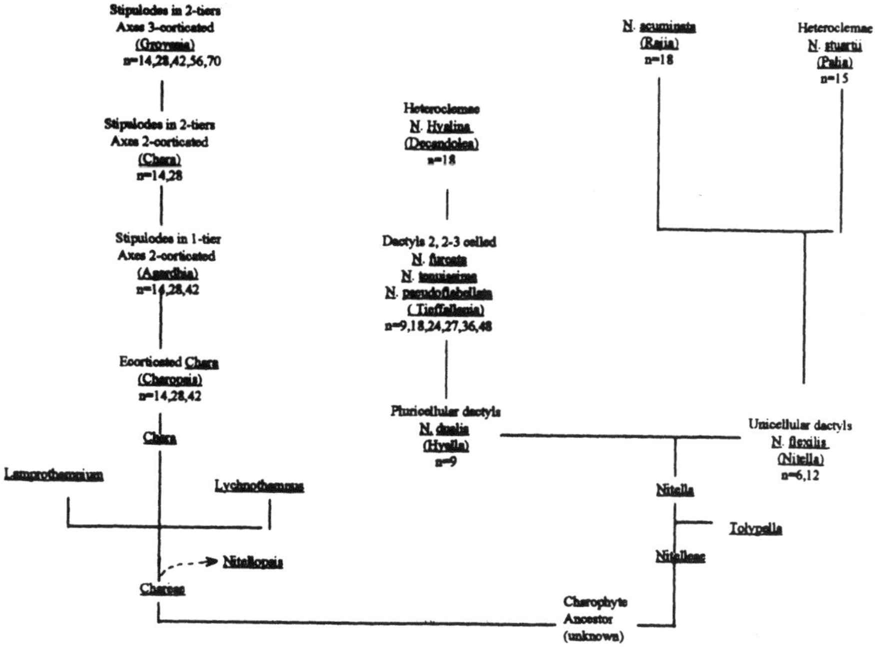


Fig. 7. Evolutionary sequence in Charophyta based on cytological findings. (After Sarma *et al.* 1970; reproduced from Bhatnagar, 1988)

(i) Existence of third tribe *Tolypella* apart from the tribes Chareae and Nitelleae,

(ii) Base numbers for the tribes Chareae, *Tolypella* and Nitelleae are respectively x = 7; x = 5, 8, 11 and x = 3, 6.

(iii) *Lamprothamnium* and *Lychnothamnus* are of ancient nature and

(iv) *Nitellopsis* originated quite early and independently from the stock of tribe Chareae. It seems approaching towards extinction.

As is evident from Fig. 4 the chromosome numbers for only 41 out of a total of 81 species belonging to the seven genera of Characeae family have been reported so far; thus covering only 50% of known species. This information obviously, is not adequate for proper understanding of the evolutionary trends in the family. Further, one may have to keep in mind the observation of Briggs and Walter (1986) and Merxmuller (1970) that "some, perhaps many, chromosome numbers published in the literature are incorrect". However, based on presently available cytological and karyotypic data including those on 52 taxa investigated by present authors, a discussion on aspects like the basic chromosome constitution, karyomorphological details and the role of polyploidy and aneuploidy in speciation and evolution of the family is being presented.

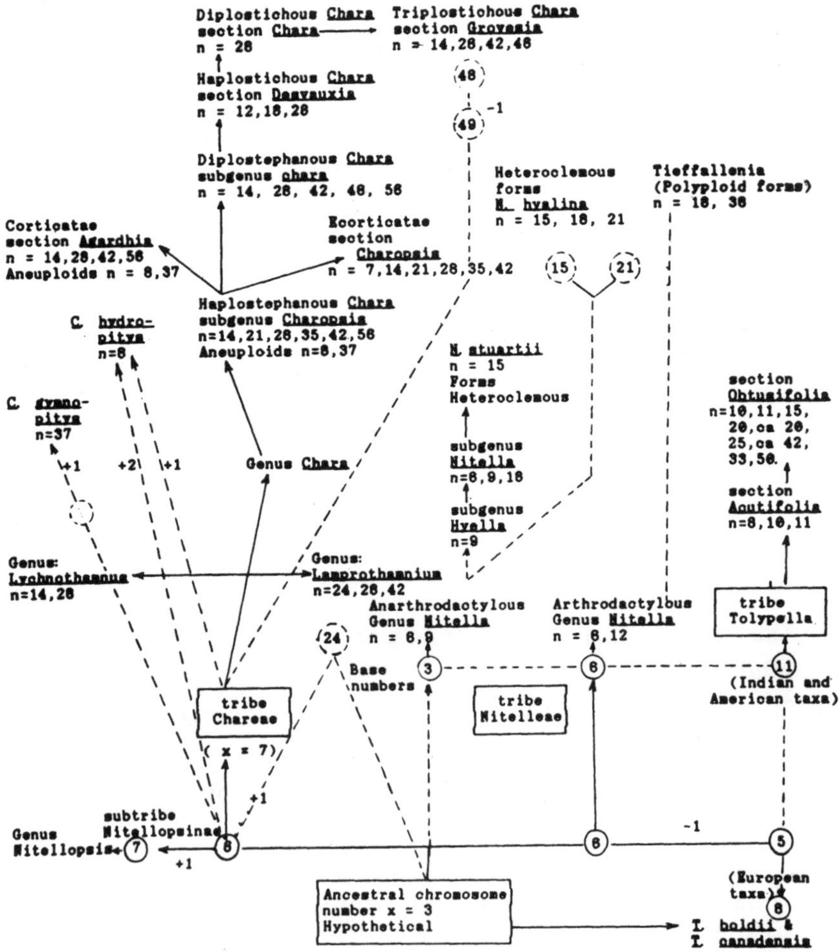


Fig. 8. A diagrammatic representation of evolution, interrelationship and phylogeny of Charophyta (After Bhatnagar, 1988)

## BASIC CHROMOSOME NUMBERS

The chromosome counts reported so far in different genera of the family Characeae are shown in Fig. 9. The patterns emerging from these counts are discussed below.

*Tribe Nitelleae*: *Nitella* possesses x = 3, 8, 14, 17, 19 and 29 where x = 3 is hypothetical and x = 19 and 29 lack euploid series based on them. x = 8, 14, 17 and 29 may be explained as the first descending aneuploid derivatives from n = 9, 15, 18 and 30 respectively, whereas x = 19 may be an ascending aneuploid derivative from n = 18.

In *Tolypella* x = 3, 5, 8 and 11 have been reported where x = 3 and 5 are hypothetical chromosome numbers and x = 8 lacks a euploid series based on it. x = 5, 8 and 11 may be regarded as the descending aneuploid derivatives from n = 6, 9 and 12. This aneuploid series may be expressed through the relation (3k–1) where k represents the degree of euploidy.

*Tribe Chareae*: Five possible euploid series based on x = 6, 7, 8, 9 and 10 may be located in the genus *Chara*, where x = 6 and 10 are hypothetical basic chromosome numbers. Further, x = 19, 26 and 37 are aneuploid counts lacking euploid series based on them. With x = 7 assumed as the ancestral form, x = 6, 19 and 26 and x = 8, 9, 10 and 37 may respectively be interpreted as descending and ascending aneuploid counts at different levels of polyploidy. *Lamprothamnium* is represented by three euploid series based on x = 14, 24 and 25 where the missing count n = 48 is marked by a dot on the series based on x = 24. Further, x = 24 and 25 may have derived through successive stages of chromosome reduction.

*Lychnothamnus* possesses a euploid series based on hypothetical x = 7. The only aneuploid count reported in the genus is x = 60, probably obtained through progressive aneuploid decrease from n = 63. *Nitellopsis* is associated with x = 14 and 15 with a euploid series based on x = 14.

A persual of Fig. 9 indicates that all the genera of Characeae family are multibasic although *Lychnothamnus* and *Nitellopsis* have only one euploid series. The array of basic chromosome constitution can be arbitrarily classified into three main groups such as (i) x = 3-6, (ii) x = 7-14 and (iii) x = 15 and above. Basic numbers belonging to x = 3-6 are seen to be widely occurring in the tribe Nitelleae whereas the tribe Chareae is dominated by x = 7-14. The third group x = 15 and above has a very low frequency and is associated with poorly developed euploid series.

The dominant euploid series for *Lamprothamnium* and *Nitellopsis* in Fig. 9 are based on x = 14, whereas those in *Chara* and *Lychnothamnus* appear to be based on x = 7, being hypothetical for *Lychnothamnus*. Therefore, evolution of Characeae might have occurred along the following two routes:

(i) x = 7 as the ancestral chromosome number for each of the four genera, implying parallel evolution for them. This would further imply assumption of hypothetical basic number for *Lamprothamnium* and *Nitellopsis* as well as alternate missing chromosome counts in the dominant euploid series on them. Alternatively, it may be argued that the original constitution x = 7 may not have flourished for some reasons, and its diploid n = 14 might have progressed through polyploidization due to conducive environment for x = 14.

(ii) Parallel evolution of *Chara* and *Lychnothamnus* at x = 7 and bifurcation of *Lamprothamnium* and *Nitellopsis* from *Chara* at n = 14 (2x).

The dominant euploid series in *Nitella* and *Tolypella* appear to have originated from hypothetical x = 3 and x = 3 and 5, respectively. Therefore x = 3 may be considered as the ancestral chromosome constitution for the tribe Nitelleae, thus implying that x = 5 for *Tolypella* have originated through a combination of euploid and aneuploid developments.

x = 7 for Characeae is in conformity with the view held by Kasaki (1964), Guerlesquin (1970), Sarma *et al.* (1970), Sawa (1971) and Bhatnagar (1988). But as far as Nitelleae is concerned they suggest x = 6 for the tribe; additionally x = 9 and x = 3, respectively have also been suggested by Sarma *et al.* (1970) and Bhatnagar (1988). x = 6 and x = 9 as ancestral chromosome numbers for the tribe Nitelleae may require aneuploids at different levels of euploidy for covering the range of chromosome counts reported in the tribe. Thus x = 3 appears to be more logical for Nitelleae.

Another approach for deciding the ancestral chromosome number is based on the preponderance of a chromosome count in the tribe. Accordingly, the most preponderant chromosome count is considered to be the ancestral chromosome number. This criterion is supported by Solbrig *et al.* (1964) and Abraham and Mathew (1988) in angiosperms, and Subramanian (1983) in Charophyta. A perusal of Fig. 4 reveals the overwhelming preponderance of n = 14 followed by n = 28 in Chareae and n = 18 followed by n = 12 and n = 9 in Nitelleae. Accordingly, this approach suggests x = 14 and x = 18 as the respective ancestral chromosome constitution for the tribe Chareae and Nitelleae, which confirm the findings of Subramanian and Ganesan (1985a, b). The basic chromosome numbers for different genera as shown in Fig. 9 may be explained through progressive polyploidization from the original constitution.

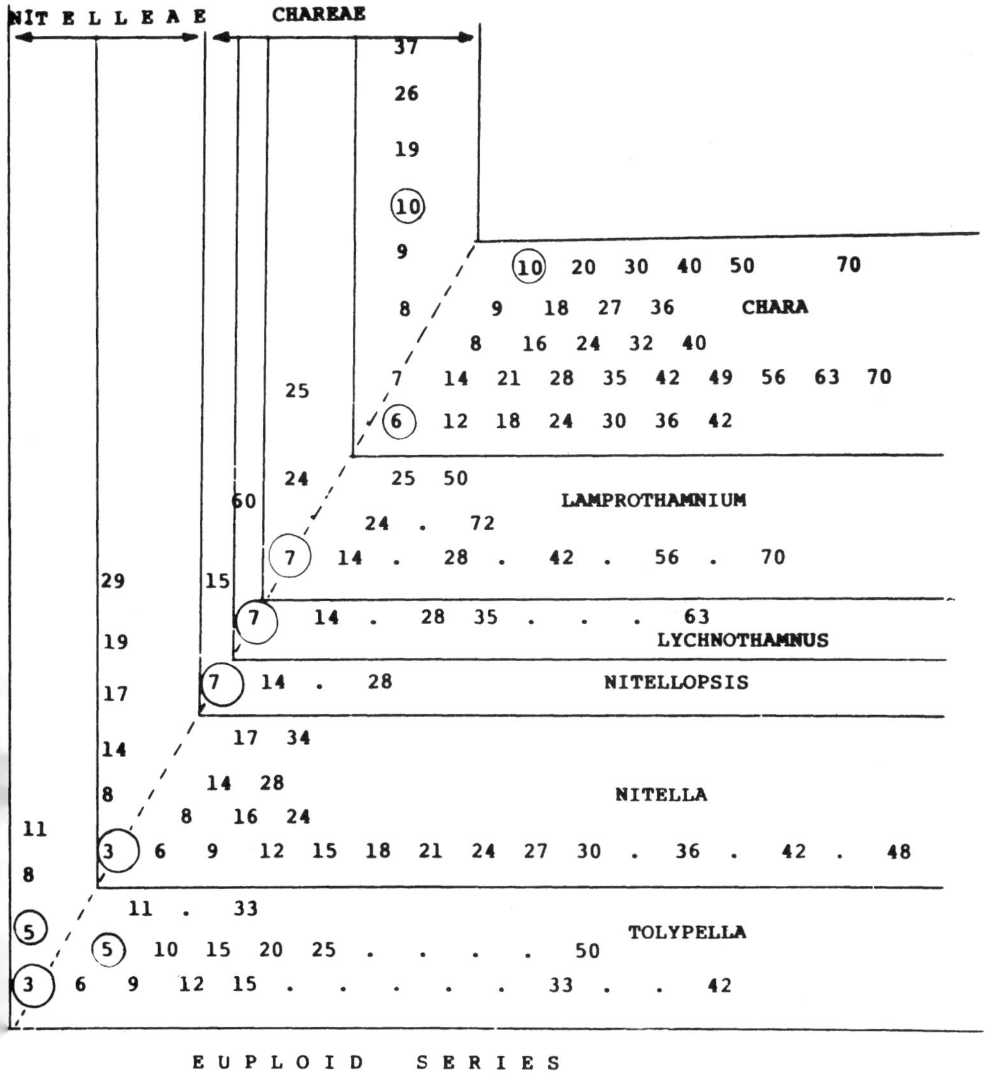


Fig. 9. Chromosome numbers in Charophyceae—Euploid and Aneuploid series: 1. The probable ancestral chromosome number for different genera are located on the dotted lines. 2. The encircled chromosome numbers are hypothetical ones. 3. Missing chromosome counts in a euploid series are represented by dots.

## POLYPLOIDY

Polyploidy is the most wide-spread cytogenetic process that has contributed to species formation ad evolution in plants. Higher tolerance of polyploids to adverse environmental conditions and their ability to increase the chances of fertilisation by breaking reproductive barriers are mainly responsible for this contribution (Stebbins, 1971). This abrupt speciation is associated with either an increase in chromosome number in integral multiple (euploidy) or in a loss or gain of generally one chromosome at a time (aneuploidy). Fig. 9 shows that degree and extent of polyploidy in the tribe Chareae and Nitelleae are respectively, highest in *Chara* (10x) and *Nitella* (16x). These two genera record maximum number of euploid series as well. Thus *Chara* and *Nitella* may be considered as the most advanced genus of their respective tribes. The inhibited polyploidisation of Nitellopsis may be an indicator of its probable movement towards extinction. The difference in degree and extent of polyploidy in these genera may be due to variation in growth habit and breeding system (Stebbins, 1971).

*Chara* and *Nitella* are associated with relatively higher number of aneuploids suggesting that these two genera in characeae family have relatively higher rate of incidence of Karyological events such as unequal translocation between nonhomologous chromosomes, fragmentation of centric fragments and unequal separation of chromosomes into daughter cells during mitosis. According to Stebbins (1971) aneuploidy is known to occur most often at the diploid level in angiospoerms. No such trend is discernible in Charophyta.

## KARYOMORPHOLOGY

The initial phase of karyomorphological studies was essentially confined to investigation of size of chromosomes and the position of centromeres in the chromosomes of a complement, and accordingly the chromosomes were classified. A chromosome is termed as large (5–10μ), medium (3–5μ), small (1–3μ) or minute (0.25–1μ) as suggested by Sarma (1959). Similarly, a chromosome is considered thick (more than 1μ wide), moderately thick (1μ wide) or slender (less than 1μ wide). The chromosomes were further categorised on the basis of ratio of the length of their long arm (l) to that of short arm (s), expressed as $r = l/s$, as per the system proposed by Levan (1964). Accordingly, chromosomes with absolute median position ($r = 1$) are designated as 'M' type, with median position ($r = 1–1.7$) as 'm' type, with submedian position ($r = 1.7–3.0$) as 'sm' type, with sub-terminal position ($r = 3.0–7.0$) as 'st' type and with terminal position ($r$ = more than 7) as "t" type. The position of centromere is generally determined on the basis of shape and bending of the chromosome as the centromeres in Charophytes are not clearly demarcated; thus resulting in large variation in karyotypes from worker to worker. The consequential confusion may hopefully be cleared by using Giemoa C-banding technique for centromeric assessment by heterochromatin differentiation as proposed by Bhatnagar (1989).

Recently, karyomorphological studies of Charophytes have also been directed towards assessment of the nature and extent of asymmetry of karyotype as per the criteria proposed by Stebbins (1971) who has classified the karyotype asymmetry under 12 categories (1A–4C) taking into account both the degree of difference between the largest and smallest chromosomes of the complement and position of centromere on the chromosome. According to the above criteria the route followed by a taxon in course of its evolution from the most primitive stage (4A) to the most advanced one (1C) is (4A-3A-2A-1A-4B-3B-2B-1B-4C-3C-2C-1C). The present authors have made similar studies on 52 taxa belonging to Characeae family which were classified according to the scheme proposed by Stebbins (1971). The results obtained in terms of their grouping is shown in Table-3 and the karyotypic types of the taxa concerned have been indicated on the hypothetical evolutionary sequence proposed by Wood and Imahori (1965) for the tribes Nitelleae and Chareae as shown in Fig. 10 A and 10 B.

**Table 3. Types of karyotype obtained in 52 taxa of characeae investigated by the present authors**

| Proportion of Chromosomes with arm ratio (r) less than 1.7 : 1 | Types of Karyotype & Number of taxa Investigated in Nitelleae (N) and Chareae (C) | | | | | |
|---|---|---|---|---|---|---|
| | L/S < 2:1 | | 2:1 < L/S < 4:1 | | L/S > 4:1 | |
| | N | C | N | C | N | C |
| 1 | 4A | | 4B | | 4C | |
| | ~ | 1 | 2 | ~ | ~ | ~ |
| 0.5-0.99 | 3A | | 3B | | 3C | |
| | 3 | 5 | 12 | 6 | 1 | ~ |
| 0.01-0.5 | 2A | | 2B | | 2C | |
| | 2 | 5 | 9 | 2 | ~ | 1 |
| 0 | 1A | | 1B | | 1C | |
| | ~ | 1 | ~ | 1 | ~ | 1 |

L—Length of the longest chromosome
S—Length of the shortest chromosome

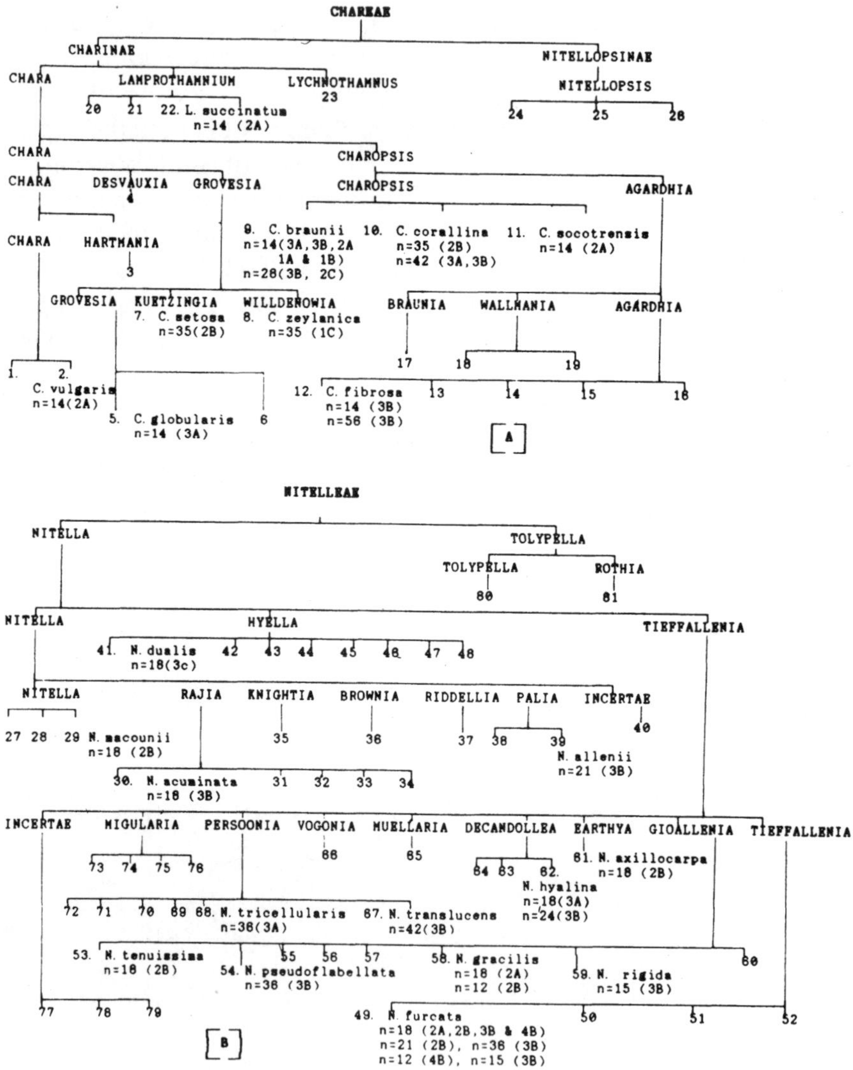


Fig. 10. A & B. Chromosome counts and Karyotype category obtained by present authors for certain members of tribes Chareae and Nitelleae with reference to evolutionary scheme proposed by Wood and Imahori (1965).

Number on the sequence are species number as considered by Wood and Imahori (1965).

Change of basic chromosome numbers and processes which transform a symmetrical karyotype to an asymmetrical one are the two major factors known to contribute to the evolution of karyotype of a taxon within a group (Abraham and Mathew, 1988). Obviously, a co-relation should be expected to exist between the chromosome numbers, the karyotypic type and the morphological specialisation of a taxon. A perusal of Fig. 10 A and 10 B fails to indicate any such co-relation; may be, due to meagre information available on karyotypes and chromosome numbers in the tribe Chareae and Nitelleae. A definite conclusion on the pattern of such a co-relation may be drawn only when sufficient data are available. In case of herbaceous genera Stebbins (1971) has recognised four different patterns of such co-relation within a genus such as (i) increasing asymmetry of the karyotype associated with decreasing chromosome number and increasing morphological specialisation (ii) increasing karyotype asymmetry associated with increasing chromosome number and increasing specialisation, (iii) two different basic chromosome numbers associated with little difference in the asymmetry of the karyotype and (iv) genera having constant basic chromosome numbers but possessing variation in karyotype asymmetry (Abraham and Mathew, 1988). This observation is presently applicable in case of Characeae family as well.

The absence of clear pattern between morphological characteristics, number of chromosomes and karyotype may be due to many factors such as variation in environmental conditions, geographical location, developmental stages etc. whose effects are yet to be clearly understood. According to Briggs and Walters (1984) there is strong evidence for the proposition that the phenotype and behaviour of the plant are determined by interaction between genotype and environment; different genotypes react differently to a given set of environmental conditions and plants of identical genotype produce different phenotypes under contrasting environmental conditions. Therefore, the pattern of relationship between different characters of a taxon as well as the interrelationship among these characters within the members of Characeae family shall be discovered not by examining these characters in isolation but also by considering the influencing factors such as ecological, geographical, developmental process, etc.

## BIOCHEMICAL SCENE

Study of the variation in chemical constituents, especially secondary plant products is being increasingly regarded by many botanists as of great value in investigating ancestry. Initially the study remained confined to patterns of variation in unknown substances in extracts of plants, but recently many chemotaxonomists interested in aspects of ancestry have attempted to identify these chemical compounds. Presently, information about the chemotaxo-nomical studies of Indian Charophyta is very scarce.

Publication of "Biochemical Systematics" by Alston and Turner (1963) has generated interest in comparative biochemical data for supplementing morphological characteristics of different taxa in order to assess the affinities of species. The problems of interrelationship among species of same genus and among species and genera of particular family have been examined by Subramanian and associates by studying variations in chromatographic pattern of amino acids.

Chemotaxonomical studies of four species of *Nitella*, six species of *Chara;* one species each of *Lychnothamnus* and *Nitellopsis* have been made by Subramanian and Ganesan (1985a, b) with special reference to protein amino acids and comparative ratio of chlorophyll a and b. Their findings are, (i) a total of 15 amino acids are present in various combinations, (ii) *Nitella* and *Nitellopsis* are chemically more specialised than *Chara* and *Lychnothamnus*, as the former group have larger number of amino acids in comparision to the latter, (iii) *Nitellopsis* seems to be related to *Nitella* and *Chara* to *Lychnothamnus*, (iv) chlorophyll estimation more or less leads to similar conclusion as above and (v) the chemical specialisation of Charophyta is as much as found in advance angiospermic taxa such as families Compositae, Rubiaceae, Acanthaceae etc.

Glycolate oxidase appears to be a essential enzyme for photorespiration in all the higher plants and the Charophytes. Frederick *et al.* (1973) and Stewart and Mattox (1975) report glycolate oxidase as the key enzyme of glycolate metabolism in Charophyta. Reviewing the cytological studies in green algae with regard to origin of land plants Stewart and Mattox (1975) report that Charophycean group is characterised by persistent interzonal spindles during cytokinesis, and glycolate oxidase is present as a key enzyme of glycolate metabolism. According to Frederick *et al.* (1973) glycolate oxidase has been found in those green algae which also has persistent interzonal microtubules at telophase similar to vascular plants. Bhatnagar and Saxena (1990) and Gupta and Bhatnagar (1992) have used glycolate oxidase as a biochemical marker for tracing the phylogenetic alliances between Charophytes and land plants while investigating eleven species of *Chara*, one species of *Tolypella* and two species of *Nitella*. They report minor difference in the optical density of these taxa, indicating different levels of glycolate oxidase present in these taxa, which according to them represent different levels of evolution. On the basis of their findings they suggest that higher plants are closer to Charophyceae and feels that the Charophytes may be on the evolutionary line of land plant as possible ancestor.

Study of the chloroplast envelope in algae is believed to be an important means for a better understanding of the different aspects of the organisation in different groups, but very little attention has been given to this aspect until now.

Reviewing the scene in this regard Marcelle Lefort-Tran (1989) refers to different number of membranes for the chloroplastic envelope and different organisation having been observed in different groups of algae and observes that "these differences reinforce the idea that the contemporary diversity of algal chloroplast might reflect a polyphylectic origin through endosymbiosis," whereas, "there is much evidence concerning a common origin for chloroplast of higher plants".

Sakano (1989) has tried to explore interphylogenetic relations among different genera of Charophyceae by using Iso-aspargine as a tool, and their findings are (i) *Nitella* species of subsection *Nitella* (anarthrodactylous) are primitive than *Nitella* species of sub-sections *Hyella* and *Tieffallenia* and (ii) the anarthrodactylous *Nitella*, like that of *Chara*, *Lamprothamnium* and *Nitellopsis*, possesses Iso-aspargine and hence is closer to Chareae.

Erich Kessler (1989) in context of *Chlorella*, observes that "Chemotaxonomy, using a wide variety of biochemical and physiological characters, can serve as a sound basis for the taxonomy of morphologically ill-defined or highly variable micro algae. Studies at the level of DNA, DNA hybridization and DNA-RNA hybridisation, might provide the ultimate information for the elucidation of evolutionary relationships among different taxa (cf. Erich Kessler, 1984)".

DNA replication in antheridial filaments of four species of *Chara* with different DNA-C value (genome size), heterochromatin content and cell cycle duration have been studied by Zanusz (1991) by using fibre autoradiographic technique wherein he explains the S-phase timing differences on the basis of different number of DNA replic on families that function serially during nuclear DNA synthesis.

Comparative susceptibility of Charophytes may also provide a clue on their genetic complexities and degree of resistance towards mutagens, irrespective of their ploidy levels and chromosome morphology.

## PERSPECTIVE

The interest in phylogeny and its relation to taxonomy has grown appreciably in recent years with the development of chemotaxonomy. According to Smith (1976) much of this interest lies in the possibility that certain kinds of chemical evidence may be reliable guide to phylogenetic relationship of living species. Further recent discoveries in Cytology often tend to have a bearing on taxonomy. Naturally a host of new information is being added to the literature at an increasing rate. But despite this the situation regarding evolutionary trend in Characeae and the origin of Charophytes continued to be confusing at the moment. What appears to be settled on the basis of one type of Character, at times, gets unsettled with the addition of new characters. Stebbins' (1971) reference of four patterns of evolution in regard to herbaceous genera, as referred earlier is an interesting indication in this regard. In view of conflicting interpretations on the basis of different characters some adjustment of the existing taxonomy may be necessary to accomodate new, uncomfortable data, but always in such a manner as to retain a practical frame work of morphologically based taxa.

The emergence of computers on the scene with its vastly superior means of information storage and retrieval and the consequent impact of mathematics on taxonomy in recent years, have added a new dimension to taxonomy 'Numerical taxonomy' subsequently established by Sokal and Sneath (1963) as an important branch of classification theory and practice, involves a set of defined and repeatable procedures to bring out a measure of similarity between groups of organism in terms of their phenotypic characters and their subsequent ranking on the basis of these similarities. Numerical taxonomy implies and objectively satisfactory taxonomic procedure involving all available characters of phenotype would produce, not just a single natural classification as in the Omega taxonomy, but any number of natural classifications according to the procedure adopted, and its growth represents a general tendency towards quantitative mathematical approach to biology.

A large amount of new information, significant for study of variation in plants, have been provided by chemotaxonomy and cytotaxonomy through the use of sophisticated techniques or apparatus, but, strictly speaking, these remain 'observational' and not 'experimental'. Stace (1980) refers to this aspect of the term 'experimental taxonomy' in his introductory discussion of the scope of taxonomy and recommends the term 'biosystematics' for studies "largely concerned with genetical, cytological and ecological aspects of taxonomy (which) must involve studies in the field and experimental garden". Research in areas broadly related to aspects of biosystematics, in recent years, have been undergoing a shift in emphasis from a concern with definition of categories to an examination of the detailed pattern of variation and the processes underlying these patterns. According to Briggs and Walters (1984) "Modern discussion about relation between classification and evolution centres mainly around two separate, but obviously related areas, which concern on the one hand the nature of the evolutionary process and on the other the relation between phenotypic resemblance (in its widest sense) and evolutionary history. As Darwin saw, the species we recognise are varying both in space and in time, and an understanding of the causes of that variation is a major concern of biologists today".

## REFERENCES

Adanson, M. 1763. Families des plantes. Paris. **2**: 537.

Agardh, C.A. 1824. Systema Algarum. Lund. pp. 123-130.

Ahmad, S. and Sinha, J.P. 1973. Chromosome numbers in some taxa of *Chara. Proc. Ind. Sci. Cong. Assoc.* p. 302.

Alston, R.E. and Turner, B.L. 1962. *Proc. Nat. Acad. Sci.* **48**(2): 132-137.

Alston, R.E. and Turner, B.L. 1963. Biochemical Systematics. Prentice Hall. In New York.

Bandyopadhyay, S. and Chatterjee, P. 1989. *Chara fibrosa* var. *fibrosa* form *Keukensis*—A new record for India and its chromosome number. *Phykos.* **28**(1 & 2): 178-180.

Bell, P. and Woodcock, C. 1968. The diversity of green plants. Edward Arnold Publishers Ltd.

Bennett, A.F. 1878. On structure and affinities of Characeae. *J. Bot. Lond.* **16**: 202-206.

Bharati, S.G. 1990. Morphological and Cytological studies on two taxa of *Chara fibrosa Ag. Ex. Bruz., Em RDW. J. Indian Bot. Soc.* **69**: 315-318.

Bharati, S.G. and Chennaveeraiah, M.S. 1975. Cytotaxonomical studies on *Chara zeylanica* complex. *Proc. Inter. Symp. Taxo. Algae.* Madras. p. 8.

Bharati, S.G. and Chennaveeraiah, M.S. 1983. *Lychnothamnus barbatus* var. nov. *Cytologia.* **48**: 843-846.

Bhatnagar, S.K. 1981. Cytotaxonomy of Charophytes of Rohilkhand division and effects of different chemicals on selected taxa. Doctoral thesis. Rohilkhand University, Bareilly.

Bhatnagar, S.K. 1983. Concept of basic chromosome numbers in Charophyta. A Review *Crypto. Algol.* **6**: 111-116.

Bhatnagar, S.K. 1985. On *Tolypella intricata* (Trent. ex. Roth) Leonh. em. R.D.W., Characeae, from Indian subcontinent—A karyomorphological Approach. *Cryptogamie Algologie.* **6**(3): 185-190.

Bhatnagar, S.K. 1987. Phylogeny and interrelationships of the genus *Tolypella*: A multidisciplinary outlook. *Phykos.* **26**: 140-148.

Bhatnagar, S.K. 1988. A karyological approach to phylogeny and interrelationship of Charophyta. *Cytologia.* **53**: 723-729.

Bhatnagar, S.K. 1989. Cytological perspective of Charophyta. I. *Chara globularis* complex. *Phykos.* **28**: 156-165.

Bhatnagar, S.K. 1989. Cytological perspectives of Charophyta. II. *Nitella furcata* complex. *Phykos.* **28**: 166-177.

Bhatnagar, S.K. 1990. A tentativ reappraisal of cytotaxonomy and systematic position of the genus *Tolypella. Vegetos.* **3**(1): 119-122.

Bhatnagar, S.K. and Gangwar, L.P. 1985. On *Tolypella intricata* (Trent. ex Roth.) Leonh. em. R.D.W. (Characeae) from Indian sub-continent. A karyomorphological approach. *Crypt. Algol.* **6**: 185-190.

Bhatnagar, S.K. and Johri, M. 1986. The genus *Tolypella* (Div. Charophyta) from Indian sub-continent—A cytotaxonomic approach. *Cytologia.* **51**: 413-419.

Bhatnagar, S.K. and Johri, M. 1991. Preliminary investigations on cytology and cytotaxonomy of Charophytes from Rohilkhand Division. *J. Indian Bot. Soc.* **70**: 55-58.

Bhatnagar, S.K. and Saxena, A. 1990. Glycolate oxidase in Charophyta—A Phylogenetic marker. *Vegetos.* **3**(1): 239-242.

Bhattacharyya, S.S. 1972. Cytological studies of some Libyan Charophytes. *The Libyan J. Sci.* **2**: 1-7.

Bhattacharyya, G.N. 1990. Chromosomal evolution in the family Araceae. (In Abstracts: Inter. Symp. Chromosome Research, Calcutta. p. 7).

Blackman, F.F. 1900. The primitive algae and flagellata. An account of modern work bearing on the evolution of algae. *Ann. Bot.* **14**: 647-658.

Bold, H.C. 1967. Morphology of plants. A Harper International Edition.

Bold, H.C. and Wynne, M.J. 1978. Introduction to Algae. Structure and Reproduction. Prentice Hall. Inc. England.

Braun, A. 1849. Characeae Indian orientalis et insularum maris pacifici; or characters and observations on the Characeae of the East Indian Continent, Hooker's *Jour. Bot. & Kew Gard. Misc.* **1**: 292-301.

Braun, A. 1857. Uber Parthenogenesis bei Pflanzen. *Abh. K. Akad. Wiss.* Berlin. **1856**: 311-376.

1 pl. (Characeae, pp. 338-351; *Lychnothamnus* and *Tolypella* of doubtful rank, footnote. p. 338).

Braun, A. 1876. *Chara succinata.* Sitz. Berliner Akad. Sci. 1876 (Fide Ascherson. **1878**: 257).

Briggs, D. and Walters, M. 1984. Plant variation and evolution. Cambridge University Press, Cambridge.

Chapman, V.J. and Chapman, D.J. 1973. The algae (2nd edn) Macmillan, London.

Chatterjee, P. 1970. *Lychnothamnus barbatus* (Meyen. Leonh.) in West Bengal and its chromosome number. *Bull. Bot. Soc. Bengal, India.* **24**: 131-133.

Chatterjee, P. 1971. An analysis of the structure and behaviour of chromosomes as an aid in the taxonomy of certain higher and lower groups of plants and the cytological effects of certain chemical and physical agents in the latter. Ph.D. Thesis, University of Calcutta.

Chatterjee, P. 1975. Some additions to the Charophytes of West Bengal. *Bull. Bot. Soc. Bengal India.* **29**: 105.

Chatterjee, P. 1976a. Cytotaxonomical studies of West Bengal Charophyta: Karyotypic analysis in *Chara braunii. Hydrobiologia.* **49**: 171-174.

Chatterjee, P. 1976b. Cytotaxonomic studies on *Chara socotrensis* f. *nuda* (Pal) R.D.W. from West Bengal, India. *Cytologia.* **41**: 659-663.

Chatterjee, P. 1979a. Karyological investigation of *Chara wallichii A. Br. Cell & Chromosome Newsletter.* **2**: 21-23.

Chatterjee, P. 1979b. *Chara fibrosa* var. *fibrosa* f. *longicorollata*: A new record for India and its Cytology. *Curr. Sci.* **48**: 545-547.

Chaudhary, B.R. 1990. Algal karyology, Prospects and problems. In: *Inter. Symp.* Chromosome Research, Calcutta. Abstracts. pp. 15.

Chennaveeraiah, M.S. and Bharati, S.G. 1974. Morphological and cytological observations on *Chara gymnopitys A. Br. Cytologia.* **39**: 443-451.

Church, A.H. 1919. Thallasiophyta and the subaerial transmigration. Oxford University Press, London.

Corillion, R. 1955. Les Charophycees de France et d Europe Occidentale. These, Toulose et aussi. *Bull. Soc. Sci. Bretag.* **32**: 1.

Corillion, R. 1960. *Tolypella salina* sp. nov. Charophycees nouvelle des macais de croix-de-vie (Vendee). *Revue Algologigue.* **5**: 198-207.

Corillion, R., Gillet, C. and Guerlesquin, M. 1959. Criteres cytologiques, anatomiques et ecologiques en faveur du maintien dun genre Charopsis chez les Charophycees. *Bull. Soc. Sci. Bretag.* **34**: 63.

Corillion, R. and Guerlesquin, M. 1959. Premieres observation cytotaxonomiques sur genre *Tolypella* (Charophycees). *Bull. Soc. et Sci. Angres.* **2**: 167-170.

Corillion, R. and Guerlesquin, M. 1972. Recherches sur les Charophycees d' Afrique occidentale. *Travaux et Memoirs.* **25**: 1-169.

Das, R.N. and Mazumdar, P.K. 1984. Chromosome number of *Nitella* from two districts (Purulia & Bankura) of West Bengal. Proc. 71st Ind. Sci. Cong. pp. 15.

Debski, B. 1897. Beobachtungen uber Kernteilung bei *Chara fragilis. Jahrb. Wiss. Bot.* **30**: 227-248.

Delay, C. 1949. Observations cytologiques sur les Characees, I: L. evolution du noyan pendant la spermiogenese de *Chara vulgaris L. Rev. Cytol. Biol. Veg.* **11**: 315.

Delay, C. 1953. Nombres chromosomiques chez les cryptogames. *Rev. Cytol. Biol. Veg.* **14**: 62.

Desikachary, T.V. and Sundaralingam, V.S. 1962. Affinities and interrelationship of the Characeae. *Phycologia.* **2**(1): 9-16.

Dittmer, H.J. 1968. Phylogeny and forms in the plant Kingdom. pp. 34-94.

Erich, K. 1989. Comparative physiology and biochemistry of *Chlorella* species as the basis for their taxonomy and for their utilisation in Research Biotechnology. In: Phycotalk I Ed. H.D. Kumar, Varanasi. pp. 142-145.

Ernst, A. 1918. Bastardierung als Ursache der Apogamic im Pflazenreich. Jena.

Frederick, S.E., Gruber, P.L. and Tolbert, N.E. 1973. The occurrence of glycolate dehydrogenase and glycolate oxidase in green plants. An evolutionary survey. *Plant physiol.* **52**: 318-323.

Fritsch, F.E. 1935. The structure and reproduction of the algae. Vol. I. Cambridge University Press. Charales, pp. 447-469.

Fott, B. 1974. The phylogeny of eukaryotic algae, *Taxon.* **23**: 446-461.

Gangulee, H.C. and Kar, A.K. 1982. College Botany Vol. 2, th Edn. pp. 134-143.

Geitler, L. 1948. Uber die Teilungs rhythmen in den spermatogenen Faden der Characeen. *Osterr Bot. Zeitschr.* **95**: 147-162.

Gillet, C. 1959. Nombres chromosomiques de plusieurs especes de Charophycees (Generes *Nitella* et *Chara*). *Rev. Cytol Biol. Veg.* **20**: 229-234.

Gonclaves, D.A. and Cunha, G.A. 1941-42. Contribuicao para estudo des carofitos portugueses. *Trab. Inst. Bot. Lisboa.* **6**: 136.

Goetz, G. 1899. Uber die Entwickelung der Eiknospe bei den Characeen. *Bot. Zeit.* **57**(1): 1-13.

Graham, L.E. 1984. *Coleochaete* and the origin of land plants. *Amer. J. Bot.* **71**: 603-608.

Graham, L.E. and Mc Bride, G.E. 1974. Multilayered structure in motile cells of *Trentepohlia aurea* and the evolutionary implications. *J. Phycol.* **10**: 8.

Graham, L.E. and Mc Bride, G.E. 1979. The occurrence and phylogenetic significance of a multilayered structure in *Coleochaete* spermatozoids. *Amer. J. Bot.* **66**(8): 887-894.

Grambast, L. 1974. Phylogeny of the Charophyta. *Taxon.* **23**: 463-481.

Griffin, III D.G. 1965. The taxonomy of *Chara zeylanica* Klein ex Willd. M.S. dactylographic 1.

Griffin, D.G. and Proctor, V.W. 1964. A population study of *Chara zeylanica* in Texas Oklahoma and New-Mexico. *Amer. Jour. Bot.* **51**: 120-124.

Groves, J. 1916. On the name *Lamprothamnus* Braun. *Jour. Bot.* **54**: 336-337.

Groves, J. and Allen, G.O. 1927. On some Indian Charophyta. *Jour. Bot.* **65**: 335-339.

Groves, J. and Bullock-Webster, G.R. 1920-24. The British Charophyta, I. Nitelleae: 1-141 & II. Chareae: 1-129, Vol. I & II. Ray Society, London.

Guerlesquin, M. 1961. Contribution al'etude chromosomique des Charophycees d Europe occidentale et d Afrique du Nord-1. *Rev. Gen. Bot.* **68**: 360-372.

Guerlesquin, M. 1963. Contribution a l'etude chromosomique des Charophycees d' Europe Occidentale et d' Afrique du Nord. *Ibid.* **70**: 354-370.

Guerlesquin, M. 1964. Contribution a l'etude chromosomique des Charophycees d'Italie Peninsulaire. *Ibid.* **71**: 283-292.

Guerlesquin, M. 1967. Recherches caryotypiques et cytotaxonomiques chez les Charophycees d' Europe Occidentales et d' Afrique du Nord. Ph.D. Thesis (France).

Guerlesquin, M. 1972. Cytologie et nombres chromosomiques. In Recherches sur les Charophycees d' Afrique occidentale: Systematique, phytogeographic, ecology et cytology (ed R. Corillion and M. Guerlesquin). *Bull. Soc. Sci. Bretag.* **27**: 125.

Guerlesquin, M. 1977. Observations chrosomiques sur les Characees due sud Tunisien. *Bull. Soc. Phycol.* France. **22**: 60-69.

Guerlesquin, M. 1984. Nombres chromosomiques et ploidie chez les Charophytes. *Cryptogamie-Algologie.* **5**(2-3): 115-126.

Gupta, P. and Bhatnagar, S.K. 1992. Occurrence of Glycolate oxidase in the genus *Nitella* (Div. Charophyta) from India. *Vegetos.* **5**(1 & 2): 82-86.

Hotchkiss, A.T. 1958. Some chromosome numbers in Kentucky Characeae. Trans Kentucky *Acad. Sci.* **19**: 14-18.

Hotchkiss, A.T. 1963. A first report of chromosome number in the genus *Lychnothamnus* (Rupr) Leonh and comparision with the other Charophyte genera. *Proc. Linn. Soc.* New South Wales. **88**: 368-372.

Hotchkiss, A.T. 1964. Chromosome numbers and relationships in *Chara leptopitys A. Br. Proc. Linn. Soc.* New South Wales. **89**: 191-198.

Hotchkiss, A.T. 1965a. Chromosome numbers in Characeae from the South Pacific. *Pacific Sci.* **19**: 31-37.

Hotchkiss, A.T. 1965b. In a Revision of the Characeae (ed R.D. Wood and K. Imahori). *J. Cramer.* **6**: 93.

Hotchkiss, A.T. 1966. A new and revised base chromosome number for the Genus *Tolypella. Bull. Torrey Bot. Club.* **93**: 426-432.

Hotchkiss, A.T. 1966a. Chromosome numbers in *Lamprothamnium. Proc. Linn. Soc.* New South Wales. **91**: 118.

Hy, F. 1889. Sur les modes de ramification et de cortication dans la famille des Characees, et les caracteres qu'ils peuvent fournir a la classification *Bull. Soc. Bot.* France. **36**: 393-398.

Imahori, K. and Kato, T. 1961. Notes on chromosome numbers of Charophytes in Fukui Perfecture, Japan, 1. *Sci. Reports.* **10**: 39-48.

Iyengar, M.O.P. 1951. Chlorophyta in Smith's Manual of Phycology. Chronica Botanica, Walth, Mass.

Iyengar, M.O.P. 1958. *Nitella terrestris* sp. nov. a terrestrial Charophyte from South India. *Bull. Bot. Soc. Beng.* **12**: 85-90.

Jawale, A.K. and Patel, R.J. 1985. Morphology and Cytology of *Nitella polycarpa* Pal; forma *gujratensis* f. nov. from India. *Jour. Ind. Bot. Soc.* **64**: 317-322.

Kanahori, T. 1971. Cytotaxonomical research on the Characeae: Karyotype of the section Nitella of the genus *Nitella. Bot. Mag.* Tokyo. **84**: 327.

Kanahori, T. 1974. The karyotypes of two species of Characeae. Chromosome *Inf. Service.* **16**: 29.

Karling, J.S. 1928. Nuclear and cell division in antheridial filaments of Characeae. *Ibid.* **55**: 17-39.

Kasaki, H. 1964. The Charophyta from the lakes of Japan. *J. Hattori Bot. Lab. Bryol.* **27**: 217.

Khan, M. and Sarma, Y.S.R.K. 1967a. Studies on cytotaxonomy of Indian Charophyta I. *Chara. Phykos.* **6**: 36-47.

Khan, M. and Sarma, Y.S.R.K. 1967b. Studies on cytotaxonomy of Indian Charophyta II. *Nitella. Phykos.* **6**: 48-61.

Khan, M. and Sarma, Y.S.R.K. 1967c. Some observations on the cytology of Indian Charophyta. *Phykos.* **6**: 62-74.

Khan, M. and Sarma, Y.S.R.K. 1980. New chromosome numbers for Indian Charophyta. *Cell* and *Chr. Newsletter.* **3**: 24.

Khan, M. and Sarma, Y.S.R.K. 1981-82. Cytogeographic study of Charophyta with particular reference to India (from Recent advances in Cryptogamic Botany, Ed. D.C. Bharadwaj, Publisher: Paleobotanical Society, Lucknow).

Khan, M. and Sarma, Y.S.R.K. 1984. Cytogeography and Cytosystematics of Charophyta. In: Systematics of green algae. Ed. D.E.G. Irvine and D.M. John, Academic Press, London.

Klein, R.M. and Cronquist, A.A. 1967. A consideration of the evolutionary and taxonomic significance of some biochemical, micromorphological and physiological characters in the thallophytes. *Q. Rev. Biol.* **42**: 105-296.

Kumar, H.D. and Singh, H.N. 1982. A text book on Algae. pp. 76-97.

Labh, L. and Verma, B.N. 1984. Cytological studies in three species of *Tolypella. Cytologia.* **49**: 717-723.

Labh, L. and Verma, B.N. 1985a. First report of chromosome number in four taxa belonging to genus *Chara Linn. Cytologia.* **50**: 21-24.

Labh, L. and Verma, B.N. 1985b. New records of chromosome numbers for the Charophytes. *Cytologia.* **50**: 55-58.

Labh, L. and Verma, B.N. 1986. New counts for *Chara fibrosa* complex. *Cytologia.* **51**: 185-191.

Lefort-Tran, M. 1989. Ultrastructure of the chloroplast envelope in algae: Functional and phylogenetic implications. In: Phycotalk Vol. I. Ed. H.D. Kumar. pp. 61-92.

Leonhardi, H.V. 1863. Ueber die bohmischen Characeen. *Lotos.* **13**: 55-111-148.

Levan, A., Fredga, K. and Sandberg, A.A. 1964. Nomenclature of centromeric position on chromosomes. *Heriditas.* **52**: 201-220.

Lindenbein, W. 1927. Beitrag zur Cytologie der Charales. *Planta.* **4**: 436-466.

Linnaeus, C. 1774. Systema Vegetabilium, secundum classes, ordines, genera, species, cum characteribus et differentiis. ed. 13. Gottingae.

Mac Donald, M.B. and Hotchkiss, A.T. 1956. An estipulodic form of *Chara australis* R. Br. (Protochara australis Womersley and Ophel) Proc. *Linn. Soc.* New South Wales. **80**(3): 279-284.

Mathew, A. and Mathew, P.M. 1987. Cytological studies on the South Indian Compositae. In Glimpses in plant research III. Ed. P.K.K. Nair: Today and Tomorrow's printers and Publishers New Delhi, India. pp. 3-177.

Mattox, K.R. and Stewart, K.D. 1977. Cell division in the scally green flagellate *heteromatrix angulata* and its bearing on the origin of Chlorophyceae. AM. *J. Bot.* **64**: 931-945.

Mattox, K.R. and Stewart, K.D. 1984. Classification of the green algae. A concept based on comparative cytology. In Systematics of the green algae (Ed. D.E.G. Irvine and D.M. John). Academic Press, London. pp. 29-72.

Mendes, E.J. 1946. Mitosis in the spermatogenous threads of *Chara vulgaris* var. *longibracteata* Kuetz. Portugal *Acta. Biol.* **1**: 251.

Merxmuller, M. 1970. Biosystematics still alive? Provocation of biosystematics. *Taxon.* **9**: 140-150.

Migula, W. 1890-1897. Die Characeen Deutschlands, Oesterreich und der Schweiz. In: L. Rabenhorst, Kryptogamen-Flora Von Deutschland, Oesterreich und der Schweiz. 2 *Aufl.* **5**: 1-765.

Mirsaidov, T.I. 1971. Nombre chromosomique de quelques especes de Characees d' ouzbekistan (in Russian), *Uzbekbiol. Zh.* **15**: 59.

Mitra, D., Mitra, J.N. and Chaudhuri, S.K. 1991. Studies in Botany, Vol. 1.

Moutschen, J.M., Dahmen, M. and Gillet, C. 1956. Sur le modifications induties par les hyrazide malique et isonicotinique dans les antheridies de *Chara vulgaris*. *La Cellule*. **58**: 85.

Muhldorf, F. 1930. Berichtigungen und Erganzungen unserne Kenntnisse weber der Morphologie und Histologie pflanzlicher spermien. *Bot. Generalis*. **6**: 457-482.

Mukherjee, S. 1978. Polyploid chromosome numbers in the genus *Nitella*. *Curr. Sci.* **47**: 386.

Mukherjee, S. and Noor, M.N. 1980. On a new Charophyte from India. *Hydrobiologia*. **68**(1): 81-84.

Nair, P.K.K. 1987. Glimpses in Plant Research. Today and Tomorrows Printers and Publishers, New Delhi.

Noor, M.N. 1968. A new record of chromosome number in *Nitella mucronata* (A. Br.) Miquel from India. *Sci. Cult.* **34**: 214-215.

Noor, M.N. 1969-70. A preliminary report on the chromosome numbers of some Indian Characeae. *J. Ranchi Univ.* **6 & 7**: 233.

Noor, M.N. and Mukherjee, S. 1974. An investigation into the karyology of *Nitella superba* from India. Proc. 2nd *Inter. Symp. Taxo. Algae*, Madras Univ. p. 28.

Noor, M.N. and Mukherjee, S. 1975. On the aneuploid chromosome number in *Chara hydropitys* Reich from India. *Cytologia*. **40**: 227-232.

Noor, M.N. and Mukherjee, S. 1977. Some new records of chromosome numbers in Indian Charophyta. *Cytologia*. **42**: 227-232.

Noor, M.N. and Purak, I. 1985. Effects of streptomycin on the karyology of *Chara braunii*. Proc. 72nd Indian Sci. Cong. p. 8.

Noor, M.N. and Purak, I. 1989. Effects of two antibiotics on nuclear division of *Chara braunii* Gm. (Characeae). *Cryptogamie Algologie*. **10**(2): 143-152.

Oehlkers, F. 1916. Beitrag zur kenntnis der kerntheelungen bei den Charazeen Ber Deutsch *Bot. Gesell.* **34**: 223-227.

Oltmann, F. 1904. Morphologie und Biologie der Algen. 1. Jena. (Charales). pp. 328-347.

Pal, B.P., Kundu, B.C., Sundaralingam, V.S. and Venkataraman, G.S. 1962. Charophyta. I.C.A.R., New Delhi.

Papenfus, G.F. 1946. Proposed names for phyla of algae. *Bull. Torrey Bot. Club*. (Phylum Charophycophyta). **73**(3): 217-218.

Papenfus, G.F. 1955. Classification of the algae. In century of progress in the Natural Sciences, San Francisco (Charophycophyta, pp. 129-133). pp. 115-224.

Pascher, A. 1925. Die Susswasser—Flora Deutschlands, Osterreichs und der Schweiz. Vol. II. Charophyta-Charales. pp. 207-250.

Patel, R.J. and Jawale, A.K. 1979. *Chara pseudohydropitys* (Imah) R.D.W. A new record for India. *Geobios*. **6**: 181-183.

Patel, R.J. and Jawale, A.K. 1978. On *Lychnothamnus barbatus* (Meyen) Leonh. from Gujrat, India. *Phykos*. **10**(1 & 2): 51-52.

Peshkov, M.A., Mirsaidov, T.I. and Tikhomirov, L.A. 1973. Etude caryologique de quelques especes d algues du genre *Chara* de Uzbekistan. *Akad. Nauk Ser. Biol.* **5**: 672.

Peshkov, M.A., Mirsaidov, T.I. and Tikhomirov, L.A. 1974. Etude des nombres chromosomiques et des caryotypes approximatits de queiques algues du genre Chara en 1 Uzbekistan. *Akad. Nauk. SSSR Ser. Biol.* **13**: 114.

Pekarek, J. 1932. Ein vergessenes Objekt fur das Studium der Kern und Zellteilunsvorgange im Leben, *Planta*. **16**: 788-800.

Pickett-Heaps, J.D. 1975-76. Cell division in eukaryotic algae. *Bio. Sci.* **26**: 445-450.

Pickett-Heaps, J.D. and Marchant, H.J. 1972. The phylogeny of the green algae: A new proposal. *Cytobios.* **6**: 255-264.

Prasad, P.K. and Verma, B.N. 1985. Aneuploid count for *Chara setosa* Klein ex. Willd. *Cytologia.* **50**: 241-245.

Prescott, G.W. 1969. The Algae: A Review. Nelson, London. pp. 37-45.

Proctor, V.W. 1971. Taxonomic significance of monoecism and dioecism in the genus *Chara. Phycologia.* **10**: 299.

Pundhir, H.S. 1994. Studies on the karyomorphology of *Chara hornemannii* f. *longifolia* (ROB) RDW: A new record for India. *J. Mendel.* **11**(3-4): 157-160.

Pundhir, H.S., Gautam, A. and Vidyavati. 1993. Comparative study and new chromosome counts for *Nitella hyalina* var. *hyalina* f. *indica* (A. Br.) RDW from India. *Phykos.* pp. 169-174.

Pundhir, H.S. and Gautam, A. 1994. New chromosome count for *Chara hornemannii* f. *longifolia* (Rob.) RDW: A new record from India. *Cytologia.* **59**: 305-307.

Pundhir, H.S. and Vidyavati. 1993. Taxonomic enumeration and chromosome counts of some Charophytes from district Aligarh, Uttar Pradesh. *J. Mendel.* **10**: 83-85.

Pundhir, H.S. and Vidyavati. 1994. New karyomorphological observations on Charophytes from Uttar Pradesh, India. *The Nucleus.* **37**(1 & 2): 39-45.

Purak, I. 1986. Cytotaxonomic and mutagenic studies of some members of Charophyceae and other green algae. Ph.D. Thesis, Ranchi University.

Purak, I. 1994. Morphological and Cytological studies of *Chara fibrosa* form *tylacantha* (Nordst.) R.D.W. Proc. 81st Indian *Sci. Cong.* p. 114.

Ramjee. 1969. Contributions to the systematics, cytology and cytotaxonomy of Indian Charophyta. Ph.D. thesis. Banaras Hindu University, India.

Ramjee and Bhatnagar, S.K. 1978. Studies on Charophytes from Rohilkhand division-1. Moradabad: Taxonomic enumeration and chromosome counts. *Phykos.* **17**(1 & 2): 87-92.

Ramjee and Bhatnagar, S.K. 1978a. Significance of a new chromosome number in *Nitella mirabilis* (Nordst. ex. Gr.) em RDW. *Hydrobiologia.* **57**: 99-101.

Ramjee and Bhatnagar, S.K. 1978b. Cytotaxonomic studies on *Nitella furcata* subsp. *flagellifera* f. *patula* Gr. ex Allen and comparision with other taxa of N. furcata complex. *Caryologia.* **31**(4): 457-462.

Ramjee and Sarma, Y.S.R.K. 1971. Some observations on the morphology and Cytology of Indian Charophyta. *Hydrobiologia.* **37**(2): 367-382.

Ray, S. and Chatterjee, P. 1986. Karyotype analysis in *Chara braunii* GMF *braunii* (Characeae). *The Nucleus.* **29**: 33-36.

Ray, S. and Chatterjee, P. 1987. Karyomorphological investigation of three populations of *Chara corallina* var and f. *corallina* in West Bengal. *Cytologia.* **52**: 455-458.

Ray, S. and Chatterjee, P. 1988. Cytotaxonomical study of some members of *Chara fibrosa* complex. *Cryptogamie Algologie.* **9**: 65-72.

Riker, A.J. 1921. Chondrisomes in *Chara. Bull. Torrey Bot. Club.* **48**: 141.

Richard, L.C.M. 1815. Characeal, Validaing Prior Publication missing. p. 38.

Rodrigues, M. 1945. Sobre O localization da meiose no ciclo da vida das Characeae. *Bot. Soc. Broteriana.* **19**: 609.

Round, F.E. 1963. The biology of the Algae. Edward Arnold Publishers, London.

Round, F.E. 1963. The taxonomy of Chlorophyta. *Br. Phycol. Bull.* **2**(4): 224-235.

Round, F.E. 1971. The taxonomy of Chlorophyta II Br. *Phycol. J.* **6**(2): 235-264.

Round, F.E. 1984. The systematics of Chlorophyta. A historical review leading to some modern concept (Taxonomy of the Chlorophyta III). In: Systematics of green algae. Ed. D.E.G. Irvine and D.M. John. Academic Press, London.

Ruprecht, F.J. 1845. Distribution Cryptogamarum Vascularium in Imperio Rossico. Peteropoli. Characeae, pp. 7-18.

Sachs, J.V. 1875. Text book of Botany. Oxford. pp. 292-305.

Sarma, Y.S.R.K. 1964. Cytology in relation to systematics of algae with particular reference to Chlorophyceae. *Nucleus.* **7**: 127-136.

Sarma, Y.S.R.K. 1968. Cytology and cytotaxonomy of Indian Charophyta. A resume. Proc. Inter. Sem. Chromosome structure and function. *The Nucleus.* **11**: 128-137.

Sarma, Y.S.R.K. 1982. Chromosome numbers in Algae. *The Nucleus.* **25**: 66-108.

Sarma, Y.S.R.K. 1973. Algal Karyology in India. In: Advancing Frontiers in Cytogenetics. (A collection of papers in honour of Prof. P.N. Mehra). pp. 266-285.

Sarma, Y.S.R.K. and Khan, M. 1964. Chromosome numbers in some Indian species of *Nitella. Chromosoma* (Berl.). **15**: 246-247.

Sakano, K.H. 1989. Iao-aspargine as a marker amino acid in the taxonomy of Characeae. *Bot. Mag.* Tokyo. **102**(1066): 235-244.

Sarma, Y.S.R.K. and Khan, M. 1965a. Chromosome numbers in some Indian species of *Chara. Phycologia.* **4**: 173-176.

Sarma, Y.S.R.K. and Khan, M. 1965b. Some new observations on the karyology of *Chara zeylanica* Klein ex Willd. *Curr. Sci.* **34**: 293-294.

Sarma, Y.S.R.K. and Khan, M. 1965c. A preliminary report of the survey of chromosome numbers of Indian Charophyta. *Nucleus.* **8**: 33-38.

Sarma, Y.S.R.K. and Khan, M. 1966. A new form of *Lychnothamnus barbatus* Leonh. f. *iyengarii* f. nov. *Phykos.* **5**(1 & 2): 195-197.

Sarma, Y.S.R.K. and Khan, M. 1967. On two new Charophytes from India. *Hydrobiologia.* **30**: 405-410.

Sarma, Y.S.R.K. and Khan, M. 1967a. Dioecism and monoecism as taxonomic criteria in Charophyta. *Curr. Sci.* **36**: 245-246.

Sarma, Y.S.R.K., Khan, M. and Ramjee. 1970. A cytological approach to phylogeny interrelationship and evolution in Charophyta. *Indian Biologist.* **2**: 11-19.

Sarma, Y.S.R.K. and Ramjee. 1967. A new chromosome number in *Nitella tenuissima* (Desv.) Kutz. em. R.D.W. *Caryologia.* **20**: 61-63.

Sarma, Y.S.R.K. and Ramjee. 1971. Chromosome number in three taxa of Indian Charophyta. *J. Cytol. Gen.* **4**: 32-36.

Sarma, Y.S.R.K. and Ramjee. 1971a. Significance of chromosome numbers in Charophyta—A discussion. *Caryologia.* **24**: 391-401.

Sarma, T.A. and Bala, V. 1984. A new chromosome count in four taxa of *Chara* from Punjab. *J. Cytol.* and *Genet.* **19**: 27-32.

Sasaki, M. 1961. Cytological studies in Charophyta with special reference to synchronus mitosis in antheridial filament. St. Paul. *Rev. Sci. Nat. Sci.* **8 & 9**: 1.

Sato, D. 1959. The protokaryotype and phylogeny in plants. Sci. Pap. Coll. *Gen. Edu. Um.* Tokyo. **9**: 303-327.

Sawa, T. 1965. Cytotaxonomy of the Characeae: Karyotype analysis of *Nitella opaca* and *Nitella flexilis*. *Am. J. Bot.* **52**: 962-970.

Sawa, T. 1971. Chromosome numbers in the genus *Tolypella*. *J. Phycol Suppl.* p. 7.

Sawa, T. 1974. New chromosome number for the genus *Tolypella* (Characeae). *Bull. Torrey. Bot. Club.* **1-1**: 21-26.

Schottlander, P. 1892. Beitrage zur Kenntnis des Zellkernes und der Sexualzellen bei Kryptogamen. *Beitr. Z. Biol. d. Pflanzen.* **6**: 267-304.

Schmucker, T. 1927. Uber Bildunganomalien bei *Chara. Planta.* **4**: 780-787.

Sharma, O.P. 1986. Text book of algae. pp. 121-260.

Silva, P.C. 1952. A review of nomenclature conservation in algae from the point of view of the type method. *Univ. Calf. Publ. Bot.* **25**(4): 241-324.

Silva, P.C. 1962. In physiology and biochemistry of algae, Ed. A.R. Lewin, New York.

Silva, P.C. 1982. Chlorophyta. In: Synopsis and classification of living organisms. (S.P. Parker ed.) Vol. I. Mc. Graw. Hill. New York.

Sinha, J.P. and Noor, M.N. 1967. Chromosome numbers in some members of Chlorophyceae of Chotanagpur (India). *Phykos.* **4**: 74-75.

Sinha, J.P. and Noor, M.N. 1971. Studies on the karyology of some members of Charophyta of Bihar. *Phykos.* **10**(1 & 2): 112-117.

Sinha, J.P. and Sinha, S. 1989. Study of mitotic chromosomal counts in some members of Charales of Chhotanagpur (Bihar). *Biojournal.* **1**(2): 69-72.

Sinha, J.P. and Verma, B.N. 1970. Cytological analysis of the Charophytes of Bihar. *Phykos.* **9**(2): 92-99.

Sinha, J.P. and Verma, S. 1969. Cytological studies on three species of *Chara* Linn. Proc. 56th Ind. Sci. Cong. III. p. 366.

Sinha, S. 1974. Cytotaxonomic and cultural studies of common members of Charales and other green algae of Chotanagpur plateau. Ph.D. Thesis, Ranchi University, Ranchi.

Sinha, S. and Purak, I. 1994. Morphological investigation of *Nitella furcata* complex. *Neo Botanica.* **2**(2): 87-91.

Sinha, S. and Purak, I. 1995. Karyomorphological analysis of *Nitella furcata* complex. 82nd Ind. *Sci. Cong. Proc.* p. 82.

Sinha, S. and Sinha, J.P. 1992. Karyomorphological studies on three species of *Chara. J. Appl. Biol.* **2**: 7-11.

Smith, G.M. 1938. Cryptogamic Botany 1. Algae and Fungi. N.Y. Charophyceae. pp. 127-135.

Smith, G.M. 1950. The fresh-water algae of the united States. Second edition. New York. Charophyceae. pp. 336-347.

Smith, G.M. 1951. Manual of Phycology. Waltham.

Smith, G.M. 1955. Cryptogamic Botany. Algae and Fungi. ed. 2. New York. Charophyceae. pp. 121-131.

Smith, J.E. 1800. Flora Britannica. Vol. I. Londini. *Chara.* pp. 4-7.

Smith, P.M. 1976. The chemotaxonomy of plants. London. Arnold.

Sokal, R.R. and Sneath, P.H.A. 1963. Principles of numerical taxonomy. London and San Fransico: Freeman.

Solbrig, O. *et al.* 1964. Chromosome numbers in Compositae. V. Asteraceae. II. *Am. J. Bot.* **56**: 348-353.

Strasburger, E. 1912. A text book of Botany (Revised by W.H. Lang, 4th Ed. London), Characeae, pp. 422-425.

Stace, C.A. 1980. Plant taxonomy and Biosystematics. London. Arnold.

Steil, W.N. 1941. Antherozoids of certain Characeae. *Bot. Gaz.* **102**: 623-628.

Strasburger, E. 1908. Einiges ueber Characeen und Amitose. Wiesner Festschr. Vienna. (from Fritsch, 1935). pp. 24-47.

Stebbins, G.L. 1950. Variation and evolution in plants. Columbia University Press, New York.

Stebbins, G.L. 1958. Longivity, habitat and release of genetic variability in higher plants. *Cold Spring Harb. Symp. Quart. Biol.* **23**: 365-378.

Stebbins, G.L. 1971. Chromosomal evolution in higher plants. Addison Wesley Publishing Company.

Stewart, K.D. and Mattox, K.R. 1975. Comparative cytology, evolution and classification of green algae with some consideration of the origin of other organism with chlorophyll *a* and *b*. *Bot. Rev.* **41**: 104-135.

Subramanian, D. 1983. Studies on Charophytes from Tamil Nadu 1. Taxonomic enumeration and chromosome counts. *Phykos.* **22**: 18-22.

Subramanian, D. and Ganeshan, R. 1985a. Chemotaxonomical studies of Charophyta from Tamil Nadu. In: Advances in Applied Phycology, 1985. pp. 107-112.

Subramanian, D. and Ganeshan, R. 1985b. Cytotaxonomical studies of Charophyta from Tamil Nadu. In: Advances in Applied Phycology, 1985. pp. 184-191.

Subrahmanyam, B.V.S. and Chowdary, Y.B.K. 1992. Karyology of *Nitella* from South India. *Cytologia.* **57**: 209-212.

Subrahmanyam, B.V.S. and Chowdary, Y.B.K. 1992. Karyological studies on some Indian Charophyta. *Cytologia.* **57**: 409-415.

Subrahmanyam, B.V.S., Chowdary, Y.B.K. and Sarma, Y.S.R.K. 1991. Aneuploidy in two taxa of *Nitella* (Div. Charophyta) from Karnataka, India. *The Nucleus.* **34**(3): 174-176.

Sundaralingam, V.S. 1946. The cytology and spermatogenesis in *Chara zeylanica* Wild. *J. Ind. Bot. Soc.* pp. 289-303.

Sundaralingam, V.S. and Bharatan, S. 1978. Occurrence of *Lychnothamus barbatus* Leonh. near Madras. *Phykos.* **17**(1 & 2): 123-125.

Telezynski, M. 1929. Garnitures des chromosomes et synchronisme des divisions dans les filaments d'antherozoides chez certain especes du genera Chara Vaill. *Acta. Soc. Bot. Poloniae.* **6**: 230-237.

Tilden, J.E. 1937. The algae and their life relations. Univ. Minnesota Press.

Tindall, D.R. 1967. A new species of *Nitella* (Characeae) belonging to the *Nitella flexilis* species group in north America. *J. Phycol.* **3**: 229.

Tindall, D.R. and Sawa, T. 1964. Chromosomes of the Characeae of Wood Hole (Massachusetts). *Amer. J. Bot.* **51**: 943-949.

Tindall, D.R., Sawa, T. and Hotchkiss, A.T. 1965. *Nitellopsis bulbifera* in North America. *J. Phycol.* **1**: 147.

Vaidya, B.S. 1963. *Hydrobiologia.* **21**: 188.

Vaillant, S. 1721. Caracteres de quqtorze genres de plantes. Mem. Acad. Roy. Sci. Paris. (1719) pp. 17-20.

Vasishtha, B.R. 1995. Botany for degree students. pp. 242-266.

Verma, B.N. 1967. A new chromosome count of *Chara braunii* Gmelin. *Proc.Ind. Sci. Cong.* p. 365.

Walter, E. 1929. Entwicklungsgeschichtliche und cytologische Untersuchunge an einigen Nitellen, *Arch. J. Kl-usstiftung.* **4**: 23-121.

West, G. 1910. A further contribution to a comparative study of the dominant phanerogamic and higher cryptogamic flora of aquatic habit in Scottish Lakes. *Proc. Roy Soc. Edinburgh.* **30**: 65-181.

Wildeman, E.D. 1900. Les Algues de la Flora de Buitenzorg. (Charophycees, pp. 372-377).

Williams, J.T. and Tindall, D.R. 1975. Chromosome numbers for species of Characeae from Southern Illinois. *Amer. Midl. Naturalist.* **93**: 330-338.

Womersley, H.B.S. and Ophel, I.L. 1947. Protochara, a new genus of Characeae from Western Australia. *Trans. Roy. Soc. S. Austr.* **71**(2): 311-317.

Wood, R.D. and Imahori, K. 1965. A revision of the Characeae. Monograph and Iconograph. Vertag. Von. J. Cramer Pub. New York.

Wood, R.D. and Manson, R. 1977. Characeae of New-Zealand. New Zealand *J. Bot.* **15**: 87.

Zaneveld, J. 1940. The Charophyta of Malayasia and adjacent countries. *Blumea.* **4**(1): 1-224.

Zanusz, M. 1991. DnA replication in antheridial filaments of *Chara* spp. with different genome sizes and S phase duration. *Biol. Zentralbl.* **110**(1): 23-32.

Recent Trends in Algal Taxonomy
Vol. I. Taxonomical Issues
Editors : Vidyavati and A.K. Mahato
APC Publications Pvt. Ltd., New Delhi, 1999, pp. 221-228

# *A Reappraisal on the Taxonomy of* Lamprothamnium *J. GR.*

S. Bharathan
Department of Botany, Pachaiyappa's College, Chennai-600 030

**ABSTRACT**

*Lamprothamnium* J. Gr. a leuco-halophyte which has been listed as an endangered species in certain countries has not been investigated in detail. Despite the fact that a number of taxa have been assigned to *Lamprothamnium* only two species viz. *L. papulosum* and *L. succintum*, have been studied with regard to their ecology, morphology and development. Present review on *Lamprothamnium* brings out areas investigated so far and also reveals the scope for further investigations.

## INTRODUCTION

The genera and families in algae are largely based on morphology and reproduction (Padmaja and Desikachary, 1967). However, difficulties have arisen in the demarcation of various taxa due to the prevalance of phenotypic plasticity. It is more true with regards to the family Characeae, which has six genera and eighty one species (Wood and Imahori, 1965). Among the genera of Characeae, the genus *Lamprothamnium* is the most disputed and the least investigated taxon. Reviews on the various aspects of *Lamprothamnium* have been published by Corillion (1953), Hotchkiss (1966), Daily (1967) and Blindow and Langangen (1995). The present review is an attempt to update and to identify lacunae in our knowledge on *Lamprothamnium*.

## DISTRIBUTION

*Lamprothamnium*, primarily occurs in brackish water habitats. Davis and Lipkin (1986) have reported of the growth of *Lamprothamnium* in permanently hypersaline water (salinity ranging from 0.1 to 12.5% TSC). They have observed morphological variations including pedicellate gametangia and deviation in size of plant parts.

It has been reported to occur in Europe, Africa, Denmark, S. Asia, Italy, Australia, China, Japan and Newzealand (Ophel, 1947; Imahori,

1963; Wood and Imahori, 1965 and Hotchkiss, 1966). Corillion (1953) has reported its occurrence in Baltic, Atlantic and Mediterranian areas. Similarly, Blindow and Langangen (1995) have reported from Scandinavia on sand/silt bottom in shallow waters at 1-2m depth.

In India, only one spcies, *L. succinctum* has been reported from Salsette, Bombay (Dixit, 1931), Pulicat lake near Madras (Bharathan and Sundaralingam, 1984), Chotanagpur and Ranchi in Bihar (Sinha and Verma, 1970).

There are reports of the disappearance of *Lamprothamnium* in a number of localities due to eutrophication and pollution. In Britain the alga has been included under endangered species (Moore, 1991).

## TAXONOMY

*Lamprothamnium* as typified by *L. papulosum*, is ecorticate, with stipulodes opposite to branchlets and under conjoined conditions the oogonium is below or by the side of the antheridium. The second species *L. hansenii* (Sond.) Cor. was established during 1890. Subsequently, Ophel (1962) transferred the sejoined taxon *C. succinctum* to *Lamprothamnium* as *L. succinctum*. Wood and Imahori (1965) in their Monograph on 'A revision of Characeae' have listed the following taxa under *Lamprothamnium*.

A. *L. papulosum* (Wallr.) J. Gr. em. R.D.W.
var. *papulosum* (Wallr.) J. Gr. em. R.D.W.
1. f. *papulosum*
2. f. *carrissoi* (G. da. C.) R.D.W.
3. f. *aragonense* (Prosp.) R.D.W.
4. f. *pouzolsi* (Gay ex A. Br., R.D.W.
5. f. *macropogon* (A. Br.) R.D.W.
var *toletanum* (Prosp.) R.D.W.

B. L. *hansenii* (Son.) Cor.

C. L. *succinctum* (A. Br.) R.D.W.

Although Wood and Imahori (1965) have reduced a number of taxa to the level of varieties and forma, in the present study they are treated as individual species and a key to the species is given below.

## KEY TO THE SPECIES

1. Gametangia conjoined ... 2

1. Gametangia sejoined ... 4
   2. Gametangia at the base of whorls ... *L. aragonense*
   2. Gametangia at the base of whorls ... 3
3. Plants with fox-tail appearance ... *L. papulosum*
3. Plants with 'Chara' like appearance ... *L. hansenii*
   4. Plants with fox-tail appearance ... *L. macropogon*
   4. Plants with 'Chara' like appearance ... 5
5. Gametangia at the base of whorls ... *L. succinctum*
5. Gametanga not at the base of whorls ... *L. pouzolsii*

Daily (1967) emended *Lamprothamnium* by including characters like 'exterior of stem node structure, starch grain type, and external and internal features of the lime shells'. Based on these features she has transferred three species of *Chara* and one species of *Nitellopsis* to *Lamprothamnium* and redesignated them as:

| | |
|---|---|
| *L. hornemannii* (Wallm.) | Daily |
| *L. longifolium* var. *longifolium* (Rob.) | Daily |
| *L. longifolium* var. *buckellii* (G.O. Allen) | Daily |
| *L. longifolium* var. *bulbilliferum* (C. Dont.) | Daily |

All the above taxa are dioecious and therefore the position of antheridia in relation to oogonia cannot be determined.

Proctor *et al.* (1967) have carried out interfertility tests between *C. longifolia C. buckellii* and *Nitellopsis bulbillifera.* Based on the experiment they concluded that all the three were conspecific and treated *C. hornemannii* as separate species.

Developmental morphological studies on *L. hornemannii* (Bharathan, 1980) has proved that *L. hornemannii* properly belongs to *Chara.*

Thus, *Lamprothamnium* is left with six taxa as recognized by Wood and Imahori (1965) while there are a few reports on ecology, morphology and cytology of *L. papulosum.* and *L. succinctum,* very little information is available about other taxa.

## DEVELOPMENTAL MORPHOLOGY

The developmental morphology of two species of *Lamprothamnium* viz., *L. papulosum* and *L. succinctum* has been investigated by Giesenhagen (1896), Iwasaki (1963) and Bharathan and Sundaralingam (1964). Information regarding developmental morphology of other taxa of *Lamprothamnium* is almost nil.

The apical cell with unlimited growth is responsible for the development of the thallus which consists of internodes and nodes. The later bears the branchlets, axillary branch, stipulodes, and reproductive structures. The apical cell cuts off segments distally and each of which divide into a node and an internodal cell. As in other member of the tribe Chareae, the internode without any further divisions develop directly into a long, vacuolated, multinucleate cell. The nodal initial by a halving wall divides into two cells and further cuts off 6-10 peripheral cells. Each of the peripheral cell by further divisions form the branchlets. The two central cells of the axis node divide further resulting in the formation of more than two central cells. It is interesting to note that among Chareae *Chara, Lychnothamnus*, and *Nitellopsis* have two central cells, while in *Lamprothamnium* it is more than two, as in Nitelleae.

Yet, another interesting feature which demarcates *Lamprothamnium* from other members of Chareae is the characteristic features of the basal node of the branchlet (Bharathan, 1986). Here, the basal node consists of a central cell with four peripheral cells (one adaxial, one abaxial and two laterals). The abaxial peripheral cell produces a single stipulate opposite to the branchlet. Such a feature is characteristic of *Lamprothanium.* The structure of the basal node of branchlet in *L. succinctum* is little different from that of the branchlet in *L. papulosum.* Here, the basal node consists only of four cells. Axial cortication is absent.

The nodes of the branchlets have a central cell and four or five peripheral cells. The peripheral cells divide into an outer and an inner cell. The outer cell without any further division develops into bract and the inner cell remains without any change. Since the inner cell remains undivided, the cortex formation is absent.

The developmental morphological studies have revealed that the position of the stipulode is opposite to the branchlets in *Lamprothamnium* and this feature cannot be compared to other members of the tribe Chareae where the stipulodes are alternate to branchlets.

In the development of the axillary branch *Lamprothamnium* is different from other members of Characeae. In both the species of *Lamprothamnium*, the first internodal cell functions as initial for a axillary branch, cuts off a segment by a transvers wall and the upper one behaves as a apical cell of the axillary branch.

Sex organs: While in, *L. papulosum* sex organs are conjoined and produced at the branchlet nodes, in *L. succinctum* they are sejoined and the antheridia are distributed at branchlet nodes while the oogonia occur at the base of the whorl of branchlets.

The terminal part of the main axis of the thallus gives a foxtail appearance due to short internodes, incurved branchlets and restricted appearance of the reproductive structures in the branchlets of the apical regions. Yet, another diagnostic feature is that under conjoined conditions the antheridia are seen above the oogonia, a position reverse to that of *Chara.* But under sejoined conditions this feature cannot be taken into consideration which results in difficulties in assigning taxa to *Lamprothamnium.*

The development of the reproductive structure is similar to other members of Chareae except that the oogonium develop from the lower most or abaxial cell of the antheridial node. This is the reason for the development of oogonium below the antheridium. In sejoined taxa the development oogonium takes place from the basal node of the branchlet as in *C. wallichii.*

The oogonium of *Lamprothamnium* has characteristic globose crown cells and has single sterile cell. The structure of the oospores of *Lamprothamnium* has been investigated by Leitch (1989). He has described the development and structure of the compound oosporangial wall (COW) which was formerly referred to as the oospore membrane.

In both *L. papulosum* and *L. succinctum* the development of the spherical bulbils from rhizoids have been observed (Bharathan and Sundaralingam, 1984 and Bharathan, 1987).

## CYTOLOGY

Cytological investigations have been carried out by number of authors. The basic number has been identified as 14 (Hotchkiss, 1966; Khan and Sarma, 1984). In contrast to *Chara,* high level of polyploidy has been observed in *Lamprothamnium* (*L. papulosum* n = 50, 56; *L. succinctum* n-42; *L. macropogon* n = 56—Hotchkiss, 1966). It has been reported that the level of poliploidy is higher in taxa with disjunct gametangia in *Lamprothamnium.* This is in contrast to the condition members of found among disjunct and dioecious members of *Chara* Guerlesquin (1967) has related polyploidy with saline habit.

## FOSSIL

The charophytes are one of the most ancient groups of algae. The calcified fructifications represent the taxa which have been

**Table 1. Comparative account of the morphological features of species of *Lamprothamnium***

| | *Lamprothamnium Papulosum* (Wallr.) J.Gr. em. R.D.W | *L. pouzolsii* (Gay ex A.Br) R.D.W. Comb nov | *L. aragonesnes* (Prosp.) G.da. C | *L. macropogon* (A.Br.) Ophel | *L. hansenii* (Sond.) Cor. | *L. succinctum* (A.Br.) R.D.W |
|---|---|---|---|---|---|---|
| Apical region | Foxtail | Chara like | Chara like | Foxtail | Chara like | Chara like |
| Axis diameter | 600-800μ | c. 320μ | 750μ | c. 400μ | 600-1500μ | 500-700μ |
| Gametangia | Conjoined | Sejoined | Conjoined | Sejoined | Conjoined | Sejoined |
| Oogonia position | Oogonia below or beside antheridia | solitary at lowest branchlet node | oogonia below antheridium aggregate at base of whorl | solitary or geminate at lowerst 1-2 br. nodes and clustered at base of whorl | oogonium below antherdium only at branchlet | rarely conjoined when conjoined oogonium below or beside anthe-ridium clustered at base of whorl |
| Oogonia dimension | 750-950μ x 400-575μ | 670μ x 500μ | 600-700μ X 420-500 | 840-1000μ X 450-555 | 710-975μ x 405-450 | 710-975μ x 405-450 |
| Oospore | 600-750μ x 275 x 430μ | 530-640μ x 330-350 | 600-700μ x 350-475 | 645-685μ x 360-375 | 500-550μ x 420-460 | 535-670μ x 312-345 |
| Antheridia | c. 100μ | | c-480μ | 375-425μ | 356-525μ | 356-525μ |
| Chromosome numbers | 50<br>56 | | | 56 | | 42 |

reported to occur since late silurian. The gyrogonites from early eocene have been considered as related to the genus *Lamprothamnium* based on the apical structure and are generally included in the subfamily Aclistocharoideae. (Soulie-Marsche, 1979). The fossil genus *Aclistochara* Peck emend Peck, 1957, reported from Godaveri valley of Peninsular India, has lamprothamoid condition (Feist *et al.* 1991). Feist-Castel (1973) has described the fine structure of intercellular sutures in the walls of gyrogonites of *Lamprothamnium*. The pillars and undulations in the walls have been described by him.

## CONCLUSION

*Lamprothamnium* currently is represented by six taxa (table 1). Number of additions, transfers, mergers and reduction in ranks have been carried out by number of Charaphytalogists including Corillion (1953), Daily (1967), Proctor *et al.* (1967) and Wood and Imahori (1965). But for the two taxa viz., *L. papulosum* and *L. succinctum* other taxa remains uninvestigated for their ecology, morphology, development and cytology. Till detailed investigations are carried out on the rest of the taxa these would have to be kept distinct under *Lamprothamnium*.

Already there are reports of the fast disappearance of *Lamprothamnium* due to adverse ecological conditions (Moore, 1991; Blindow and Langangen, 1995). Therefore, it is imperative on the part of algologists to take effective steps to conserve these taxa and also to carryout detailed investigation on their distribution, growth, development, physiology of tolerance to salinity and conservation of *Lamprothamnium* at a global level.

## ACKNOWLEDGEMENT

The author is grateful to Prof. V.S. Sundaralingam for his valuable suggestions and encouragement, to Prof. T.V. Desikachary for his critical comments, to Prof. B. Subramanian and Prof. M.P. Murugesan, Department of Botany, for going through the manuscript. The Author is also thankful to Principal, Pachiayappa's college, and Prof. and Head of the Department of Pachiyappa's College for providing facilities for research.

## REFERENCES

Bharathan, S. 1980. Studies on Characeae. Thesis submitted to the university of Madras for Ph.D.

Bharathan, S. 1983. Development morphology of *Nitellopsis obtusa* (Desv.) Groves. *Proc. Indian Acad. Sci.* (Plant Sci.). **92**: 373-379.

Bharathan, S. 1986. The basal node of the branchlet in the taxonomy of Chareae. *Indian Bot. Soc.* pp. 346-349.

Bharathan, S. 1987. Bulbils of some Charophyts. *Proc. Indian Acad. Sci.* **97**: 257-263.

Bharathan, S. and Sundaralingam, V.S. 1984. Developmental morphology of *Lamprothamnium papulosum* (Wallr). J. Groves and *L. succintum* (A. Br.) R.D.W. *J. Indian Bot. Soc.* **63**: 97-107.

Blindow, I. and Langangen, A. 1995. *Lamprothamnium papulosum* (Wallr.) J. Grovees a threatened charophyte in Scandinavia. *Cryptogamie Algol.* **16**: 47-55.

Corillion, R. 1953. *Lamprothamnium papulosium J. Groves. Bull. Soc. Sci.* de Bretagne. **28**: 33-41.

Daily, F.K. 1967. *Lamprothamnium* in America. *J. Phycol.* **3**: 201-207.

Daily, F.K. 1968. *L. succinctum* (Characeae) with imperfect cortex. Bull of Torrey *Bot. Club.* **96**: 656-660.

Davis, J.S. and Lipkin, Y. 1986. *Lamprothamnium* prosperity in permanently hypersaline water. *Schweiz. J. Hydrol.* **48**: 240-246.

Dixit, S.C. 1931. Some Charophyta from Salsetee. *J. Indian Bot. Soc.* **10**: 205-208.

Feist, M., Bhatia, S.B. and Yadagiri, P. 1991. On the oldest representatives of the family Characeae and its relationships with the Poracharaceae *Bull. Soc. Hot.* Fr. pp. 25-32.

Feist-Castel, M. 1973. Observations nouvells surla paroi des fructificatios chez les charophytes. *Geobios.* **6**: 13-15.

Frame, P.W. and Sawa, J. 1975. Comparative anatomy of Charophyta 11. *J. Phycol.* **11**: 202-205.

Giesenhagen, K. 1896. Untersuchungen uber die Characeen I. *Flora.* **82**: 381-433.

Grambast, L.J. 1974. Phylogeny of the Charophyta. *Taxon.* **23**: 463-481.

Hotchkiss, 1966. Chromosome members in *Lamprothamnium proc. Linn. Soc. New S. Wales.* **91**:1/8-120.

Iwasaki, N. 1963. Entwicklungs geschichtliche untersuchungen an den Characeen. *Jap. Jur. Bot.* **18**: 225-254.

Khan, M. 1980. Occurrence of *Lamprothamnium J. Gr.* (Charophyta) in India *Acta. Botanica. India.* **8**: 97-98.

Khan, M. and Sarma, Y.S.R.K. 1984. Cytogeography and cytosystematics of Charophyta in Systematics of the green algae Ed. D.E.G. Irvine and D.H. John, Academic Press. pp. 303-330.

Leitch, A.R. 1989. Formation and ultrastructure of a complex, multilayered wall around Chara and *Lamprothamnium* (Characeae) *Br. Phycol. H.* **24**: 229-236.

Moore, A.J. 1991. *Lamprothamnium papulosium*, a pioneer in the conservation of Characeae and their habitats. *Bull Soc. Bot. France.* **138**: 73-74.

Ophel, I.L. 1947. Notes on the genera *Lychnothamnus* and *Lamprothamnium* (Characeae). *Trans. Roy. Soc. S. Aust.* **71**: 318-323.

Proctor, V.W., De Donterberg, C.C.C., Hotchkiss, A.T. and Imahori, K. 1967. Conspecificity of some Charophytes. *J. Phycol.* **3**: 208-211.

Srinivasa Rao, P. and Leelavathi, C. 1979. Flora of Pulicat Lake, Madras State. Abstract International symposium on Marine algae of Indian Ocean Region. pp. 9-12.

Soulie-Marsche, L. 1979. Origin and evolution of extant charophyte genera. *Procd. Elf. Aquitaine.* **3**: 821-891.

Wood, R.D. and Imahori, K. 1964. A revision of the Characeae I. Iconograph of Characeae, *J. Cramer. Weinheim.* pp. 1-395.

Wood, R.D. and Imahori, K. 1965. A revision of the Characeae I. Monograph of Characeae. *J. Cramer, Weimbein.* pp. 1-903.

Recent Trends in Algal Taxonomy
Vol. I. Taxonomical Issues
Editors : Vidyavati and A.K. Mahato
APC Publications Pvt. Ltd., New Delhi, 1999, pp. 229-239

# *An Annotated List and Selected Bibliography of Indian Charophyta*

H.S. Pundhir* and M. Khan**
Department of Botany, Faculty of Life-Sciences
Aligarh Muslim University, Aligarh- 202 002 (U.P.), India

## ABSTRACT

The present communication deals with the composition of Charophyte flora of India. A detailed survey of available literature concerned with Indian Charophytes revealed the occurrence of about 142 taxa growing in different parts of India. A comprehensive list of all the reported taxa has been presented. The taxa are arranged following Wood and Imahori (1965) classification.

## INTRODUCTION

Charophytes are known to occur both in temperate and tropical countries of the world. India as abound with approximately all the climatic conditions facilitate the luxuriant growth, prevalency of their occurrence and great diversifications in morphological characters. All the living Charophyte genera, viz., *Chara, Nitellopsis, Lamprothamnium, Lychnothamnus, Nitella* and *Tolypella* were reported from various parts of India. Among them *Tolypella, Lamprothamnium* and *Nitellopsis* were found to restrict only in North India, whereas, the species of *Nitella, Chara* and *Lychnothamnus* showed countrywide distribution.

Voluminous work of various investigators reflected the abundant distribution of Charophytes throughout the Indian subcontinent. In early phase, most of the work on Indian Charophytes was based on taxonomy and systematics. The publication of "*Characeae Indiae Orientalis et insularum maris pacifici*" by Braun (1849) seemed to appear the first systematic account from India, in which he described 11 species. Further additions of a large number of Indian records were made by Braun and Nordstedt (1882). Groves (1924) gave the resume of available information of

* Corresponding author with permanent address: Rohina Singh Pur, Aligarh-202 121 (U.P.), India.

** Posthumously published in his memory.

Indian Charophytes upto that time in his publication of "*Notes on Indian Charophyta*". Valuable additions to the list of Indian Charophytes were made by Groves (1922), Allen (1925, 1928, 1933, 1936, 1942, 1961), Kundu (1938, 1941), Agharkar and Kundu (1937), Dixit (1931, 1935, 1940, 1942, 1966), Mukherjee (1934), Sundaralingam (1959). Publication of I.C.A.R. monograph by Pal *et. al.* (1962) brought together all the described taxa from India and its neighbouring countries upto that time. Subsequently, with the publication of this monograph, our knowledge on the systematics of Indian Charophytes is further enriched by vigorous systematic works of Sinha and Choudhary (1962), Agarkar (1963, 1967), Vaidya and Gonzalves (1963), Chacko (1966), Sarma and Khan (1966, 1967), Khan and Sarma (1967a and b), Rao and Rao (1969), Chatterjee (1975), Ramjee and Bhatnagar (1978a and b), Subramanian (1983), Anand and Langer (1988), Pundhir *et. al.* (1994) and Pundhir and Vidyavati (1994a) and others. A large number of new taxa to the list of Indian Charophytes were added by various workers. A critical survey on the literature of Indian Charophytes was carried out and revealed the occurrence of about 142 taxa in Indian flora. A comprehensive list of all these taxa has given below in which they are arranged following Wood and Imahori (1965) classification.

## GENUS-*CHARA*

1. *Chara vulgaris* var. *vulgaris* f. *vulgaris* RDW
2. *C. vulgaris* var. *vulgaris* f. *crispa* (Wallm.) RDW
3. *C. vulgaris* var. *vulgaris* f. *atrovirence* (lowe.) H. & J. Gr. RDW
4. *C. vulgaris* var. *vulgaris* f. *contraria* (A. Br. ex Kuetz.) RDW
5. *C. vulgaris* var. *vulgaris* f. *longibracteata* (Kuetz.) H. & J. Gr. RDW
6. *C. vulgaris* var. *nitelloides* (A. Br.) RDW
7. *C. vulgaris* var. *inconnexa* f. *inconnexa* RDW
8. *C. vulgaris* var. *inconnexa* f. *hippelliana* (Vilh.) RDW
9. *C. vulgaris* var. *inconnexa* f. *brionica* (Steph.) RDW
10. *C. vulgaris* var. *imperfecta* f. *dissoluta* (A. Br. ex Leonh.) RDW
11. *C. globularis* var. *globularis* f. *globularis* RDW
12. *C. globularis* var. *globularis* f. *capensis* (Meyer ex Kuetz.) RDW
13. *C. globularis* var. *virgata* f. *virgata* (Kuetz.) RDW
14. *C. globularis* var. *virgata* f. *leiopitys* (Wheld.) RDW
15. *C. globularis* var. *virgata* f. *chrysospora* (J.Gr. & Steph.) RDW

16. *C. globularis* var. *virgata* f. *fischeri* (Mig.) RDW
17. *C. globularis* var. *leptosperma* f. *leptosperma* RDW
18. *C. globularis* var. *aspera* f. *strigosa* (A.Br.) RDW
19. *C. globularis* var. *kokeilii* (A.Br.) RDW
20. *C. setosa* f. *setosa* RDW
21. *C. setosa* f. *inermis* (Zanev.) RDW
22. *C. setosa* f. *pseudobrachypus* (J.Gr. & Steph.) RDW
23. *C. zeylanica* var. *zeylanica* f. *zeylanica* RDW
24. *C. zeylanica* var. *zeylanica* f. *michauxii* (A.Br.) H. & J.Gr. RDW
25. *C. zeylanica* var. *zeylanica* f. *berteroi* (A.Br. ex Kuetz.) H. & J. Gr. RDW
26. *C. zeylanica* var. *zeylanica* f. *humboldtiana* (H. & J.Gr.) RDW
27. *C. zeylanica* var. *zeylanica* f. *elegans* (A.Br. ex T.F.A.) RDW
28. *C. zeylanica* var. *zeylanica* f. *filicaulis* (Rob.) RDW
29. *C. zeylanica* var. *sejuncta* f. *sejuncta* RDW
30. *C. zeylanica* var. *diaphana* f. *diaphana* RDW
31. *C. zeylanica* var. *diaphana* f. *oerstediana* (A.Br.) RDW
32. *C. zeylanica* var. *diaphana* f. *armata* (Meyer) RDW
33. *C. zeylanica* var. *diaphana* f. *vandalurensis* (Sund.) RDW
34. *C. braunii* f. *braunii* RDW
35. *C. braunii* f. *schweinitzii* (A.Br.) RDW
36. *C. braunii* f. *novi-mexicana* (A.Br.) RDW
37. *C. braunii* f. *coromandelina* (A.Br.) RDW
38. *C. braunii* f. *kurzii* (Zanev.) RDW
39. *C. braunii* f. *oahuensis* (Meyen) RDW
40. *C. braunii* f. *perrottetii* (A.Br.) RDW
41. *C. corallina* var. *corallina* f. corallina RDW
42. *C. corallina* var. *wallichii* (A.Br.) RDW
43. *C. socotrensis* f. *socotrensis* RDW
44. *C. socotrensis* f. *nuda* (Pal) RDW
45. *C. hornemannii* f. *longifolia* (Rod.) RDW
46. *C. fibrosa* var. *fibrosa* f. *fibrosa* RDW
47. *C. fibrosa* var. *fibrosa* f. *acanthopitys* RDW
48. *C. fibrosa* var. *fibrosa* f. *tylacantha* (Nordst.) RDW
49. *C. fibrosa* var. *fibrosa* f. *curtissii* (A. ex Rob.) RDW
50. *C. fibrosa* var. *fibrosa* f. *keukensis* (T.F.A.) RDW
51. *C. fibrosa* var. *fibrosa* f. *longicorollata* (Kasaki) RDW
52. *C. fibrosa* var. *fibrosa* f. *erythrogyna* (Griff.) RDW
53. *C. fibrosa* var. *hydropitys* f. *hydropitys* RDW
54. *C. fibrosa* var. *hydropitys* f. *mexicana* (T.F.A.) RDW
55. *C. fibrosa* var. *hydropitys* f. *schneckii* (Rob.) RDW

56. *C. pseudohydropitys* Imah.
57. *C. bharadwajae* Sarma et Khan.

**GENUS - *LAMPROTHAMNIUM***

58. *Lamprothamnium succinctum* (A. Br.) RDW

**GENUS - *LYCHNOTHAMNUS***

59. *Lychnothamnus barbatus* f. *barbatus* RDW
60. *L. barbatus* f. *iyengarii* Sarma et Khan
61. *L. barbatus* f. *gigantea* Patel et Jawale
62. *L. barbatus* var. *gigantii* Bharati et Chennaveeraiah
63. *L. barbatus* f. *mahmoodia* Pundhir et Gautam

**GENUS - *NITELLOPSIS***

64. *Nitellopsis obtusa* (Desv.) J.Gr. RDW

**GENUS - *NITELLA***

65. *Nitella flexilis* var. *flexilis* f. *flexilis* RDW
66. *N. mirabilis* var. *mirabilis* f. *mirabilis* RDW
67. *N. mirabilis* var. *inokasiraensis* (Kasaki ex Wood) RDW
68. *N. macounii* var. *macounii* f. *macounii* (T.F.A.) RDW
69. *N. acuminata* var. *acuminata* f. *acuminata* RDW
70. *N. acuminata* var. *acuminata* f. *capitulifera* (T.F.A.) RDW
71. *N. acuminata* var. *acuminata* f. *subglomerata* (A.Br.) RDW
72. *N. acuminata* var. *acuminata* f. *blankinshipii* (T.F.A.) RDW
73. *N. acuminata* var. *belangeri* f. *belangeri* (A.Br.) RDW
74. *N. acuminata* var. *greenii* RDW
75. *N. stuartii* (A.Br.) RDW
76. *N. allenii* f. *allenii* RDW
77. *N. dualis* var. *dualis* (T.F.A.) RDW
78. *N. dualis* var. *pulchella* (T.F.A.) RDW
79. *N. dualis* var. *pulchella* f. *superba* (Pal) RDW
80. *N. furcata* var. *furcata* f. *furcata* RDW
81. *N. furcata* var. *furcata* f. *nicobarica* (A.Br.) RDW
82. *N. furcata* var. *furcata* f. *burmanica* (Pal) RDW
83. *N. furcata* var. *furcata* f. *coreana* (Mor. ex Imah.) RDW
84. *N. furcata* var. *furcata* f. *roxburghii* (A.Br.) RDW
85. *N. furcata* var. *furcata* f. *polycarpa* (Pal) RDW
86. *N. furcata* var. *sieberi* f. *sieberi* RDW
87. *N. furcata* var. *sieberi* f. *microcarpa* (A.Br.) RDW

88. *N. furcata* var. *sieberi* f. *glaziovii* (Zell.) RDW
89. *N. furcata* var. *sieberi* f. *papuana* (Zanev.) RDW
90. *N. furcata* var. *mucronata* f. *mucronata* RDW
91. *N. furcata* var. *mucronata* f. *oligospira* (A.Br. ) RDW
92. *N. furcata* var. *mucronata* f. *dictyosperma* (H.& J.Gr.) RDW
93. *N. furcata* var. *mucronata* f. *wrightii* (H. & J. Gr. ) RDW
94. *N. furcata* var. *ogivalis* (J.Gr. & Steph.) RDW
95. *N. furcata* var. *inaequalis* (J.Gr.) RDW
96. *N. furcata* subsp. *flagelliformis* (A.Br.) RDW
97. *N. furcata* subsp. *flagellifera* f. *flagellifera* RDW
98. *N. furcata* subsp. *flagellifera* f. *tuberculata* (Kundu) RDW
99. *N. furcata* subsp. *flagellifera* f. *graciliformis* (J.Gr.) RDW
100. *N. furcata* subsp. *flagellifera* f. *patula* (J.Gr. and G.O.A.) RDW
101. *N. furcata* subsp. *megacarpa* f. *megacarpa* RDW
102. *N. furcata* subsp. *tumulosa* f. *tumulosa* (Zanev.) RDW
103. *N. furcata* var. *orientalis* f. *orientalis* RDW
104. *N. furcata* var. *erecta* f. *erecta* (Imah.) RDW
105. *N. tenuissima* var. *tenuissima* f. *tenuissima* RDW
106. *N. tenuissima* var. *tenuissima* f. *transilis* (T.F.A.) RDW
107. *N. tenuissima* var. *tenuissima* f. *bengalensis* (Kundu) RDW
108. *N. tenuissima* var. *tenuissima* f. *sanukensis* (Imah.) RDW
109. *N. pseudoflabellata* var. *pseudoflabellata* f. *pseudoflabellata* RDW
110. *N. pseudoflabellata* var. *elegans* f. *elegans* RDW
111. *N. pseudoflabellata* var. *elegans* f. *sejuncta* (Imah.) RDW
112. *N. pseudoflabellata* var. *mucosa* f. *stabilis* (Imah.) RDW
113. *N. pseudoflabellata* var. *leptodactyla* f. *leptodactyla* RDW
114. *N. pseudoflabellata* var. *leptodactyla* f. *megacarpa* (J. Gr.) RDW
115. *N. pseudoflabellata* var. *imperiales* f. *wattii* (J. Gr.) RDW
116. *N. pseudoflabellata* var. *imperiales* f. *sonderi* (A. Br.) RDW
117. *N. pseudoflabellata* var. *pusilla* (Imah.) RDW
118. *N. sphaerocephala* J. Gr. RDW
119. *N. gracilis* var. *gracilis* f. *gracilis* RDW
120. *N. gracilis* var. *gracilis* f. *arvernica* (Hy) RDW
121. *N. gracilis* var. *gracilis* f. *intermedia* (Nordst.) RDW
122. *N. gracilis* var. *bipartita* (Fil) RDW
123. *N. gracilis* var. *confervacea* f. *confervacea* RDW
124. *N. gracilis* var. *confervacea* f. *conformis* (Nordst.) RDW
125. *N. gracilis* var. *leptosoma* f. *globulifera* (Pal) RDW
126. *N. gracilis* var. *annandalei* (Pal) RDW
127. *N. rigida* f. *rigida* RDW

128. *N. subtilissima* f. *subtilissima* RDW
129. *N. axillocarpa* (J. Gr. & steph.) RDW
130. *N. hyalina* var. *hyalina* f. *hyalina* RDW
131. *N. hyalina* var. *hyalina* f. *brachyactis* (A. Br.) Feldm. RDW
132. *N. hyalina* var. *hyalina* f. *indica* (A. Br.) RDW
133. *N. translucens* var. *translucens* f. *translucens* RDW
134. *N. translucens* var. *axillaris* f. *axillaris* RDW
135. *N. terrestris* Iyengar
136. *N. terrestris* var. *minor* Chacko

## GENUS - *TOLYPELLA*

137. *Tolypella nidifica* var. *nidifica* f. *nidifica* RDW
138. *T. nidifica* f. *stenhammariana* (Wallm.) RDW
139. *T. nidifica* var. *glomerata* (Desv.) RDW
140. *T. intricata* var. *intricata* f. *intricata* (T.F.A.) RDW
141. *T. intricata* var. *intricata* f. *prolifera* (Ziz. ex A. Br.) RDW
142. *T. jwalaii* Labh and Verma

Above list based on the scattered information existed in the publications upto December, 1995. Work on cytology, cytotaxonomy and SEM study of Indian Charophytes is still going on in various schools of India. Thus, further additions are expected in above list. Moreover, North-Eastern belt and Western parts of India have remained yet to be explored.

## ACKNOWLEDGEMENTS

Authors are greatful to D.S.T. and C.S.I.R. for financial assistance and to Prof. (late) Y.S.R.K. Sarma for inspiration and blessings. One of us (H.S.P.) is also greatful to Prof. Vidyavati for encouragements and to Miss Aruna Gautam for her kind cooperation.

## REFERENCES

Agarkar, D.S. 1963. Systematic studies of the Charophytes of Gwalior, Madhya Pradesh. *The Vikram.* **7**: 43-51.

Agarkar, D.S. 1967. A note on *Chara vandalurensis* Sund. from Gwalior (M.P.). *Science and Culture.* **33**: 59-60.

Agharkar, S.P. and Kundu, B.C. 1937. Charophytes of West Bengal. *J. Dept. Sci. Calcutta Univ.* **1**: 1-23.

Allen, G.O. 1925. Notes on Charophytes from Gonda, U.P. *J. Bomb. Nat. Hist. Soc.* **30**: 589-599..

Allen, G.O. 1928. Charophyte notes from Saharanpur, U.P. *J. Indian Bot. Soc.* **7**: 49-69.

Allen, G.O. 1933. Charophyte notes from Agra, U.P. J. *Indian Bot. Soc.* **12**: 17-19.

Allen, G.O. 1936. Charophyte notes from Bareilly. *J. Indian Bot. Soc.* **15**: 51-54.

Allen, G.O. 1942. On the distribution of some Indian Charophytes. *Bull. Royal Bot. Gdn. Calcutta* (150th Anniv.). **2**: 183-188.

Allen, G.O. 1961. Charophyte notes from Banaras, U.P. India. *J. Indian Bot. Soc.* **40**: 234-242.

Anand, V.K. and Langer, G. 1988. Occurrence of Charophytes in Jammu region. *Phykos.* **27**: 25-27.

Bandhopadhyay, S. and P. Chatterjee. 1989. *Chara fibrosa* var. *fibrosa* f. keukensis: A new record for India and its chromosome number. *Phykos.* **28**: 178-180.

Bharati, S.G. 1990. Morphological and cytotaxonomical studies on two taxa of *Chara fibrosa* Ag. ex Bruz. em RDW. *J. Indian Bot. Soc.* **69**: 315-318.

Bharati, S.G. and M.S. Chennaveeraiah. 1980. Cytotaxonomical studies on *Chara zeylanica* complex. In: *Taxonomy of Algae* (eds. T.V. Desikachary and V.N. Rajarao). Univ. Madras Pub. pp. 553-564.

Bharati, S.G. and Chennaveeraiah, M.S. 1983. *Lychnothamnus barbatus* var. *gigantii* var. novo. (A new variety). *Cytologia.* **48**: 843-846.

Bhatnagar, S.K. 1988a. The first cytotaxonomic report in *Chara braunii* f. *novi-mexicana* (Div. Charophyta). *Curr. Sci.* **57**: 88-90.

Bhatnagar, S.K. 1988b. *Nitella hyalina* f. *hyalina* (Dc.) Ag. RDW: A new chromosome count and its cytotaxonomy. *Curr. Sci.* **57**: 343-344.

Bhatnagar, S.K. 1989a. Cytological perspectives of Charophyta. I. *Chara globularis* complex. *Phykos.* **28**: 156-165.

Bhatnagar, S.K. 1989b. Cytological perspectives of Charophyta. II. *Nitella* furcata complex. *Phykos.* **28**: 166-177.

Bhatnagar, S.K. and L.P. Gangwar. 1985. On *Tolypella intricata* (Trent. ex Roth.) Leonh. em RDW Characeae from Indian subcontinent. A karyomorphological approach. *Cryptogamie, Algologie.* **6**: 185-190.

Bhatnagar, S.K. and Johri, M. 1986. The genus *Tolypella* (Div. Charophyta) from Indian subcontinent. A cytotaxonomical approach. *Cytologia.* **51**: 413-419.

Bhatnagar, S.K. and Johri, M. 1991. Preliminary investigations on cytology and cytotaxonomy of Charophytes from Rohilkhand division, Bareilly. *J. Indian Bot. Soc.* **70**: 55-58.

Bhatnagar, S.K. and Sharma, M. 1985. A new chromosome number in the genus *Nitella* from India. *Science and Culture.* **51:** 195-196.

Braun, A. 1849. Characeae Indiae Orientalis et insularum maris pacifici. *Hooker's J. Bot. & Kew Gdn. Misc.* **1**: 292-301.

Chacko, P.M. 1966. A new variety of *Nitella terrestris* (var. minor var. novo.) from Kerala, South India. *Curr. Sci.* **35**: 630-631.

Chatterjee, P. 1970. *Lychnothamnus barbatus* (Meyen) Leonh in West Bengal and its chromosome number. *Bull. Bot. Soc. Bengal.* **24**: 131-133.

Chatterjee, P. 1975. Some additions to the Charophytes of West Bengal. *Bull. Bot. Soc. Bengal.* **29**: 105-109.

Chatterjee, P. 1976a. Cytotaxonomic studies on *Chara socotrensis* Nordst in Kuhn f. *nuda* (Pal) RDW from West Bengal, India. *Cytologia.* **41:** 659-663.

Chatterjee, P. 1976b. Cytotaxonomical studies of West Bengal Charophytes: Karyotype analysis in *Chara braunii. Hydrobiologia.* **49**: 171-174.

Chatterjee, P. 1979a. *Chara fibrosa* var. *fibrosa* f. *longicorollata*: A new record for India and its cytology. *Curr. Sci.* **48**: 545-547.

Chatterjee, P. 1979b. Karyological investigation of *Chara wallichii* A. Br. from West Bengal. *Cell & Chromosome Newsletter.* **2**: 21-23.

Chennaveeraiah, M.S. and Bharati, S.G. 1974. Morphological and cytological observations in *Chara gymnopitys* A. Br. *Cytologia.* **39**: 443-451.

Choudhary, M.C. and Sinha, J.P. 1962. The Chhotanagpur Charophytes. *J. Ranchi Univ.* **1**: 51-56.

Dixit, S.C. 1931. Some Charophytes from Salsette. *J. Indian Bot. Soc.* **10**: 205-208.

Dixit, S.C. 1935. The Charophytes of the Bombay Presidency. I. *J. Indian Bot. Soc.* **14**: 257-263.

Dixit, S.C. 1940. The Charophytes of the Bombay Presidency. II. *J. Indian Bot. Soc.* **18**: 231-239.

Dixit, S.C. 1942. The Charophytes of the Bombay Presidency. III. *J. Indian Bot. Soc.* **21**: 355-362.

Dixit, S.C. 1966. Annotation on the Charophytes of Western India. *J. Univ. Bomb.* **34**: 1-28.

Goyal, S.K. 1960. Algal flora of Jodhpur and its environs. I. Charophyta. *J. Bomb. Nat. Hist. Soc.* **59**: 447-452.

Groves, J. 1919. Notes on *Lychnothamnus*. *J. Bot.* **57**: 125-129.

Groves, J. 1922. *Nitellopsis obtusa* in Northern India. *J. Bot.* **60**: 54-55.

Groves, J. 1924. Notes on Indian Charophyta. *J. Linn. Soc.* (Bot.). **46**: 359-376.

Hate, V.N. 1909. Two species of *Chara* from Bombay Island. *J. Bomb. Nat. Hist. Soc.* **19**: 762-763.

Iyengar, M.O.P. 1958. *Nitella terrestris* sp. nov. A terrestrial Charophyte from South India. *Bull. Bot. Soc. Bengal.* **12**: 85-90.

Jawale, A.K. and Patel, R.J. 1985. Morphology and cytology of *Nitella polycarpa* Pal f. *gujaratensis* f. nov. from India. *J. Indian Bot. Soc.* **64**: 317-322.

Kant, R. and Pandey, D.C. 1985. Observations on *Nitellopsis obtusa* (Desv. in Lois.) J. Groves from Allahabad. *Phykos.* **24**: 95-97.

Khan, M. 1966. Studies on cytology and cytotaxonomy of Indian Charophyta. Ph.D. Thesis, Banaras Hindu University, Varanasi, India.

Khan, M. 1980. Occurrence of *Lamprothamnium* J. Gr. (Charophyta) in India. *Acta Botanica Indica.* **8**: 97-98.

Khan, M. and Sarma, Y.S.R.K. 1967a. Studies on cytotaxonomy of Indian Charophyta. I. *Chara. Phykos.* **6**: 36-47.

Khan, M. and Sarma, Y.S.R.K. 1967b. Studies on cytotaxonomy of Indian Charophyta. II. *Nitella. Phykos.* **6**: 48-61.

Khan, M. and Sarma, Y.S.R.K. 1967c. Some observations on the cytology of Indian Charophyta. *Phykos.* **6**: 62-74.

Khan, M. and Sarma, Y.S.R.K. 1980. New chromosome numbers for Indian Charophyta. *Cell and Chromosome Newsletter.* **3**: 24.

Kundu, B.C. 1938. A new species of poly-arthrodactylous *Nitella* with a review of allied species. *J. Indian Bot. Soc.* **16**: 263-268.

Kundu, B.C. 1941. Two new nitellas. *J. Bomb. Nat. Hist. Soc.* **12**: 843-846.

Labh, L. and Verma, B.N. 1983. *Tolypella Jwalaii*, a new species of *Tolypella* from India. *Bibliotheca Phycologica.* **66**: 379-388.

Labh, L. and Verma, B.N. 1984. Cytological studies in three species of *Tolypella. Cytologia.* **49**: 717-723.

Labh, L. and B.N. Verma. 1985a. First report of chromosome numbers in four taxa belonging to the genus *Chara* Linn. *Cytologia.* **50**: 21-24.

Labh, L. and Verma, B.N. 1985b. New record of chromosome numbers for the Charophytes. *Cytologia.* **50**: 55-58.

Labh, L. and Verma, B.N. 1986. New counts for *Chara fibrosa* complex. *Cytologia.* **51**: 185-191.

Mukherjee, S. 1978. Polyploid chromosome numbers in the genus *Nitella. Curr. Sci.* **47**: 386.

Mukherjee, S. and Noor, M.N. 1973. A first report of karyological studies in *Nitella wattii* Groves from India. *Science and Culture.* **39**: 459-460.

Mukherjee, S.K. 1934. The Charophytes of Dal lake, Kashmir. *Proc. Indian Sci. Cong. Asso.* p. 295.

Mukhopadhyay, A. and Chatterjee, P. 1980. Cytomorphology of *Chara flaccida* A. Br. from West Bengal. *Cell and Chromosome Newsletter.* **3**: 37-39.

Noor, M.N. 1968. A new record of chromosome number in *Nitella mucronata* (Br.) Miquel from India. *Science and Culture.* **34**: 214-215.

Noor, M.N. 1969-1970. A preliminary report on the chromosome numbers of some Indian Characeae. *J. Ranchi Univ.* **6 & 7**: 233.

Noor, M.N. and Mukherjee, S. 1975. On the aneuploid chromosome number in *Chara hydropitys* Reich from India. *Cytologia.* **40**: 803-807.

Noor, M.N. and Mukherjee, S. 1977. Some new records of chromosome number in Indian Charophyta. *Cytologia.* **42**: 227-232.

Pal, B.P., Kundu, B.C., Sundaralingam, V.S. and Venkataraman, G.S. 1962. *Charophyta.* I.C.A.R. Pub., New Delhi.

Patel, R.J. and Jawale, A.K. 1979a. *Chara pseudohydropitys* Imah. A new record for India. *Geobios.* **6**: 181-183.

Patel, R.J. and Jawale, A.K. 1979b. On *Lychnothamnus barbatus* (Meyen) Leonh. from Gujarat, India. *Phykos.* **18**: 51-52.

Patel, R.J. and Jawale, A.K. 1984. *Nitella furcata* (Rox. ex Bruz.) subsp. *furcata* RDW var. *sieberi* (A. Br.) RDW f. *sieberi* RDW—A new record for India. *Geobios New Reports.* **1**: 37-38.

Patel, R.J. and Jawale, A.K. 1984. On *Chara haitensis* Turpin from Gujarat. *Geophytology.* **14**: 235-237.

Patel, R.J. and Jawale, A.K. 1985. A new form of *Lychnothamnus barbatus* (Meyen) Leonh f. *gigantea* f. novo. from Gujarat, India. *J. Indian Bot. Soc.* **64**: 323-327.

Prasad, P.K. and Verma, B.N. 1985. Aneuploid count for *Chara setosa* Klein ex Willd. *Cytologia.* **50**: 241-245.

Pundhir, H.S. 1994. Studies on the karyomorphology of *Chara hornemannii* f. *longifolia* (Rob.) RDW: A new record for India. *J. Mendel.* **11**: 157-160.

Pundhir, H.S. and Gautam, A. 1993. Occurrence of *Nitella hyalina* var. *hyalina* f. *brachyactis* (A. Br.) Feldm. RDW. A new record for India. *Phykos.* **32**: 99-104.

Pundhir, H.S. and Gautam, A. 1994a. Observations on morphology and cytology of *Chara vulgaris* var. *vulgaris* f. *crispa* (Wallm.) RDW: A new record for India. In: *Current Researches in Plant Sciences* (eds. T.A. Sarma, S.S. Saini, M.L. Trivedi and M. Sharma). Pub. Bishen Singh, Mahendra Pal Singh, Dehradun, India. pp. 197-201.

Pundhir, H.S. and Gautam, A. 1994b. A new chromosme count for *Chara hornemannii* f. *longifolia* (Rob.) RDW: A new record for India. *Cytologia.* **59:** 305-307.

Pundhir, H.S. and Gautam, A. 1995. Cytomorphological studies on a new form of *Lychnothamnus barbatus* f. *mahmoodia* f. nov. *Phykos.* **34**: 39-48.

Pundhir, H.S., Gautam, A. and Vidyavati. 1993. Comparative study and new chromosome counts for *Nitella hyalina* var. *hyalina* f. *indica* (A. Br.) RDW, from India. *Phykos.* **32**: 169-174.

Pundhir, H.S., Gautam, A. and Vidyavati. 1994. Some Charophytes from Aligarh district, Uttar Pradesh. *Phykos.* **33**: 1-9.

Pundhir, H.S. and Vidyavati. 1993. Taxonomic enumeration and chromosome counts of some Charophytes from district Aligarh, Uttar Pradesh. *J. Mendel.* **10**: 83-85.

Pundhir, H.S. and Vidyavati. 1994a. New karyomorphological observations on Charophytes from Uttar Pradesh, India. *The Nucleus*. **37**: 39-45.

Pundhir, H.S. and Vidyavati. 1994b. Some new karyomorphological observations on *Chara fibrosa* var. *fibrosa* f. *longicorollata* (Kasaki) RDW. J. *Indian Bot. Soc.* **73**: 245-249.

Ramjee and Bhatnagar, S.K. 1978a. Studies on Charophytes from Rohilkhand division. I. Moradabad. Taxonomic enumeration and chromosome counts. *Phykos*. **17**: 87-93.

Ramjee and Bhatnagar, S.K. 1978b. Significance of a new chromosome number in *Nitella mirabilis* Nordst. ex Groves em RDW. *Hydrobiologia*. **53**: 99-101.

Ramjee and Bhatnagar, S.K. 1978c. Cytotaxonomic studies on *Nitella furcata* subsp. *flagellifera* f. *patula* Gr. ex Allen and comparison with the other taxa of *Nitella furcata* complex. *Caryologia*. **31**: 457-462.

Ramjee and Sarma, Y.S.R.K. 1971. Some observations on the morphology and cytology of Indian Charophyta. *Hydrobiologia*. **37**: 367-382.

Rao, K.R. and Rao, K.V.M. 1969. Algal flora of Chittor district, Andhra Pradesh. II. Charophyta. *Phykos*. **8**: 56-60.

Ray, S. and Chatterjee, P. 1983. A new technique in karyological investigations of Characeae. *Cell and Chromosome Research*. **6**: 27-29.

Ray, S. and Chatterjee, P. 1986a. *Chara vulgaris* var. *inconnexa* f. *hippelliana*. A new record for India and its karyotype. *Curr. Sci.* **55**: 662-664.

Ray, S. and Chatterjee, P. 1986b. Karyotype analysis in *Chara braunii* G.M.F. *braunii* (Characeae). *The Nucleus*. **29**: 33-36.

Ray, S. and Chatterjee, P. 1987. Karyomorphological investigations on three populations of *Chara corallina* var. and f. *corallina* in West Bengal. *Cytologia*. **52**: 455-458.

Ray, S. and Chatterjee, P. 1988a. Cytotaxonomical study of some members of *Chara fibrosa* complex. *Cryptogamie, Algologie*. **9**: 65-71.

Ray, S. and Chatterjee, P. 1988b. Cytotaxonomical study of West Bengal Charophyta: A new chromosome number count of *Nitella stuartii* and its karyotype. *Acta. Botanica Neerl.* **37**: 523-525.

Ray, S. and Chatterjee, P. 1989. A new chromosome number report of *Chara globularis* var. *leptosperma* f. *leptosperma*. *Curr. Sci.* **58**: 312-313.

Sarma, T.A. and Bala, V. 1984. New chromosome counts in four taxa of *Chara* from Punjab. *J. Cytol. & Genet.* **19**: 27-32.

Sarma, Y.S.R.K. and Khan, M. 1964. Chromosome numbers in some Indian species of *Nitella*. *Chromosoma*. **15**: 246-247.

Sarma, Y.S.R.K. and Khan, M. 1965a. Chromosome number in some Indian species of *Chara*. *Phycologia*. **4**: 173-176.

Sarma, Y.S.R.K. and Khan, M. 1965b. A preliminary report on the survey of chromosome number of Indian Charophyta. *The Nucleus*. **8**: 33-38.

Sarma, Y.S.R.K. and Khan, M. 1966. A new form of *Lychnothamnus barbatus* from India. *Phykos*. **5**: 195-197.

Sarma, Y.S.R.K. and Khan, M. 1967. On two new Charophytes from India. *Hydrobiologia*. **30**: 405-410.

Sarma, Y.S.R.K. and Ramjee. 1967. A new chromosome number in *Nitella tenuissima* (Desv.) Kuetz. em RDW. *Caryologia*. **20**: 61-63.

Sarma, Y.S.R.K. and Ramjee. 1969. Chromosome numbers in three taxa of Indian Charophyta. *J. Cytol & Genet.* **4**: 32-36.

Sinha, B.D. and Srivastava, N.K. 1980. Record of *Lychnothamnus barbatus* (Leonh) RDW in Bihar. *Phykos.* **19**: 175-176.

Sinha, J.P. and Choudhary, M.C. 1962. The Chhotanagpur Charophytes. *J. Rachi Univ.* **1**: 51-56.

Sinha, J.P. and Choudhary, M.C. 1968. A comparative study of the Charophyte flora of India and South-Eastern United States. *Phykos.* **7**: 136-138.

Sinha, J.P. and Noor, M.N. 1967. Chromosome numbers in some members of Charophyceae of Chhotanagpur plateau, India. *Phykos.* **6**: 106-109.

Sinha, J.P. and Noor, M.N. 1971. Studies on the karyology of some members of Charophyta of Bihar. *Phykos.* **10**: 112-117.

Sinha, J.P. and Verma, B.N. 1970. Cytological analysis of the Charophytes of Bihar. *Phykos.* **9**: 92-99.

Sinha, S. and Sinha, J.P. 1989. Study on mitotic chromosomal counts in some members of Charales of Chhotanagpur (Bihar). *Biojournal.* **1**: 69-72.

Sinha, S. and Sinha, J.P. 1992. Karyomorphological studies on three species of *Chara. J. Appl. Biol.* **2**: 7-11.

Subrahmanyam, B.V.S. and Chowdary, Y.B.K. 1992. Karyology of *Nitella* from South India. *Cytologia.* **57**: 209-212.

Subrahmanyam, B.V.S., Chowdary, Y.B.K. and Sarma, Y.S.R.K. 1991. Aneuploidy in two taxa of *Nitella* (Division Charophyta) from Karnataka, India. *The Nucleus.* **34**: 174-176.

Subramanian, D. 1983. Studies on Charophyta from Tamilnadu. I. Taxonomic enumeration and chromosome counts. *Phykos.* **22**: 18-22

Subramanian, D. and Ganesan, R. 1983. Cytotaxonomic studies of Charophyta from Tamilnadu. *Proc. All India Appl. Phycol. Cong.*, Kanpur. pp. 184-191.

Sundaralingam, V.S. 1946. The cytology and spermatogenesis in *Chara zeylanica* Willd. *J. Indian Bot. Soc.* (M.O.P. Iyengar Comm. Vol.). pp. 289-303.

Sundaralingam, V.S. 1959. A systematic account of the South Indian Characeae. *Proc. Indian. Acad. Sci.* **49**: 7-51.

Sundaralingam, V.S. and Bharathan, S. 1978. Occurrence of *Lychnothamnus barbatus* Leonh near Madras. *Phykos.* **17**: 123-125.

Vaidya, B.S. 1963. Systematic position of Charophytes. *Vidya J. Gujarat Univ.* **6**: 173-175.

Vaidya, B.S. and Gonzalves, E.B. 1963. A systematic enumeration of the Charophytes of Western India. *Phykos.* **2**: 33-37.

# Section - 5

Recent Trends in Algal Taxonomy
Vol. I. Taxonomical Issues
Editors : Vidyavati and A.K. Mahato
APC Publications Pvt. Ltd., New Delhi, 1999. pp. 243-252

# *Recent and Emerging Trends in Chrysophycean Taxonomy*

Daniel E. Wujek
Department of Biology, Central Michigan University
Mt. Pleasant, Michigan-48859 U.S.A.

**ABSTRACT**

This paper is a survey of the current trends in taxonomy and classification of chrysophytes. The genera are distinguished by level of cellular organization: unicellular or colonial (either motile or flagellate), amoeboid, palmelloid, filamentous or thalloid. While light and electron microscopy have been extensively used in the past and continue to be used, molecular perspectives and newer LM techniques will help in delineating what presently are called the Chrysophyceae. As further knowledge of Chrysophyceae cell and molecular structure is gained, the existing classifications will remain dynamic.

## General Features of the Cell

The Chrysophyceae or golden-brown algae cannot easily be defined because they are diverse and often distantly related to one another. First recognized as a distinct group by Pascher (1914), they obtain their energy and nutrients by photosynthesis and/or heterotrophy. In the latter, they take up complex organic molecules either in soluble form (osmotrophy) or as particulate matter, *e.g.*, bacteria or other protists (phagotrophy). Two groups of golden-brown algae, Pedinellophyceae (Kristiansen, 1990) and Synurophyceae (Andersen, 1987), are sometimes separated from the Chrysophyceae as separate classes, because of their special cellular features. For discussion purposes they are included here as part of the Chrysophyceae. Preisig (1995) and Moestrup (1995) present historical and modern interpretations of chrysophyte classification.

The chrysophytes, or chrysomonads as they are sometimes called, have vegetative or reproductive cells that are biflagellate. Unlike the Phaeophyceae (brown algae) which have lateral insertion of the flagella, the Chrysophyceae have two flagella that are inserted near the apex of the cell. One is a short smooth flagellum that points backwards along the cell. The other is a long pleuronematic (tinsel or flimmer bearing two rows of tripartite mastigonemes) flagellum. Species are unicellular or colonial; non-motile forms can be planktonic or sessile. However, a relatively small number have a simple multicellular organization.

The chloroplasts are golden-brown due to the accessory pigment fucoxanthin which masks the chlorophylls a and c. The DNA within the chloroplast is generally arranged in a ring.

Silica-walled cysts (stomatocysts) are formed endogenously and are species specific. In several genera, siliceous body scales are present.

### Size and Distribution of the Class

The Chrysophyceae are almost entirely freshwater inhabitants, although a few species are brackish or marine (Chrysomeridales, Sarcinochrysidales, and an order of uncertain taxonomic position, Parmales, the later two orders are placed by some researchers into their own respective families). They comprise nearly 200 genera and over 1000 species. They were long thought to be found chiefly in temperate areas, but studies on tropical lakes and ponds have been reporting a considerable chrysophycean flora.

In general they are associated with water bodies which are low in productivity and that has a pH below 7.5. Their ability to compete for phosphorus when it is limiting (oligotrophic waters) tends to make them dominant under these conditions. Their scarcity in more productive water bodies may be brought about by zooplankton grazing or competition with other phytoplankton.

### The Systematics and Cellular Organization of the Chrysophyceae

While in most cases light microscopy (LM) can be utilized in distinguishing chrysophycean genera, the chrysophyte genera that are covered with silica scales and other silica structures have been studied more intensively than the others. Kristiansen (1995) presents an overview of this group. It is within this group of algae that a second class, the Synurophyceae, is recongnized. Within the synurophytes, all taxa are silica-scale covered. However, by far the largest number of genera within the Chrysophyceae (*sensu stricto*) are without scales, with those that do possess them are placed within the family paraphysomonadaceae. This family contains only four genera, one of which, *Paraphysomonas*, is colourless.

By far, the most important scaled synurophyte genera are in the family Mallomonadaceae: *Mallomonas* (> 100 species) and *Synura* (ca. 20 species). The symmetry of the scales in this family (Figs. 1-2) is very different from those in the Paraphyasomadaceae (Fig.3). The scales are usually heteropolar, with the proximal and distal ends of different ornamentation. The distal end often bears a spine. Within the Chrysophyceae, *sensu stricto*, the scales are homopolar, radially symmetrical and often with a central spine.

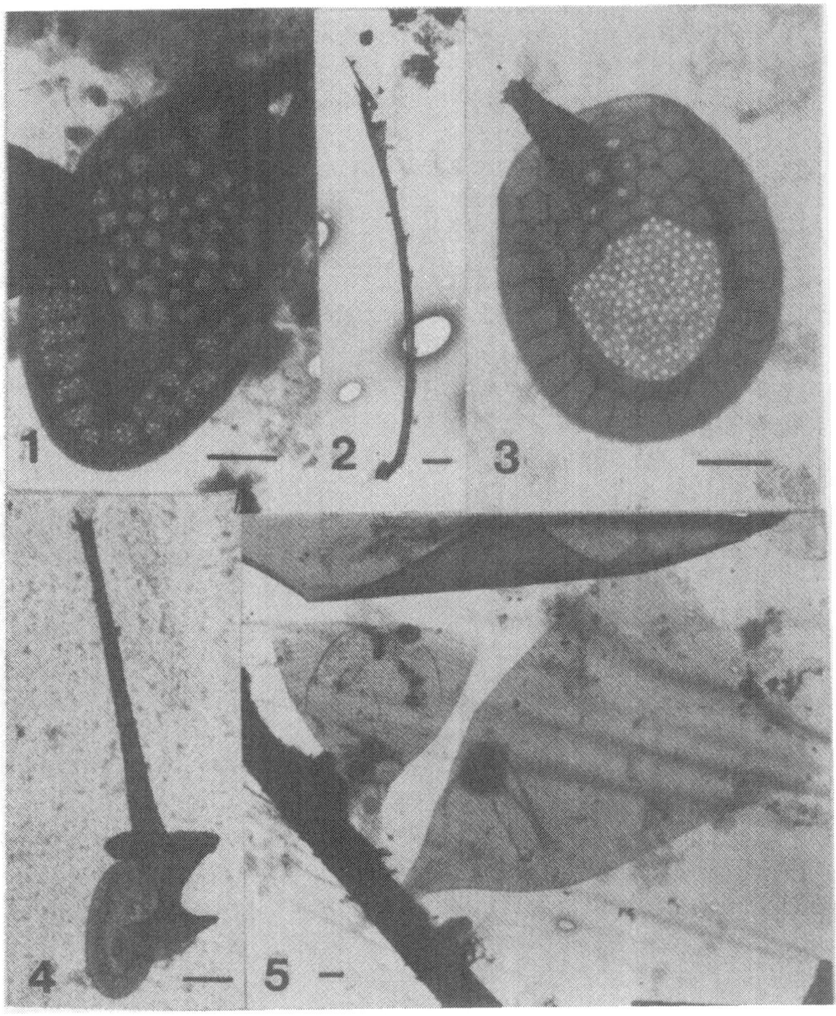


Figs. 1-5. Scales and bristles from chrysophytes: 1-2. *Mallomonas crassisquama*. 1. Body scale. 2. Helmet bristle. 3. Spine scale of *Synura uvella*. 4. Scale and bristle scales of *Chrysosphaerella brevispina*. 5. Loricas of *Biocosoeca* sp. Scale bars = 1 μm

Frequently attached to some of the scales of *Mallomonas* are hinged silica bristles. Based on both the scale and bristle structure, Asmund and Kristiansen (1986) divided the genus into 16 sections. The bristle consists of a sheet rolled up into a tube (Wujek and Kristiansen, 1978). Attached at the apical end of a scale by an as yet undetermined glue, it may lay attached to a flattened area of the scale (*Mallomonas caudata*) or, most often, the bristle's foot fits into the dome, a special differentiated part of the scale. In bristles of the primitive type (craspedont), serrations occur along the edge of the former slit. In the more derived type (notacanthic), the teeth are hollow. In some of the sections the bristle is distally adorned with structures such as hooks of 'Greek' helments (Fig. 4). Both bristle type may be infrequently observed on the same cell (section *Heterospinae*). Most often they terminate with a sharp point or are bifurcate. All of these features are of taxonomic interest. Kristiansen and Vigna (1994) have reported for *Synura spinosa* that the tubular scales of this species may represent an early stage of the evolution toward *Mallomonas* bristles.

During the first half of the current century, many scaled chrysophytes were described. As early as 1899, Iwanoff stated that scales were species specific. He was the first to use scale morphology when he described *Mallomonas caudata*. These early descriptions were all made with LM. Some of the early workers included Bourrelly (1957), Matvienko (1965) and Starmach (1980). Because of the fine structure of the silica scales (Figs. 1-4), electron microscopy (EM) is now required to identify most species; however, for a few species light microscopy can still be utilized (Wee, 1982). Indeed, when describing a new taxon, details from EM (TEM and/or SEM) are now necessary to illustrate the organism. To cite just one example of the confrontation between the taxonomy based on LM and on EM. Prowse (1962) described the two-celled colonial genus *Chrysodidymus* comprising two species. Scales from material collected in Quebec, Canade, by Puytorac *et al.* (1972) were the first EM observations attributed to the genus. Nygaard (1978) used similar scales from material collected in Greenland in describing a new species of *Synura, S. microcrepis.* Nevertheless, Wujek and Wee (1983) gave proof from LM and EM observations that the cells and scales they observed were identical to Prowse's illustrations and Puytorac's and Nygaard's micrographs. Kristiansen (1979, 1996, 1995) gives further discussion of these problems.

The question now is how minute details of the siliceous structures should be used in taxonomy. The use of EM has resulted in many varieties and forms being described based on ultrastructural details. Examples include the *Synura petersenii* complex and the *Mallomonas mangofera* group. These small morphological differences must be combined with other characters, e.g, ecological (Siver, 1989a), to support their use in taxonomy. The role of environmental factors influencing scale ornamentation needs to be known; certainly this will require experimental studies.

A characteristic feature of the chrysophytes is their ability to produce resting stages from the endogenously formed siliceous cysts also called statospores or stomatocysts. Cysts of the bicosoecids are believed to be organic. Cienkowski (1870) was the first to describe organisms that formed cysts. The shape of the cyst is oval or round. The cyst wall may be ornamented or smooth. A pore from which the germling alga (vegetative cell or zoospore) emerges can be plugged with a silicified stopper. The pore may possess a collar. Sandgren (1983) has demonstrated that in *Dinobryon cylindricum*, spines on the cyst walls and/or collar lengths varied when clones representing the same population were exposed to different temperatures.

The majority of chrysophyte cysts remain to be described. They have been classifed only to morphotype (Duff *et al.*, 1995). Because of this, they have been unimportant in taxonomy. Only in some of the non-silicified genera such as *Uroglena* and *Uroglenopsis* is cyst morphology necessary for species delineation (Bourrelly, 1963).

Because the chrysophytes are primarily planktonic and presumed to have narrow environmental requirements, they can be good indicators of water quality and paleoenvironmental conditions. The chrysophytes appear seasonally, thus the scales and cysts may cause distinct layers in the sediments that can be used to reconstruct past limnological conditions (Duff and Smol 1995a, 1995b; Smol, 1980, 1988, 1995; Siver, 1988a, 1988b, 1989b, 1995).

A lorica surrounds the protoplast in the families Bicosoecaceae, Chrysococcaceae and Dinobryaceae (Fig. 5). Its shape in the LM is a character of great importance and is a basis for distinction at both the genus and species level. The lorica has been shown to be an interwoven case of fibrils (Belcher, 1963; Karim and Round, 1967; Kristiansen, 1969, 1972) in most genera.

The shape and arrangement of the lorica àre the basis for species delineation. However, in *Epipyxis*, the lorica is composed of distinct imbricate scales. Staining for LM or phase contrast microscopy allows one to observe the arrangement of the scales and the flaring of the apical scales, all important in species determintation. EM of whole mounts is not effective unless shadow cast preparations are made; even here, generally only the fibrillar nature of the scale can be ascertained.

In some genera, e.g. *Chrysococcus*, the lorica may be stained with deposits of manganese or iron (Belcher, 1969). In a recent study of *Chrysocrinus* (Schoonoord and Ellis-Adams, 1984), these deposits were shown to be caused by oxidation resulting from photosynthesis.

Unusual scales also consisting of an interwoven fibrillar nature have been observed in the colonial genus *Lepidochrysis* Ikävalko *et al.*, 1994). Fibrillar scales are also observed on *Chromulina*, *Phaeaster* and *Syncrypta* (see Kristiansen, 1986, for review). Members of the Pedinallales also have cellulosic scales, some of which are spined.

Organic scales of taxonomic value are observed in other chrysophytes. In *Sphaleromantis*, the cell is covered with flower pot-shaped scales. Scales of varying morphologies have been reported from both the cell and flagella in *Chrysolepidomonas* (Peters and Andersen, 1993).

Flagellar scales have been reported for other genera, including *Chromulina*, *Mallomonas*, and *Synura*. To date, the scattered occurrence of these organic flagellar scales has not led to their use in taxonomy. Certainly this is an area in need of further investigation.

**Molecular Approaches**

Perhaps not a readily available tool for many taxonomists, molecular techniques are beginning to resolve some chrysophyte relationships above the ranh of species. Comparative molecular studies on chrysophytes have initially focused on the chloroplast genome, and the methods primarily were limited to mapping genes and restriction endonuclease sites, DNA cloning, and molecular probes. Subsequently, amplification of specific DNA regions via polymerase chain reaction (PCA) and advancements in nucleic acid sequencing methods, including direct sequencing of double-stranded DNA and automated sequencing followed. This allowed nucleotide sequencing to rapidly replace the previous molecular approaches in comparative studies. The increasing availability of DNA sequences via electronic data bases further enlarged data sets and also contributed to the need for increasingly sophisticated analytical methods (see especially Wee, 1996; Wee *et al.*, 1996a. 1996b).

Initial reviews of molecular biology to chrysophyte biology include Olsen (1990) and Coleman and Goff (9191). As "the use of automated sequencers and electronic databases facilitate access to DNA sequence data, future chrysophyte studies may become less limited by data acquisition, and computational and theoretical considerations of data analysis may become more important" (Wee *et al.*, 1996b). Species diversity that is not reflected in morphological species concepts and the parallels in morphological diversity that LM and EM have provided is expected to be either confirmed or reveal new relationships through use of molecular techniques.

It is at the higher taxonomic levels (family and above) that molecular techniques have contributed significantly. Molecular and structural analyses have demonstrated the affinity between both the autotrophic and heterotrophic chrysophytes and fungal-like heterotrophic protists (the oomycetes) (Leipe *et al.*, 1994). These comparative studies place the Chrysophyceae into a complex evolutionary assemblage now frequently called the stramenopiles. Cladistic analyses of traditional and molecular data of a large and biological diverse assemblage of organisms (Saunders *et al.*, 1995), including the organisms discussed above, places them into the chromophyte algae; thus uniting them with all taxa possessing a highly reduced flagellar apparatus. Saunders and co-workers study suggests that the traditional classification of diatoms with the chrysophytes and xanthophytes into the division Chrysophyta may be unsatisfactory and that a different classification should be employed. Wee (1996) lists many terms and acronyms associated with molecular data.

Unfortunately much of the information presented here is biased by my interest in involving EM in taxonomy. Certainly the extensive literature on chrysophyte taxonomy that has been published in the past quarter century has been extensively EM based. It is hoped with implementation of newer LM techniques (fluorescence, Nomarski, laser confocal, image enhancement) and molecular techniques, these techniques will become equally useful in delineating this most interesting group of algae.

## ACKNOWLEDGEMENTS

I am grateful to Dr. J.L. Wee for providing copies of manuscripts that are in press and CMU Graphic Productions for their assistance in preparing the micrographs.

## REFERENCES

Andersen, R.A. 1987. Synurophyceae classis nov., a new class of algae. *Amer. J. Bot.* **74**: 337-353.

Asmund, B. and Kristiansen, J. 1986. The genus *Mallomonas*. *Opera Botanica.* **85**: 1-128.

Belcher, J.H. 1968. Lorica construction in *Pseudokephyrion pseudospirale* Bourrelly. *Brit. Phycol. Bull.* **3**: 495-499.

Belcher, J.H. 1969. A morphological study of the phytoflagellate *Chrysococcus rufescens* Klebs in culture. *Brit. J. Phycol.* **4**: 105-117.

Bourrelly, P. 1957. Reserches sur les Chrysophycées. *Rev. Algol. Mém. Hor-Sér.* **1**: 1-412.

Bourrelly, P. 1963. Loricae and cysts in the Chrysophyceae. *Ann. N.Y. Acad. Sci.* **108**: 421-429.

Cienkowski, C. 1870. Über Palmellaceen und einige Flagellaten. *Arch. Mikrosk. Anat.* **6:** 421-438.

Coleman, A.W. and Goff, L.J. 1991. DNA analysis of eukaryotic algal species. *J. Phycol.* **27**: 463-473.

Duff, K.E. and Smol, J.P. 1995a. Chrysophycean cysts assembledges and their relationship to water chemistry to 71 Adirondack Park (New York, USA) lakes. *Arch. Hydrobiol.* **134**: 307-336.

Duff, K.E. and Smol, J.P. 1995b. The relationship of chrysophycean stomatocysts to environmental variables in freshwater lakes in British Columbia. *Can. J. Bot.* **73**: 1097-1111.

Duff, K.E., Zeeb, B.A. and Smol, J.P. 1995. Atlas of chrysophycean cysts. Developments in Hydrobiology 99. Series editor H.J. Dumont. Kluwer Academic Publishers, Dordrecht, Netherlands, 189p.

Ikävalko, J., Kristiansen, J. and Thomsen, H.A. 1994. A revision of the taxonomic position of *Syncrypta glomifera* (Chrysophyceae), establishment of a new genus *Lepidochrysis*, and observations on the occurrence of L. *glomifera* comb. nov. in brackish waters. *Nord. J. Bot.* **14**: 339-344.

Iwanoff, L. 1899. Beitrag zur Kenntniss der Morphologie und Systematik der Chrysomonaden. *Bull. Ac. Imp. Sc. St. Petersburg.* **11**: 247-262.

Karim, A.G. and Round, F.E. 1967. Microfibrils in the lorica of the freshwater alga *Dinobryon*. *N. Phytol.* **66**: 409-412.

Kristiansen, J. 1969. Lorica structure in *Chrysolykos* (Chrysophyceae). *Bot. Tidsskr.* **64**: 162-168.

Kristiansen, J. 1972. Studies on the lorica structure in Chrysophyceae. *Sven. Bot. tidsskr.* **66**: 184-190.

Kristiansen, J. 1979. Problems in classification and identification of Synurophyceae (Chrysophyceae). *Schweiz. Z. Hydrol.* **40**: 310-319.

Kristiansen, J. 1986. The ultrastructural basis of chrysophyte systematics and phylogeny. *CRC Critical Reviews in Plant Science.* **4**: 149-211.

Kristiansen, J. 1990. Phylum Chrysophyta. In: *Handbook of Protoctista* Eds. L. Margulis, J.O. Corliss, M. Melkonian and D.J. Chapman, Jones and Bartlett, Boston, pp. 438-453.

Kristiansen, J. 1995. Silica structure in the taxonomy and identification of scaled chrysophytes. *Beiheft Nova Hedwigia*. **112**: 355-365.

Kristiansen, J. and Vigna, M.A. Tubular scales in *Synura* and the possible origin of bristles in *Mallomonas* (Synurophyceae). *Phycologia*. **33**: 67-70.

Leipe, D.D., Wainright, P.O., Gunderson, J.H., Porter, D., Patterson, D.J., Valois, F., Himmerich, S. and Sogin, M.L. 1994. The stramenopiles from a molecular perspective: 16S-like rRNA sequences from *Labyrinthuloides minuta* and *Cafetera roenbergensis*. *Phycologia*. **35**: 369-377.

Matvienko, O.M. 1965. *Zolotisti vodorosti, Chrysophyta*. Viznac, Prisnovod, Vodorosteij Ukraj. RSR. Kijiv.

Moestrup, O. 1995. Current status of chrysophyte splinter groups': synurophytes, pedinellids, silicoflagellates. In: *Chrysophyte Algae: Ecology, Phylogeny and Development* Eds. C.D. Sandgren, J.P. Smol and J. Kristiansen, Cambridge University Press, Cambridge, pp. 75-91.

Nygaard, G. 1978. Freshwater phytoplankton from the Narssaq area, South Greenland. *Bot. Tidsskr.* **73**: 191-238.

Olsen, J.L. 1990. Nucleic acids in algal systematics. *J. Phycol.* **26**: 209-214.

Pascher, A. 1914. Über Flagellaten und Algen. *Ber. Deutsch. Bot. Ges.* **32**: 136-160.

Peters, M.C. and Andersen, R.A. 1993. The fine structure and scale formation of *Chrysolepidomonas dendrolepidota*, gen. et. sp. nov. (Chrysolepidomonadaceae fam. nov., Chrysophyceae). *J. Phycol.* **29**: 469-475.

Preisig, H.R. 1995. A modern concept of chrysophyte classification. In: *Chrysophyte Algae: Ecology, Phylogeny and Development* Eds. C.D. Sandgren, J.P. Smol and J. Kristiansen, Cambridge University Press, Cambridge, pp. 46-74.

Prowse, G.A. 1962. Further Malayan freshwater flagellata. *Gard. Bull. Singapore.* **19**: 105-140.

Puytorac, P., Mignot, J.P., Grain, J., Groliere, C.A., Bonnet, L. and Couillard, P. Premier relevé de certaines groups de protozoaires libres sur le territoire de la station de biologie de l'université de Montréal (Saint Hippolyte, Comte de Terrbonne, Québec). *Nat. Can.* **99**: 417-440.

Sandgren, C.D. 1980. Morphological variability in populations of chrysophycean resting cysts. I. Genetic (interclonal) and encystment temperature effects on morphology. *J. Phycol.* **19**: 64-70.

Saunders, G.W., Potter, D., Paskind, M.P. and Andersen, R.A. 1995. Cladistic analyses of combined traditional and molecular data sets reveal an algal lineage. *Proc. Nat. Acad. Sci.* **92**: 244-248.

Schoonoord, M.P. and A.C. Ellis-Adams. 1984. *Chrysocrinus homorarius* spec. Nov. (Stylococcaceae, Chrysophyceae): some stages of its lorical development and its distribution on the substrate. *Acta Bot. Neerl.* **33**: 399-418.

Siver, P.A. 1988a. Distribution of scaled chrysophytes in 17 Adirondack (New York) lakes with special reference to pH. *Can. J. Bot.* **66**: 1391-1403.

Siver, P.A. 1988b. Morphology and ecology of *Mallomonas galeiformis* (Chrysophyceae), a potentially useful paleolimnological indicator. *Trans. Amer. Micros. Soc.* **107**: 152-161.

Siver, P.A. 1989a. The separation of *Mallomonas acaroides* v. *acaroides* and v. *muskokana* (Synurophyceae) along a pH gradient. *Beihfte Nova Hedwigia*. **95**: 111-117.

Siver, P.A. 1989b. The distribution of scaled chrysophytes along a pH gradient. *Can. J. Bot.* **67**: 2120-2130.

Siver, P.A. 1995. The distribution of chrysophytes along environmental gradients: their use as biological indicators. In: *Chrysophyte Algae: Ecology, Phylogeny and Development* Eds. C.D. Sandgren, J.P. Smol and J. Kristiansen, Cambridge University Press, Cambridge, pp. 232-268.

Smol, J.P. 1980. Fossil synuracean (Chrysophyceae) scales in lake sediments: a new group of paleoindicators. *Can. J. Bot.* **58**: 458-465.

Smol, J.P. 1988. Chrysophycean microfossils in paleolimnological studies. *Paleogeography, Paleoclimatology, Paleoecology.* **62**: 287-297.

Smol, J.P. 1995. Application of chrysophytes to problems in paleoecology. In: *Chrysophyte Algae: Ecology, Phylogeny and Development* Eds. C.D. Sandgren, J.P. Smol and J. Kristiansen, Cambridge University Press, Cambridge, pp. 303-329.

Starmach, K. 1980. Chrysophyta I—Zlotowiciowce. In: *Flora Stodkowodona Polski* Vol. 5. Ed. K. Starmach and J. Sieminska. Panstwowe Wydawnictwo Naukowe, Warszawa, pp. 1-775.

Wee, J.L. 1982. Studies on the Synuraceae (Chrysophyceae) of Iowa. *Bioliotheca Phycol.* **18**: 1-183.

Wee, J.L. 1996. Molecular investigations of heterokont comparative biology: recent and emerging trends. In: *Proceedings of the 4th International Chrysophyte Symposium* Ed. J. Kristiansen (In press).

Wee, J.L., Hinchey, J.M., Nguyen, K.X., Kores, P. and Hurley, D.L. 1996a. Assessment of the phylogenetic utility of psbA and ITS/5.8S nucleotide sequences for the Synurophyceae. In: *Proceedings of the 4th International Chrysophyte Symposium* Ed. J. Kristiansen (In press).

Wee, J.L., Hinchey, J.M., Nguyen, K.X., Kores, P. and Hurley, D.L. 1996b. Investigating the comparative biology of the heterokonts with nucleic acids. *J. Euk. Microbiol.* **43**: (In press).

Wujek, D.E. and Kristiansen, J. 1978. Observations on bristle and scale-production in *Mallomonas caudata. Arch. Protistenk.* **120**: 213-221.

Wujek, D.E. and Wee, J.L. 1983. *Chrysodidymus* in the United States. *Trans. Amer. Micros. Soc.* **102**: 77-80.

Recent Trends in Algal Taxonomy
Vol. I. Taxonomical Issues
Editors : Vidyavati and A.K. Mahato
APC Publications Pvt. Ltd., New Delhi, 1999, pp. 253-272

# *A Taxonomic Appraisal of the Genera* Spirulina *and* Arthrospira

N. Jeeji Bai
Shri A.M.M. Murugappa Chattiar Research Centre
Taramani, Chennai-600 113

## *SPIRULINA*—BACKGROUND AND RECOGNISED SPECIES

The genus *Spirulina* was first described by Turpin in 1827. The monograph of Gomont (1892) which is the starting point for nomenclature of Oscillatoriales, lists 9 species of this genus. The original species of Turpin, S. *oscillarioides*, which should be the type species, is listed under "questionable species" by Gomont and he does not designate another type. Geitler (1932) lists 20 species two of which, S. *agilissima* (Lagerh.) Kirchn and S. *tenuior* (Lagerh.) Kirchn. he does not consider as members of Cyanophyceae. S. *nordstedtii* Gom., S. *versicolor* Cohn and S. *rosea* Crouan, all ex Gomont (1892) are not listed by Geitler (1932) and he also does not indicate the type species. Elenkin (1949) lists 13 species of which S. *pseudovacuolata* Uterm. is not listed by others. Desikachary (1959) lists 8 species and gives as lectotype S. *major* Kütz. based on the earlier lectotype designation by Geitler (1942). Drouet (1968) has retained S. *oscillarioides* Turpin as the type. It is generally accepted at present that the type species of *Spirulina* is S. *major* Kütz. ex Gom.

Members of this genus occur in different habitats: running and stagnant fresh water bodies, thermal springs, marine and brakish water environements. Some of the species occurring in sublittoral zones exhibit chromatic adaptation with a predominance of the red phycoerythrin pigment.

This genus was initially considered to be unicellular because of the absence of obvious cross walls in the "trichomes". Nevertheless, it has been generally classified as a member of Oscillariaceae in many of the major monographs.

The earliest clear-cut demonstration of the presence of crosswalls in *Spirulina* through special staining techniques has come from Zuelzer (1912), Dobell (1912), Schmid (1920), Figini (1925) and Crow (1927). Based on this finding by earlier workers and his own confirmatory observations Gardner (1917) has provided the first amended description of *Spirulina*; however, he retains the two genera as distinct as he considers it to be "practically advantageous" His modified diagnosis of the genus *Spirulina*, however, does not include the visible or non-visible nature of cross-walls "trichomes multicellular, cylindrical, evaginate, loosely or tightly coiled into a more or less regular spiral, apex of trichome usually not tapering, terminal cell rounded without calyptra, cell contents homogeneous or granular"

Following the demonstration of cross walls in *Spirulina* many major treatises on blue-green algae have treated the genus *Arthrospira* as a subgenus of *Spirulina* (Geitler, 1925; 1932; Fremy, 1930; Elenkin, 1949) because the earlier name *Spirulina* Turpin (1827) has priority over the later name *Arthrospira* Stizenberger (1852). Bourrelly (1970) goes a step further and includes both these genera in *Oscillatoria*, some species of which also have partially coiled trichomes (see also Jeeji Bai & Seshadri, 1980; Hoffmann, 1991; Deore, 1992 for comments). It is noteworthy that Drouet (1968) retains the two genera although on very different grounds.

The first ultrastructural study of S. *major* by Holmgren *et al.*, (1971) followed by that of Guglielmi and Cohen-Bazire (1982) has unequivocally shown the presence of cross-walls. However, the latter authors have shown certain additional distinctive ultrastructural features in *Spirulina*. Anagnostidis and Komarek (1988) have, therefore, included in their recent revision of the group some of these ultrastructural features, besides non-visibility of cross walls in light microscope, closeness of spirals, length-breadth ratio of cells and absence of calyptra (see also Gardner, 1917) as diacritical features of *Spirulina*.

Rippka *et al.*, (1979) have shown a significant difference in the GC ratios of *Spirulina* and *Arthrospira* as a justification for their retention. Subsequently, Guglielmi *et al.*, (1993) have provided a list of ultrastructural and biochemical characteristics to separate the two genera. Recently, a study of the 16S RNA by Nelissen *et al.*, (1994) has shown that the two genera have very different genotypes. There is thus a general consensus of opinion among phycologists at present that the two genera are indeed quite distinct.

## *ARTHROSPIRA*—BACKGROUND AND SPECIES DESCRIBED

The genus *Arthrospira*, first created by Stizenberger (1852) is based on the species of Hassall (1845), originally described as *Spirilllum jenneri* Hass. and hence named by Kützing (1849) as *Spirulina jenneri* (Hass.) Kütz. and transferred to *Arthrospira* by Stizenberger as *A. jenneri* (Hass.) Stiz. Stizenberger (1854) described another species, *A. baryana* Stiz. which is listed by Gomont (1892) as a synonym of *A. jenneri*. He and others following him do not designate the type. Desikachary (1959) gives *A. jenneri* (Hass.) Stiz. as the type species.

The generic diagnosis of *Arthrospira* in Desikachary (1959) includes all the essential characteristics: "Trichomes multicelluler, cylindrical without sheath, loosely and regularly coiled, coils usually of relatively large diameter, trichomes usually with fewer coils as compared to *Spirulina*, cross walls distinct, apices slightly or not at all tapering, terminal cell rounded, calyptra absent"

The members of this genus occur in running and standing fresh water bodies, brackish waters and in purely marine habitats. Unlike members of *Spirulina* they do not occur in thermal springs. A majority of the freshwater species are planktonic with the presence of gas vacuoles.

Gomont [1892, so also Geitler (1932), Fremy (1930) and others] does not include the feature "multicellular" in his generic diagnosis of the genus *Arthrospira*, probably because the term "trichome" entails "multicellularity" as an obligatory condition. He lists only 3 species: *A. jenneri* (Hass.) Stiz., *A. platensis* (Nordst.) Gom. and *A. miniata* (Hauck) Gom. Geitler (1932) lists 9 additional species under the unified genus *Spirulina*: *A. spirulinoides* Ghose, *A. breviarticulata* Setch., *A. okensis* Meyer, *A. massartii* Kuff., *A. tenuis* Bruhl et Biswas, *A. margaritae* Gom. ex Fremy and *A. curta* Lemm. Geitler's combination, *S. maxima* (Setch. et Gard.) Geitler is an invalid name as it is a later homonym of *S. maxima* Bernard and hence J. de Toni (1936) proposes a new name, *S. geitleri* J. de Toni for this species. Now that the genus *Arthrospira* is widely recognised as quite distinct from *Spirulina*, *A. maxima* Setch. et Gard. stands as a valid name.

Desikachary (1959) lists 7 species out of which *A. khannae* Drouet et Strickl. (in Drouet, 1942) is a new addition. Drouet's (1968) index includes 13 additional species: *A. aeruginea* (Buell) J. deToni, *A. amethystina* (Buell) J.de Toni, *A. ardissonii* (Cohn)

Forti, *A. balkrishnanii* Kamat (1963), *A. desikacharyensis* Vasishta (1962), *A. joshii* Vasishta (1963), *A. braunii* (Kütz.) Forti, *A. exaquiensis* (Fremy) J. de Toni, A. *funiformis* Vauk., *A. laxissima* Gom., *A. leopoliensis* Racid, *A. nägelii* (Brügg.) Forti and *A. setchellii* (Setch.) J. de Toni.

Gomont (1892) considers his own species, *A. laxissima*, as a synonym of *A. miniata*; J. de Toni (1938) considers the two species of Buell collected from the same locality at the same time and described in the same paper, as synonymous differing only in colour. Geitler (1932) and Drouet (1968) include both these under *A. jenneri*. *A. ardissonii* (Cohn) Forti, a marine form, is considered by Geitler (1932) to be insufficiently described. These species as well as *A. braunii*, *A. exaquiensis*, *A. nagelii* and *A. setchellii* have probably not been observed again as they are not listed in any of the later compilations.

In India, only *A. balkrishnanii*, *A. desikacharyensis* and *A. joshii* have been added after 1959. Drouet (1968) considers *A. balakrishnanii* and *A. joshii* to be synonyms of *A. jenneri* and excludes *A. desikacharyensis*. *A. spirulinoides* Ghose has not again been reported except for a form, *tenuis* Singh (1939). *A. tenuis* has been reported again by Biswas himself (1942).

*A. breviarticulata* Setch. et Gard. in Gardner (1918) has been included in the earlier described species, *A. funiformis* Vauk (1915) as a synonym by Komarek (1956). Hindak (1988) has shifted the species *A. okensis* Meyer to the genus *Romeria* Koczw. under a new combination *R. okensis* (Meyer) Hind. on the basis of the absence of movement and in doing so he has endorsed the view of Geitler (1932) who says that "the alga reminds one of *Romeria* except for trichome length". Anagnostidis and Komàrek (1988) have shifted *A. curta* Lemm. to the genus *Borzia* Cohn under a new combination *B. curta* (Lemm.) Anag. et Kom.

The marine species, *A. miniata* (Hauck) Gom. has often been recorded from different localities in Europe and recently also from Papua New Guinea by Hoffmann (1991).

## THE COMMONLY DISCUSSED SPECIES OF *ARTHROSPIRA*

Elenkin (1949) lists 7 species of *Arthrospira* under the name *Spirulina* of which *S. fusiformis* Woronichin (1934) has not been listed by others. This species of Woronichin has again been recognised for the first time, isolated in unialgal culture and described from a pond in Madurai by Jeeji Bai and Seshadri (1980). They have made a detailed study of the coiling pattern in this alga and have distinguished three different "morphotypes", namely 'S', 'C' and 'H' according to the number, degree and pattern of the resulting coils (Figs. 1-4) and these have been shown to be reversible according to the culture conditions. Somewhat similar patterns have also been reported by Hindàk (1985) from a natural sample from lake Bogoria, Africa. Even straight variants have been reported by Lewin (1980) and Jeeji Bai (1985). These straight morphotypes (Fig. 5), initially considered to be irreversible mutants have been found to be reversible (unpublished observations). Jeeji Bai and Sheshadri (1980) have considered another isolate studied by them from lake Nakuru, Kenya, the alga described as *A. platensis* by Rich (1931) with the classical morphological variants (figured in all the major monographs ) and the form *granulata* of *A. platensis* described by Desikachary (1959) to belong to this species. Fott and Karim (1973) have described an alga from Sudan, which they initially attributed to S. geitleri (= *A. maxima*) and later shifted to *S. fusiformis*. This species, originally described as *S. fusiformis* by Woronichin has recently been validated under the genus *Arthrospira* by Komàrek and Lund (1990) with a new combination, *A. fusiformis* (Woron.) Kom. et Lund.

In recent times, *A. platensis* (Nord.) Gom. and *A. maxima* Setch. et Gard. (= *S. geitleri*) have been most discussed because strains attributed to these two species have been mass cultured for human nutrition and other value-added products. In such studies, there has often been an erroneous or even a lack of naming of the strains investigated. In order to remove the confusion in identification and proper designation, 35 strains and natural samples of *Arthrospira* from different geographical locations have been studied by Desikachary and Jeeji Bai (1992, 1996). These studies are based mainly on light microscopic observations so that the diagnostic characteristics recognised by them can be employed effectively by taxonomists, ecologists and experimentalists.

As pointed out earlier, the generic description of *Arthrospira* does not include the presence of a calyptra. However, Fott and Karim (1973) demonstrate the emblance of a calyptra in the form of apical thickening in *A. fusiformis* (under their original name, *S. geitleri*). Jeeji Bai and Seshadri (1980) and Jeeji Bai (1985) have first brought out clearly the presence of a calyptra in the

Indian (Figs. 5,8) and Kenyan isolates of this species. Hindàk (1985) also mentions an inconspicuous, broadly rounded calyptra in this species from lake Bogoria, Africa. The first confirmatory demonstration of calyptra in the genus *Arthrospira* through SEM pictures (Figs. 14-16) has come from the recent studies of Desikachary and Jeeji Bai (1992; 1996). Based on this feature, namely the presence of a calyptra, they have created a new species, A. *indica*, as distinct from the original Russian species, A. *fusiformis*, which does not have a calyptra. Should the original holotype of A. *fusiformis* (or a sample from the type locality) be shown to possess this feature, A. *indica* will become a later homonym of A. *fusiformis*. It is noteworthy that under natural conditions only an inconspicuous thickening of the end cell is seen and clear-cut calyptra seems to develop only in culture conditions.

## GROUPING OF STRAINS AND NATURAL SAMPLES AND THEIR ASSIGNMENT TO KNOWN SPECIES

Desikachary and Jeeji Bai (1996) have recognised four distinct groups among the many strains and natural samples studied by them. The groups, their species affiliation and important characteristics are given below:

**Group 1**: A. *indica* Desik. & Jeeji Bai: Trichomes form coils which are distinctly narrower (50%) towards the ends giving a spindle-like (fusiform) appearance; during elongation the trichome forms a dumb-bell like condition with slightly narrower median coils; fragmentation occurs in the median region resulting in two "spindles". The trichomes are on an average 8-9 µm broad with a gradual but distinct attenuation of ends with up to 50% reduction in trichome diameter; cells are always less than 1/2 as long as broad; end cell is both capitate and calyptrate with a conical calyptra. The MCRC is isolate has been designated as the type (Figs. 1-5, 8, 14, 16).

A. *platensis f. granulata* Desik. has been included in this species after studying it afresh. A natural sample from a temple tank in Madurai, and isolates from the Lonar lake in Maharashtra (Courtesy, CSV, Wardha) Sambhar lake in Rajasthan (Courtesy, Prof. H. D. Kumar) and the Adyar river estuary in Madras have also been attributed to this species after a detailed study (Figs. 6, 7, 9, 15).

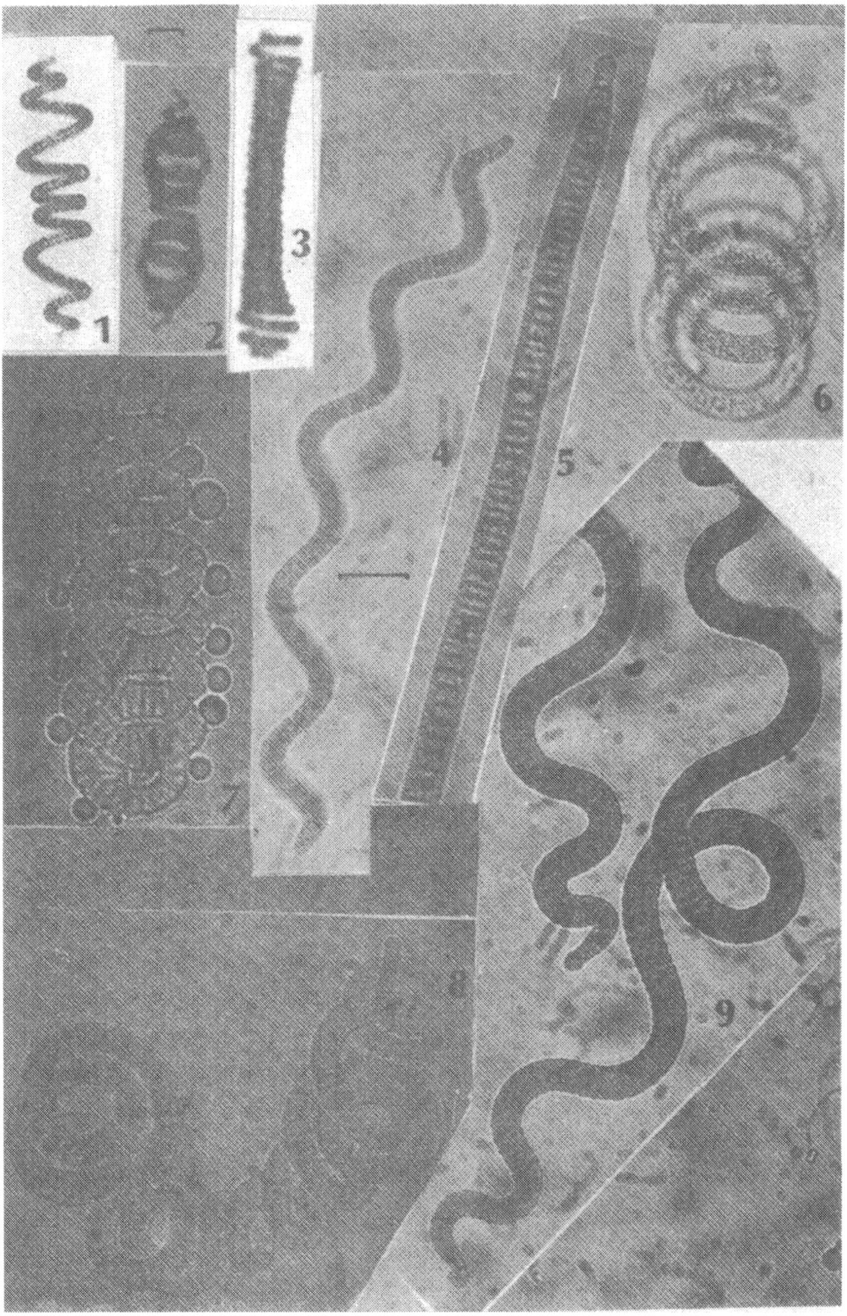


Figs. 1-9. *A. indica*; 1-5, 8; *MCRC* strain; 1, 3, 4: Morphotypes C, H, S; 2, 8: dumb-bell condition; 5: straight variant. 6: Lonar isolate showing fusiform coil; 7. Type of *A. platensis* var *granulata* Desik. with typical spindle; 9: Sambhar-lake isolate showing loose spindles; note capitate and calyptrate end cells in 5 to 9.

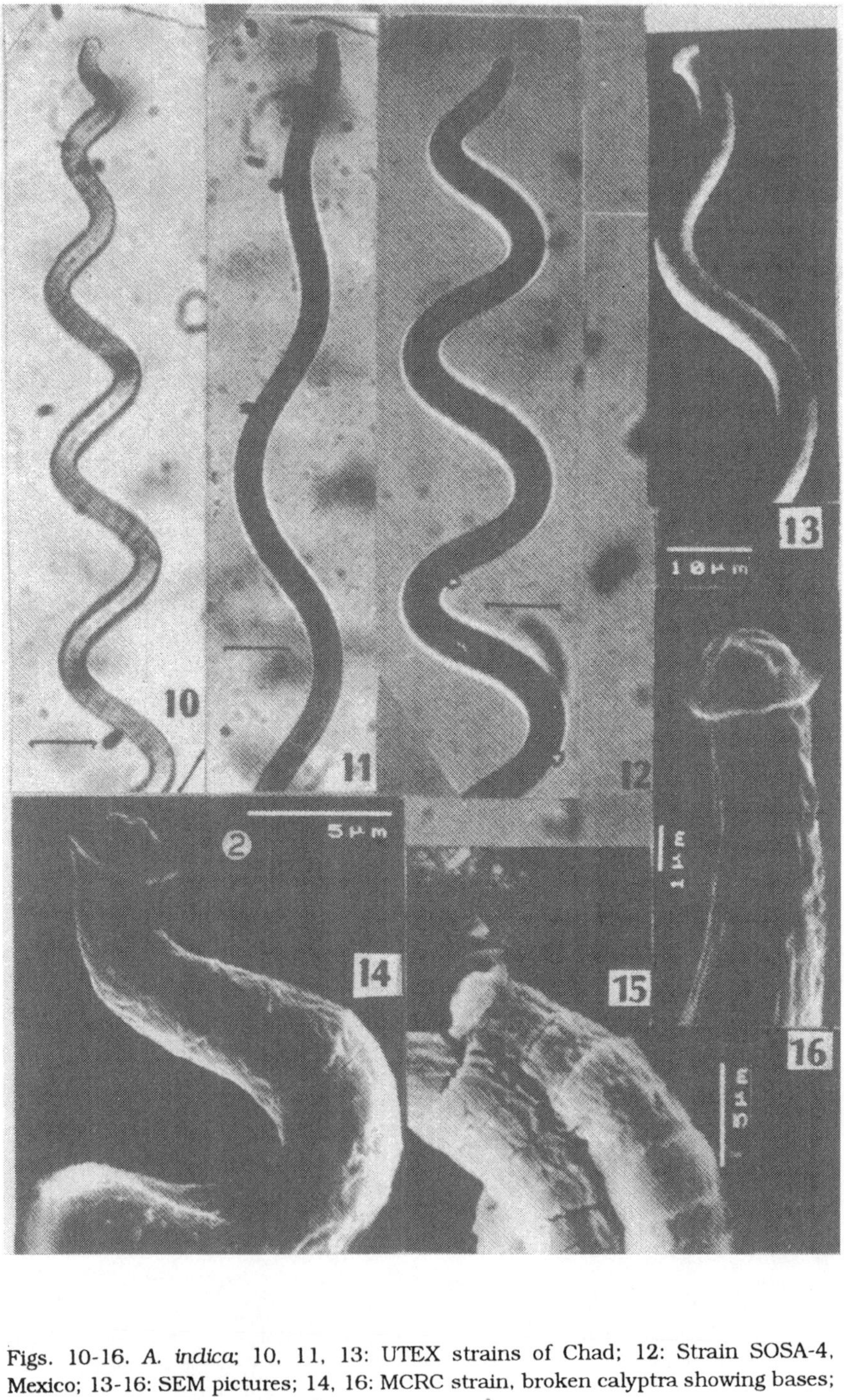


Figs. 10-16. *A. indica*; 10, 11, 13: UTEX strains of Chad; 12: Strain SOSA-4, Mexico; 13-16: SEM pictures; 14, 16: MCRC strain, broken calyptra showing bases; 15: Madurai sample, intact calyptra.

Two authentic strains from Chad obtained from the culture collections of Göttingen, Germany and UTEX, USA (Courtesy, Prof. R.C. Starr) which have usually been named A. *platensis*, have also been shown to belong to this species by a critical study of their morphology (Figs. 10, 11, 13). Another strain, Sosa 4, from lake Sosa Texcoco, Mexico (Courtesy Dr. Tomaselli, Italy) named as A. maxima, is also considered to belong to this species (Fig. 12).

**Group 2**: A. *maxima* Setch. et Gard.: Trichomes form regular coils which are uniform in diameter, occasionally the terminal coil alone may be narrower; dumb-bell or spindle like appearance never occurs during elongation and breaking of coils; trichomes have an average width of 7-8 μm and tapering towards ends is mostly sudden with a maximum of 30% reduction in diameter; cells are often 1/2 as long as broad or sometimes even longer; end cells non-capitate but calyptrate with a flatty conical calyptra.

The isolated strains and natural material from lake Huacachina, Peru (Courtesy, Dr.E. Hegewald), a sample from a lake in Pollachi (Courtesy, Dr. Sankaran), and a pond sample from Pune (Courtesy, Dr. L.T., Deore) have all been attributed to this species after a careful study (Figs. 17-22).

**Group 3**: A. *massartii* Kufferath: Trichomes form regular coils which are uniform in diameter, end coils are never narrower, dumb-bell like appearance never occurs; trichomes are usually 6 μm broad, not tapering towards the ends; cells are always less than 1/2 as long broad, end cells non capitate, often non calyptrate; calyptra when present hood like and discoid.

The samples from ponds in Punjab (Courtesy Dr. S.A. Sarma) Allahabad and Lucknow (Courtesy, Prof. G.L. Tiwari) and sample from another pond in Pune (Courtesy, Dr. L.T. Deore) have been attributed to this species (Figs. 23-27). According to Desikachary (1959) this species is not recorded from India, however, he describes a new variety, A. *massariti* var. *indica* based on the presence of a calyptra (Fig. 28). All the above referred samples from India belong to this variety as they all have calyptra.

**Group 4**: A. *platensis* (Nord.) Gom.: Trichomes form regular coils with end coils sometimes 30-50% narrower than the median coils; dumb-bell shaped condition however, never occurs during elongation; trichomes are on average 5-7 μm broad; gradually and distinctly attenuated (upto 50%) towards the ends; cells are often more than 1/2 as long as broad; end cells often subcapitate to distinctly capitate but always non-calyptrate.

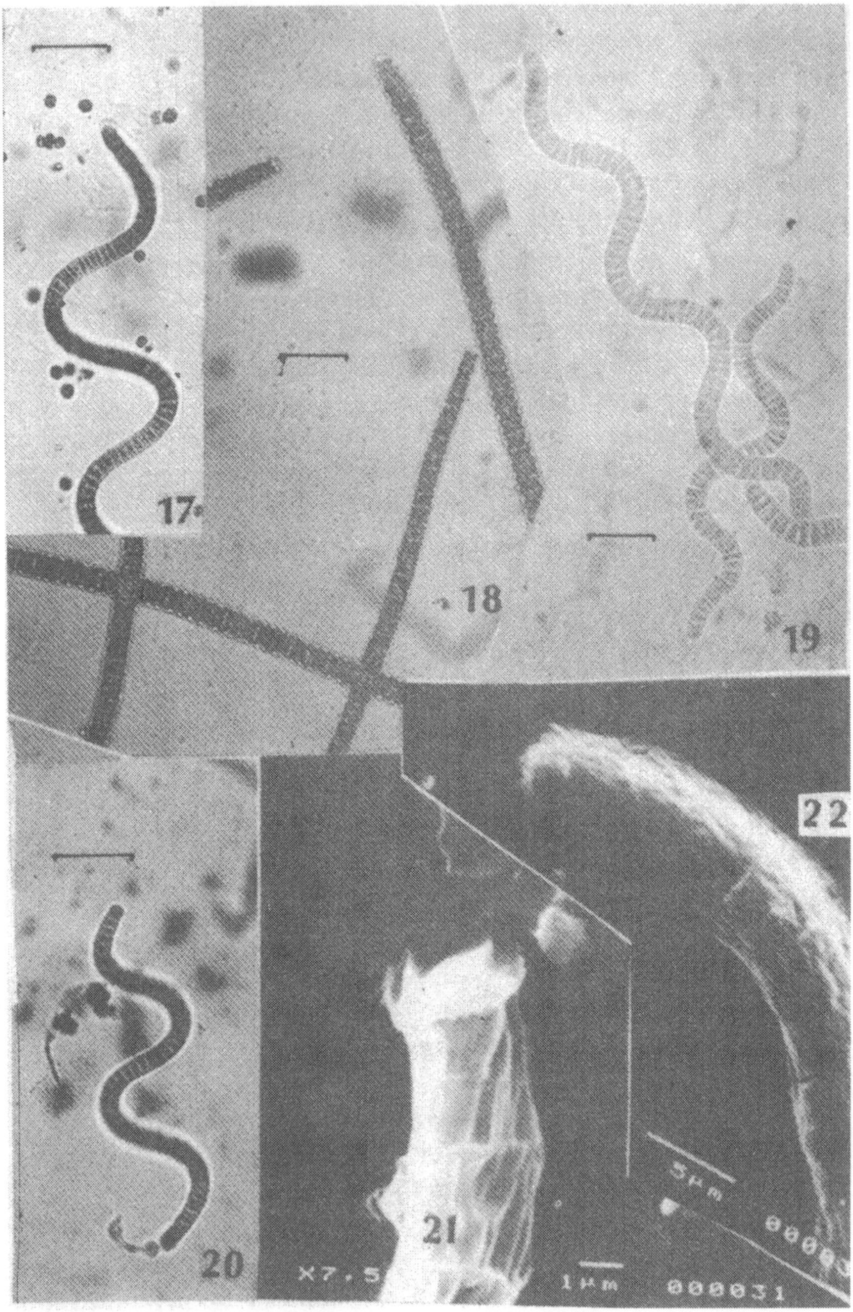


Figs. 17-22. *A. maxima*; 17: Pune sample, end view of calyptra, 19, 22 Peru sample; 19: regular loose coil; 22: SEM flatly conical calyptra; 18: Peru isolate, SAG 257-80, straight condition with calyptra; 20, 21: Pollachi sample, regular coil & SEM of calyptrate end cell.

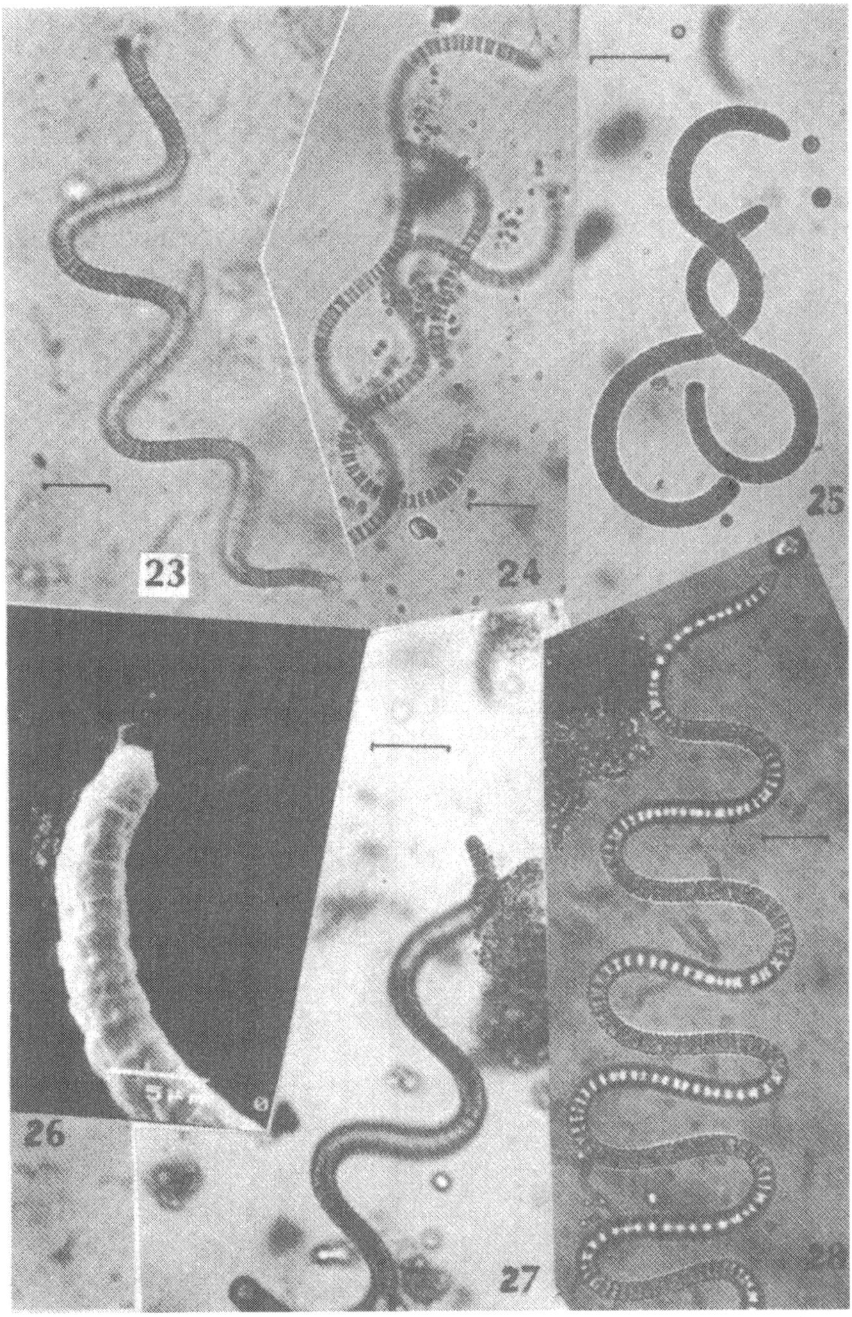


Figs. 23-28. *A. massartii*; 23, 24, 26: Allahabad, Lucknow samples, showing loose coils and SEM showing hood-like calyptra; 25: Pune sample, loose coils with calyptra; 27: Punjab sample, simple end cell; 28: Type (?) *A. massartii* var. *indica* Desik. showing regular coils with a suggestion of a calyptra.

The large number of strains from different institutions in India, such as CFTRI, Mysore (Courtesy Dr. L.V. Venkataraman), NEERI, Nagpur (Courtesy Dr. Chowdary). IARI, New Delhi (Courtesy, Dr. B.D. Kaushik), University of Rajasthan (Courtesy, Dr. P. Srivastava) and Berhampur University (Courtesy, Dr. S.B. Padhi), a strains from NIES, Japan (received from BHU, Courtesy, Dr. H.D. Kumar) and 3 strains from USA, 2 Californian isolates from UTEX collection and 1 Chinese isolate from CCMP collection, USA (Courtesy, Dr. Andersen) are all considered to belong to this group (Figs. 29-35). The figures of the type material from the British Museum studied by Komàrek and Lund (1990) show a slight dilation (subcapitate condition) of the end cells, as observed by us also in the isotype material from Belgium received from Dr. P. Compére.

The sample from the college pond, Khargone (Courtesy, Dr. S.K. Mahajan) showing very narrow trichomes with flatly conical calyptra on non-attenuate, non-capitate end cells is probably very distinct and may deserve a special status (Figs. 39, 40).

The type material of A. *platensis* var. *californica* Gardner from the Herbarium of University of California is retained as a distinct variety in view of the much broader trichomes.

A description is given below for the validation of the taxon:

A. *platensis* var. *californica* (= *Spirulina platensis* var. *californica* Gardner Ms).

Trichomes freely suspended (planktic), regularly coiled, with 5 to 12 turns, end coils sometimes slightly narrower, coils (22) 33-44 µm broad, 33-77 µm distant from each other, trichomes unconstricted, (5.6) 7.7 -9.8 µm broad, distinctly attenuated at the ends (20-35%), cells generally shorter than broad, 3.4-4.2 (5.6) µm long, end cell often slightly capitate, calyptra not evident.

***Holotype***

University of California Herbarium specimen, Accession No. 510691 (N.L. Gardner No. 7932), Figs. 36-38, 41; determined by Dr. Francis Drouet, March 1958, as *Arthrospira platensis* (Nord.) Gom. (see his handwritten note on herbarium specimen).

***Type Locality***

Lake Elsinor, Riverside County, collected by N.L. Gardner on 20th March 1936.

This material is now being described as a new variety as

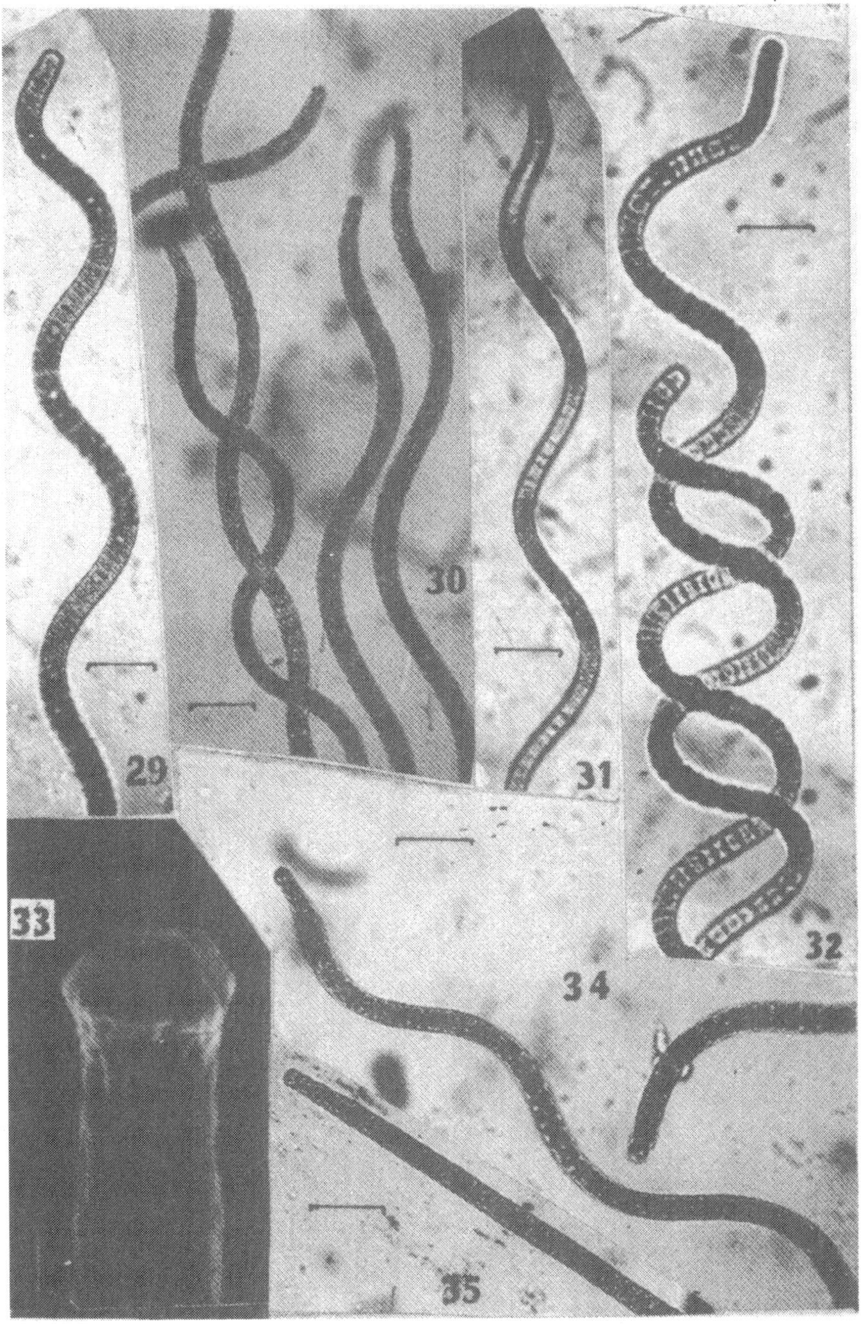


Figs. 29-35. *A. platensis*; 29: UTEX-strain; 30: CFTRI strain; 31: CCMP-strain, 32: NIES-strain, 34: Berhampur strain, 35: NEERI strain; all except 35 showing loose coils with degrees of sub-capitate to distinctly capitate condition of end cells. 33: SEM sub-capitate end cell of NEERI strain.

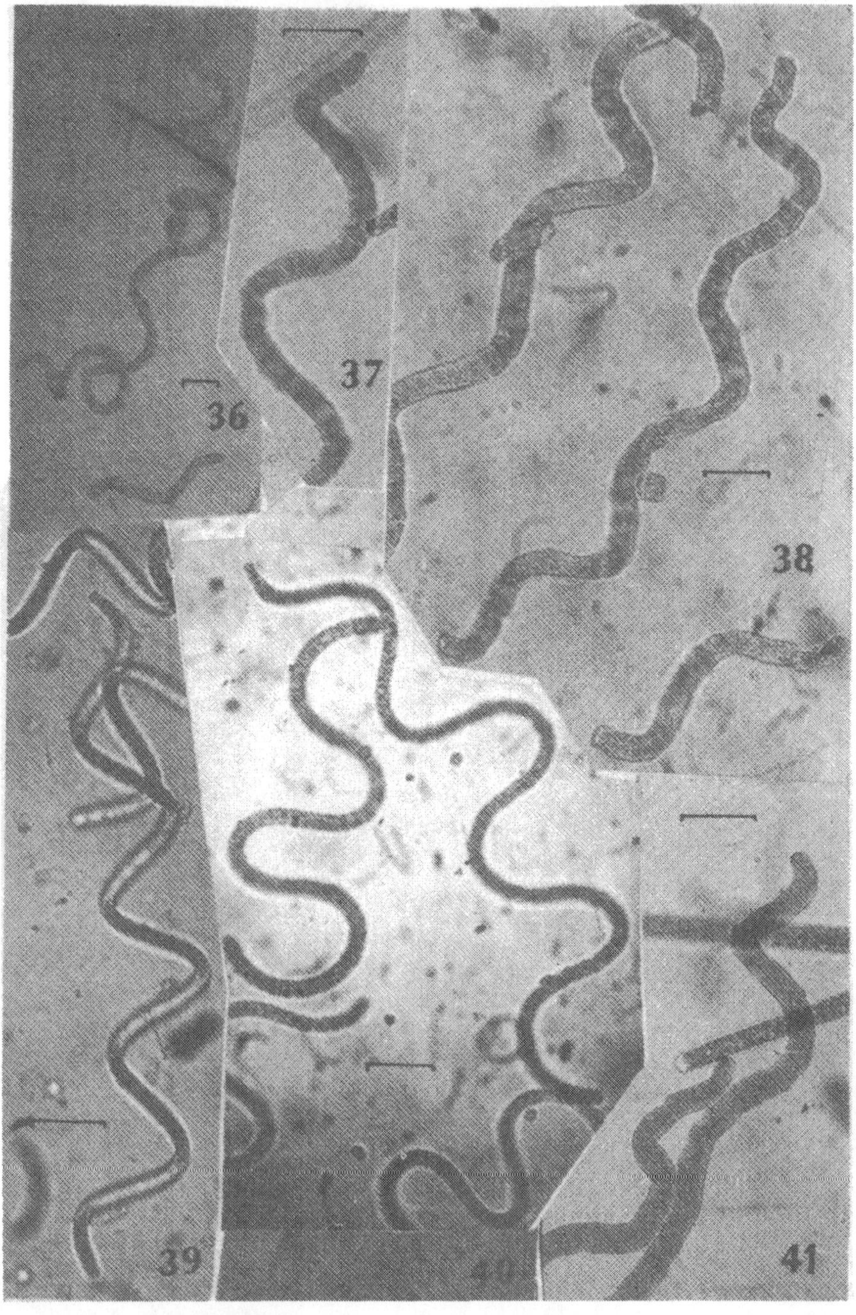


Figs.36-41. Herbarium specimen of type of *A. plantensis* var. *californica*, 36: regular loose coil; 37, 38, 41: trichomes with subcapitate end cells (see broken bit in 38). Figs. 39, 40: Pond sample from Khargone; showing trichomes with non-capitate, but calyptrate ends.

originally concluded by Gardner, as it appears different from the rest of the material of A. *platensis* studied here.

Desikachary and Jeeji Bai (1996) have recognised the following characters as useful in separating the taxa of *Arthrospira*:

1. Coiling pattern-regular (cylindrical) or spindle/dumb-bell shaped.
2. Degree of trichome attenuation towards the end
3. Average trichome width
4. Average length of cells in relation to breadth
5. End cell morphology-simple, capitate, calyptrate of both
6. Shape of calyptra, conical or hood-like.

## OTHER SPECIES OF *ARTHROSPIRA* KNOWN FROM INDIA AND THEIR AFFINITIES

Two of the species of *Arthrospira*, A. *khannae* Drouet et Strickl. and A. *gomontiana* Setch. have probably not been reported again. The isotype specimen of the former and the type specimen of the latter obtained from the herbarium of the University of California (Courtesy, Dr. P.C. Silva), have been examined and they appear to be distinct (Figs. 46-49). A. *gomontiana* var. *crassa* described by Gupta (1956) probably belongs to A. *khannae* as it has trichomes which are 4.0-4.5 μm broad with cells that are somewhat shorter than broad (1.5-3 μm long).

The type species, A. *jenneri*, has been observed and reported frequently from running and stagnant waters from different parts of the world. It can occur as free floating, attached or along with other algae. We have examined the fixed sample of a strain, UTEX 2179, received from Prof. R.C. Starr (gratefully acknowledged) which has been identified as A. *jenneri* by Dr. F. Drouet. As per the notes provided, the source is soil from a dried up pothole in a limestone outcrop near a rock quarry about a mile from a ferry across the Cumberland river in Tennessee, U.S.A. The strain is described here: Trichomes unconstricted, forming regular, lax to very tight coils, contents homogeneous or granular, coils 11.2-16.8 μm broad, touching each other when tightly coiled or 14-20 μm distant from each other when lax; trichomes 4.2-5.6 μm broad, cells 2.0-3.5 μm long, end cell broadly rounded, simple (Figs. 42-44).

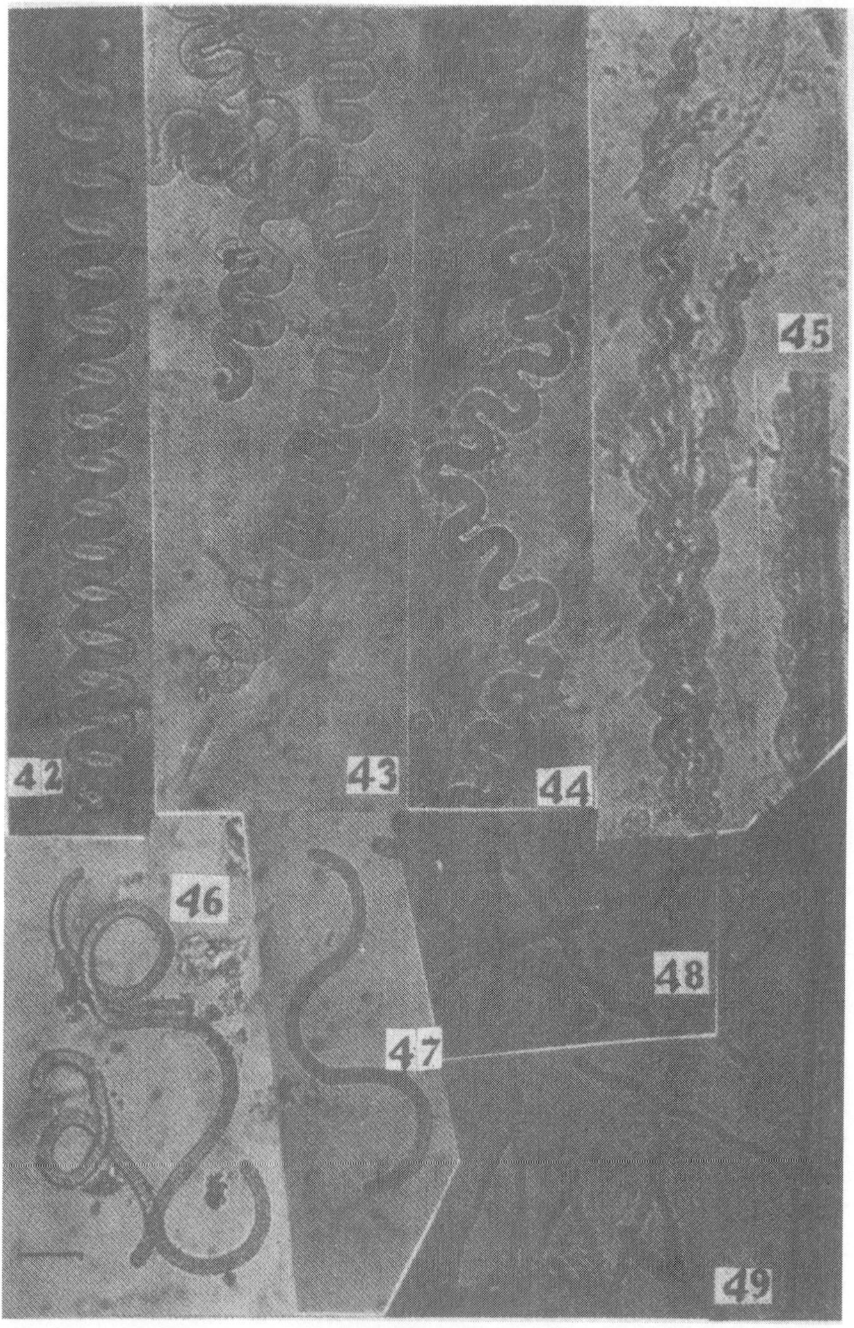

Figs.42-49. *A. jenneri*, UTEX-strain; 44: very close, tight coils, 43, 44: somewhat loose coils with rounded end cells; Fig. 45: *A. jenneri* fa. *purpurea*, herbarium specimen of type; 46, 47: *A. khannae*, herbarium specimen of isotype; 48, 49: *A. gomontiana* herbarium specimen of type.

A herbarium specimen, the isotype of A. *jenneri f. purpurea* Collins received from University of California, has also been examined. The notes on the specimen by F.S. Collins says: "Scattered filaments among A. *gomotiana* Setch. and *Oscillatoria* sp. forming a gelatinous floating mass, Silver Lake, Wilmington, Massachusetts, 4th October 1908. The filaments, even while fresh, were of a purple colour, and the form may be worthy of record as forma *purpurea* n.f." The description based on a direct examination of the herbarium specimen is as follows:

Trichomes forming very long regular, lax coils, coils are 7.0-9.8 (-11.2) µm broad, 16.8-22.4 (28.0) µm distant from each other, trichomes are 3.0-4.2 (5.0) µm broad, cells 1/3 as long as broad 2.0-2.8 µm long, end cell simple, broadly rounded (Fig. 45).

It is noteworthy that A. *jenneri* also forms very close to tight coils in laboratory cultures showing a lot of variation in this feature as seen in the UTEX strain (Figs. 42-44). As mentioned earlier A. *indica* also shows similar variation in degree of coiling (Figs. 1-4).

A. *spirulinoides* Ghose (1923) is an instance where the coils are described as touching each other (Geitler, 1932; Desikachary, 1959, Pl. 36 Fig. 9). Ghose's alga agrees well with the description of A. *jenneri* in all characters (5-6 µm broad and 4-12 µm long cells and 12-15 µm broad spirals) except in closeness of coils; the difference in cell length may be an error arising out of the coiling pattern of trichomes. Drouet (1968) also includes it as a synonym of A. *jenneri* Ghose (1919) has also described A. *jenneri* from Lahore with 4-6 µ m broad trichomes with 10-16 µm broad spirals which are 8 µm distant from each other A. *jenneri* has been described from different freshwater habitats as distinct varieties and forms (see Elenkin, 1949) and appears therefore to be cosmopolitan in occurrence with considerable morphological variability.

A. *balkrishnanii* is probably also synonym of A. *jenneri*, as it differs from it only in the somewhat narrower trichomes (see also notes in Desikachary, 1959, p. 192) and lax and wavy spirals.

A. *spirulinoides f. tenuis* (Singh, R.N.) Desik. is probably a closely coiled form of A. *tenuis* Bruhl et Biswas. as their dimensions agree in all respects.

## SUMMARY OF THE RESULTS PRESENTED

This paper has sought to bring together all available information on the historical background of *Spirulina* and *Arthrospira*. The different species of *Arthrospira* known so far and their affinities have been discussed. The detailed studies carried out earlier on different strains and natural samples of *Arthrospira* resulting in their grouping into four distinct entities and their assignment to one new and 3 existing species have been briefly dealt with. Suitable illustrations have been provided wherever necessary, A new variety *californica* of A. *platensis* has been validated.

## ACKNOWLEDGEMENTS

This study represents the culmination of a series of papers on morphological variation and taxonomy of the genus *Arthrospira* supported by the good office of Dr. C.V. Seshadri, the former director of MCRC, for which the author is very grateful. Special thanks are also due to Prof. T.V. Desikachary, University Botany Laboratory, Madras for going through the manuscript critically and making valuable suggestions. The author thanks all the persons named in the text for sending strains and samples studied here.

## REFERENCES

Anagnostidis, K. and Komàrek, J. 1988. Modern approach to the classification system of Cyanophytes, 3-Oscillatoriales. *Algolog. Studies*. **50-53**: 327-472.

Biswas, K. 1942. The role of common algal communities of the river Hoogly on the drinking water, Calcutta. 150th Ann. Vol. *Roy. Bot. Gdn.* pp. 189-206.

Bourrelly, P. 1970. Les algues d'eau douce. Vol. III: Lea algues bleues et rouges, les Englèniens, Peridiniens et Cryptomonadines. N. Boubèe et co., Paris. pp. 512.

Crow, W.B. 1927. The generic characters of *Arthrospira* and *Spirulina* Turp. *Trans. Amer. Micros. Soc.* **46**: 139-148.

De Toni, J. 1936. Not. *Nomencl. Alg.* **3**: 2.

De Toni, J. 1938. Diagn. Alg. Nov. I, Myxophyceae, Cent. IV, Brixiae, Typis Morcellanis. pp. 15.

Deore, L.T. 1992. On the occurrence of *Spirulina* Turp. em. Gardner (Family Oscilloriaceae) from Dhule, north Maharashtra. In Seshadri, C.V. & Jeeji Bai, N. (Eds.) *Spirulina* ETTA National Symposium, MCRC, Madras. pp. 49-51.

Desikachary, T.V. 1959. Cyanophyta. ICAR, New Delhi. pp. 686.

Desikachary, T.V. and Jeeji Bai, N. 1996. Taxonomic Studies in *Spirulina* I. In Seshadri, C.V. & Jeeji Bai, N. (Eds) *Spirulina* ETTA National Symposium, MCRC, Madras. pp. 12-21.

Desikachary, T.V. and Jeeji Bai, N. 1996. Taxonomic Studies in *Spirulina* II. The identification of *Arthrospira* ("*Spirulina*") strains and natural samples of different geographical origins. *Algolog. Studies* (In press).

Dobell, C.C. 1912. Research on Spirochaets and related organisms. *Arch. Protistenk.* **26**: 117-240.

Drouet, F. 1942. Studies in Myxophyceae I. Field Mus. *Nat. Hist. Bot.* **20**(6): 125-141.

Drouet, F. 1968. Revision of the classification of the Oscillatoriaceae. Monogr. 15, *Acad. Nat. Sci. Philadelphia.* pp. 370.

Elenkin, A.A. 1949. Monographia Algarum Cyanophycearum aquidulcium et terrestrium in Finibus URSS Inventarum. *Pars specialis. Moskua.* **2**: 985-1908.

Figini, G.P. 1925. Osservazioni intorno al genere *Spirulina* Turp. *La Nuova Notarisia.* pp. 31-49.

Fott, B. and Karim, A.G.A. 1973. *Spirulina* plankton community in a lake in Jebel Marra, Sudan. *Arch. Protistenk.* **115**: 408-418.

Fremy, P. 1930. Les Myxophycèes de l'Afrique èquatoriale Francaise. *Arch. Bot. Caen.* 3, Mèm no. 2. pp. 508.

Gardner, N.L. 1917. New pacific coast marine algae I. *Univ. Calif. Publ. Botany.* **6**(14): 377-416.

Gardner, N.L. 1918. New pacific coast marine algae III. *Univ. Calif. Pub. Botany.* **6**(17): 455-486 pls. 38-41.

Geitler, L. 1925. Sýnoptische Darstellung der Cyanophyceen in morphologischer und systematischer Hinsicht. *Beih. Bot. Cbl.* II. **41**: 163-294.

Geitler, L. 1932. Cyanophyceae. In: Rabenhorst's Kryptogameflora von Deutschland, Osterreich und der Schweiz. 14 Akad. Verlag. Leipzig. pp. 1196.

Geitler, L. 1942. Schizophyta (Klasse Schizophyceae). In: Engler & Prantl. Nat. Pflanzenfam. 1b. Berlin. pp. 232.

Ghose, S.L. 1919. The Myxophyceae of Lahore. *J. Ind. Bot. Soc.* **1**(1): 8-13.

Ghose, S.L. 1923. A systematic and ecological account of collection of blue-green algae from Lahore and Simla. *J. Linn. Soc. Lond. Bot.* **46**: 333-346.

Gomont, M. 1892. Monographie des Oscillariees. *Ann. Sci. Natur.* 7 ser. Botanique, Vol. 15 & 16, Paris. (Reprint, 1972). pp. 246.

Guglielmi, G. and Cohen-Bazire, G. 1982. Structure et distribution des pores et des perforations de l'envelope de peptidoglycane chez quelques Cyanobactèries. *Protistologica.* **18**: 151-165.

Guglielmi, G., Rippka, R. and Tandeau de Marsae, N. 1993. Main properties that justify the different taxonomic position of *Spirulina* spp. and *Arthrospira* spp. among Cyanobacteria. In: *Spirulina*, Algae of Life. *Bull. Inst. Oceanogr.* Monaco, Special No. **12**: 13-23.

Gupta, A.B. 1956. A contribution to the algal flora of the Allahabad district. *J. Res.* D.A.V. College, Kanpur. **3**(1): 76-81.

Hassall, A.H. 1845. A History of the British Freshwater *Algae*. pl. 75, Fig. 5. **1**: 277.

Hindàk, F. 1985. Morphology of trichomes in Spirulina fusiformis Vornonichin from lake Bogoria, Kenya. *Algolog. Studies.* **38/39**: 201-218.

Hindàk F. 1988. Contribution to the taxonomy of some Cyanophyte genera. *Preslia, Praha.* **60**: 289-308.

Hoffmann, L. 1991. Marine Cyanophyceae of Papua New Guinea I. The genera *Arthrospira* and *Spirulina*. *Nova Hedwigia.* **52**: 141-148.

Holmgren, R.P., Hostetter, H.P. and Scholes, V.E. 1971. Ultrastructural observations of cross walls in the blue-green alga *Spirulina major. J. Phycol.* **7**(4): 309-311.

Jeeji Bai, N. 1985. Competitive exclusion or morphological transformation? A case study with *Spirulina fusiformis*. *Algolog. Studies.* **38/39**: 191-199.

Jeeji Bai, N. and Seshadri, C.V. 1980. On coiling and uncoiling of trichomes in the genus *Spirulina*. *Algolog. Studies.* **20**: 32-47.

Kamat, N.D. 1963. The algae of Kolhapur, India. *Hydrobiologia.* **22**: 209-305.

Komàrek, J. 1956. "Some interesting blue-green algae from Bulgarian coast of Black sea near Burgas". Russian paper. *Univ. Carol. Biol.* **2**(1): 91-123.

Komàrek, J. and Lund, J.W.G. 1990. What is "*Spirulina platensis*" in fact? *Algolog. Studies.* **58**: 1-13.

Kützing, F. 1849. Tabulae Phycologicae I, tab. 37, Fig. XI. p. 26.

Lewin, R.A. 1980. Uncoiled variants of *Spirulina platensis* (Cyanophyceae: Oscillatoriaceae). *Algolog. Studies.* **26**: 48-52.

Nelissen, B., Wilmotte, A., Neefs, J.M. and Dewachter, R. 1994. Phylogenetic relationships among filamentous helical cyanobacteria investigated on the basis of 16S ribosomal RNA gene sequence analysis. *Syst. Appl. Microbiol.* **17**: 206-210.

Rich, F. 1931. Notes on *Arthrospira platensis*. *Rev. Algol.* **6**: 75-79.

Rippka, R., Deruelles, J., Waterbury, J.B., Herdman, M. and Stanier, R.Y. 1979. Generic assignments, strain histories and properties of pure cultures of cyanobacteria. *J. Gen. Microbiol.* **111**: 1-61.

Schmid, G. 1920. Uber die vermeintliche Einzelligkeit der Spirulinen. *Ber. Deutsch. Bot. Ges.* **27**: 473-476.

Singh, R.N. 1939. The Myxophyceae of the United Provinces, India, IV. *Proc. Ind. Acad. Sci. B.* **9**: 63-68.

Stizenberger, E. 1854. *Spirulina* und *Arthrospira. Hedwigia.* **1**: 32.

Stizenberger, E. 1854. *Spirulina* and *Arthrospira. Hedwigia* No. tab V, figs. 3 & 4. **7**: 33.

Turpin, P.J.F. 1827. Dictionnaire d'histoire naturelle de Levrault, pl. Oscl. fig. 3. **50**: 309.

Vasishta, P.C. 1962. Three colourless Cyanophyceae from Hoshiarpur. *J. Ind. Bot. Soc.* **41**(1): 99-103.

Vasishta, P.C. 1963. Five new Cyanophyceae from Hoshiarpur. *J. Ind. Bot. Soc.* **42**(4): 574-582.

Vauk, V. 1915. Dvije nove morske alge iz Hrvatskog Primorja. *Prirod istraz Hru. Slav.* **6**: 14-15.

Woronichin, N.N. 1934. Biologie des bassins mineral des steppes Kouloundine de l'Acad. Sci. d'URSS 1931-1933, I. Leningrad. *Trav. Soviet serv. Etud. Ressour. natur. Sèr. Sibèrie.* **8**: 182-183.

Zuelzer, M. 1912. Uber *Spirochaeta plicalis* Ehr. und deren Verwandschaftsbeziehungen. *Arch. Protistenk.* **24**: 1-59.

# Section - 6

Recent Trends in Algal Taxonomy
Vol. I. Taxonomical Issues
Editors : Vidyavati and A.K. Mahato
APC Publications Pvt. Ltd., New Delhi, 1999, pp. 275-285

# *Marine Algal Taxonomy in India*

N. Kaliaperumal
Regional Centre of Central Marine Fisheries Research Institute
Marine Fisheries-623 520, Tamil Nadu

## INTRODUCTION

Marine algae or seaweeds are primitive group of plants without any true root, stem and leaves. They are non-flowering plants and belong to the division Thallophyta under the plant kingdom. They occur in the intertidal region, shallow and deep waters of the sea upto 150 m depth and also in estuaries and backwaters. Marine algae grow on rocks, dead corals, stones, pebbles, other suitable substrates and also on other plants as epiphytes. They belong to four groups of algae namely Chlorophyta, Phaeophyta, Rhodophyta and Cyanophyta based on the type of pigments present in them along with morphological and anatomical characters. The potential areas in 6100 km coastline of India for luxuriant growth of several species of green, brown and red algae are the southern coast of Tamil Nadu from Mandapam to Kanyakumari covering 21 islands in Gulf of Mannar, Gujarat coast, Lakshadweep and Andaman-Nicobar Islands. The other places in the east and west coast where rich seaweed beds occur are Bombay, Ratnagiri, Goa, Karwar, Varkala, Vizhinjam, Pulieat and Chilka Lake.

The important and commonly occurring agar yielding seaweeds in different localities of Indian coast are species of *Gelidiella, Gracilaria, Gelidium* and *Pterocladia.* Among these red algae only *Gelidiella acerosa, Gracilaria edulis, G. corticata* var. *corticata, G. foliifera* and *G. verrucosa* are available in exploitable quantities. Species of *Sargassum, Turbinaria, Cysteseira, Hormophysa, Spatoglossum, Rosenvingea* and *Chnoospora* are the important algin yielding seaweeds of Indian waters. Among these brown algae *Sargassum, Turbinaria* and *Hormophysa* grow in harvestable quantities. The carrageenan yielding red alga *Hypnea* also occur in exploitable quantities at various parts of the coastline (Kaliaperumal *et al.*, 1995).

Marine algae are one of the commercially important marine living and renewable resources. They contain different vitamins, minerabls, trace elements, protein, iodine and bioactive substances. They are the only source for the production of phytochemicals like agar, carrageenan and algin. Agar is extracted from red algae such as *Gelidiella, Gracilaria, Gelidium* and *Pterocladia.* Some other red algae viz. *Eucheuma, Chondrus, Hypnea* and *Gigartina* are used for the manufacture of carrageenan. Algin is extracted from species of *Sargassum, Turbinaria, Laminaria, Undaria, Macrocystis* and *Ascophyllum.* These phycocolloids are used as gelling, stabilizing and thickening agents in food, confectionary, pharmaceutical, dairy, textiles, paper, paint, varnish industries etc. Other chemical products such as mannitol, iodine, laminarin and fucoidin are also obtained from marine algae. Many protein rich seaweeds such as *Ulva, Enteromorpha, Caulerpa, Codium, Monostroma* (green algae); *Sargassium, Hydroclathrus, Laminaria, Undaria* and *Macrocystis* (brown algae); *Porphyra, Gracilaria, Eucheuma, Laurencia* and *Acanthophora* (red algae) are used as human food in countries like Japan, China, Korea, Malaysia, Philippines and other south-east Asian countries in the form of soup, salad, curry etc. Jelly, jam, chocolate, pickle and water can also be prepared from certain seaweeds. The food value of seaweeds depend on the minerals, trace elements, protein and vitamins present in them. Seaweeds have all essential amino acids needed in the human diet which are not available in other vegetable food materials.

Since seaweeds are cheap source of minerals and trace elements, meal can be prepared from grinding the cleaned and washed seaweeds. It can be given as supplements to the daily rations of cattle, poultry and other farm animals. It can also be mixed with fish meal and used as poultry feed. Marine algae are also used in different parts of the world as fertilizer for various land crops. In India, freshly collected and castashore seaweeds and seagrasses are used as manure for coconut plantation eithr directly or in the form of compost in the coastal areas of Tamil Nadu and Kerala. Seaweed manure has been found superior to farmyard manure. The high amount of water soluble potash, other, minerals and trace elements present in seaweeds alter the nature of soil and improve the moisture retaining capacity. The liquid seaweed fertilizer obtained from seaweed extract can be used as foliar spray for inducing faster growth and yield in leafy and fleshy vegetables, fruits, orchards and horticultural plants. The trace elements and growth hormone cytokinin present in the liquid seaweed fertiliser act as growth promoters and increase the yield by 20 to 30%. It gives successful results on potatoes, cauliflower, cabbage, brinjal, lady's finger, chillies, grapes etc. There are certain medicinal properties for seaweeds. Seaweeds rich in iodine such as *Asparagopsis taxiformis* and *Sarcenema* spp. can be used for controlling goitre disease caused by enlargement of thyroid gland (Chennubhotla & Kaliaperumal, 1987; Chennubhotla *et al.*, 1987 & 1988).

**Taxonomical Works on Marine Algae of India**

The research works on marine algae were started in India in 1930. In view of the various utilities of marine algae and establishment of several agar and algin manufacturing industries in our country, a great deal of works have been done on various aspects such as taxonomy, ecology, biology, chemistry, resources assessment, culture and utilisation of seaweeds for the past more than six decades. The taxonomical works carried out in marine algae of India are dealt within this paper.

Our knowledge on marine algal flora of India is mainly due to taxonomical works carried out by Boergesen during 1930 to 1938. Based on the collections of M.O.P. Iyengar and his own collections, Boergesen has published a series of papers on the green, brown and red algae of the northern parts of west coast of India (Boergesen, 1930, 1931, 1932a, b; 1933a & b, 1934a & b and 1935) and brown and red algae of south India (Boergesen, 1937a & b and 1938a & b). The algal flora of Krusadai Island was published by Iyengar (1927).

After the valuable contribution of Boergesen much work has been done on the taxonomy of Indian marine algae during the last five decades. A general review of the marine algae of the west coast was published by Biswas (1945). Srinivasan (1946) studied the marine algal flora of Mahabalipuram. Parija and Parija (1946) studied the algal vegetation of Chilka Lake. Chacko *et al.* (1955) listed 103 species of marine algae from Krusadai Island.

Studies were made on the flora occurring on the pearl and chank beds of Tuticorin during 1958 to 1959. Totally 80 species of algae (17 green algae, 16 brown algae and 47 red algae) and one seagrass were reported (Varma, 1960). Srinivasan (1960a, b & c) has given a detailed account of the marine algae of the east and west coast of India based on the then available reports. According to the etimates given by him, 162 genera and 413 species of marine algae were known from Indian waters. The algal resources of Pamban area was studied by Varma and Krishna Rao (1962). Taylor (1964) has described the Indian species of *Turbinaria*. The checklist of Indian marine algae has been published by Dixit (1964, 1968 and 1970) and Srinivasan (1965). The list published by Dixit consists of 411 species which also includes records from Pakistan and Sri Lanka. Misra (1966) has prepared a monograph of brown algae occurring along the Indian coast. The Phaeophyceae of the west coast of India was also reported (Misra, 1967). Srinivasan (1966) has published an account of Indian species of *Sargassum*. The marine algae occurring in the Gulf of Mannar of Tuticorin was studied by Mahadevan and Nagappan Nayar (1967).

A total number of 180 algae (56 green algae, 33 brown algae, 90 red algae and 1 blue-green alga) were reported from mandapam area (Umamaheswara Rao, 1969). The composition of marine algae of Gopnath has been published by Sreenivasa Rao and Kale (1963 and 1969) and that of Gulf of Kutch by Gopalakrishnan (1969). The species of *Ulva* from Indian waters is published by Krishnamurthy and Joshi (1969). The icones of Indian marine algae was published by Srinivasan (1969 and 1973). An annoted list of 80 algae growing along the Visakhapatnam coast had been given by Umamaheswara Rao and Sreeramulu (1970). Gopalakrishnan (1970) has reported 64 species of algae from the collection grounds of Dwarka, Okha, Adatra, Hanumandandi and Balapur from Okah coast. A systematic account of 10 taxa of Indian Gelidiales has been given by Sreenivasa Rao (1970). A check list of 520 species of Indian marine algae has been published by Krishnamurthy and Joshi (1970). The addition to the algal flora of Gulf of Mannar and Palk Bay from Mandapam area was given by Umamaheswara Rao (1970).

The distribution of marine organisms on the Jaleshwar shore, Veraval was studied by Subbaramaiah (1971). Joshi and Krishnamurthy (1971) have listed 13 species of *Enteromorpha* from India. Kale (1966, 1967 & 1968) while studying algal flora of Saurashtra salt works described *Enteromorpha* species. Sixty one species of algae (28 green algae, 13 brown algae and 20 red algae) and 4 species of seagrasses were recorded in the coralreef areas of Gulf of Mannar and Palk Bay during 1968 (Umamaheswara Rao, 1972a). The description of 17 species and 2 varieties of *Gracilaria* and 2 species of *Gracilariopsis* and also their habitats and distribution in India are given by Umamaheswara Rao (1972b). Seaweed resources of Goa coast was studied by Untawale and Dhargalkar (1975). The Central Marine Fisheries Research Institute carried out detailed survey on the marine algal distribution and resources along the Tamil Nadu coast for a period of five years during 1971-76 in collaboration with the Central Salt and Marine Chemicals Research Institute, Bhavanagar and Department of Fisheries, Government of Tamil Nadu. The area covered was from Athankarai to Rameswaram in the Palk Bay (45 km distance) and from Mandapam to Colachel (13 km distance) and adjoining 21 islands in the Gulf of Mannar from high water mark to 4 m depth (Anon, 1978). The Central Salt & Marine Chemicals Research Institute conducted survey on the marine alga distribution of Pamban (Subbaramaiah *et al.*, 1977).

The marine algal resources and their distribution of Maharashtra coast was studied during 1976 (Chauhan, 1978). A total number of 36 algal species (11 green algae, 11 brown algae and 14 red algae) have been recorded. Agadi and Untawale (1978) have reported 50 algal species from the intertidal areas along the Goa coast. The marine algal flora of Goa coast was studied by Agadi (1983) and Proto-Nove region in Tamil Nadu by Kannan and Krishnamurthy (1978). The seaweed resources of Andhra Pradesh was studied by Umamaheswara Rao (1978). Ragothaman (1979) and Ragothaman and Haranath Babu (1992) studied the algae growing at Devka, Golvad, and Daman. Qasim and Wafar (1979) have recorded 72 seaweeds from Ratnagiri, Malwan and Redi along the west coast. Marine algal resources of Maharashtra coast was given by Untawale *et al.* (1979). An account of 46 species of marine algae occurring at Tiruchendur in Tamil Nadu coast was given by Krishnamurthy (1980). Sarma and Khan (1980) have published a check list of Indian fresh water and marine algae.

A taxonomic account of Indian Ectocarpales and Ralfsiales has been given by Balakrishnan and Kinker (1981). An annotated list of 44 species of algae collected from ten localities along the southern Kerala coast including Asthakmudi Lake and Kanyakumari was published by Balakrishnan Nair *et al.* (1982). The Central Marine Fisheries Research Institute conducted survey of marine algal distribution and their resources in Andaman-Nicobar Islands 1978 and altogether 55 species of algae (16 green algae, 17 brown algae and 22 red algae) were recorded (Gopinathan and Panigrahy, 1983). The marine algae of southwest coast of India were reported by Sobha and Nair (1983). Untawale *et al.* (1983) enumerated 624 species of marine algae and their distribution along the maritime states of India. Survey of marine algal resources and their distribution in Andhra Pradesh was made by the Central Salt & Marine Chemicals Research Institute (Subba Rao *et al.*, 1983 and Anon, 1984). Ninety eight species of algae (30 green algae, 23 brown algae, 44 red algae and 1 blue-green alga) were recorded from six localities along the south

Tamil Nadu coast namely Tuticorin, Mandpad, Tiruchendur, Idinthakarai, Kovalam and Muttam (Kaliaperumal and Pandian, 1984). The species of *Porphyra* from Indian region were reported by Krishnamurthy and Baluswami (1984). The list of Indian marine algae was published by Balakrishnan Nair *et al.* (1982 & 1986). The Central Marine Fisheries Research Institute and Central Salt & Marine Chemicals Research Institute have jointly surveyed the seaweed distribution and resources in deep waters (5 to 22 m depth) of Tamil Nadu from Dhanushkodi to Kanyakumari during 1986-1991. A total number of 100 algal species—20 green algae, 18 brown algae, 61 red algae and 1 blue-green alga occurred in deep waters (Anon, 1988, 1989, 1993a & b; Kaliaperumal, 1991, Kaliaperumal *et al.*, 1995 & 1996).

The key for identification for economically important seaweeds was given by Umamaheswara Rao (1987a). Survey of marine algae occurring at Pulicat, Kovalam, Muttukadu and Mahabalipuram along Madras coast was made during 1987 (Chennubhotla and Kaliaperumal, 1987). A total of 22 species (6 green algae, 3 brown algae, 10 red algae and 3 blue-green algae) were reported. The algae of Indian estuaries was given by Umamaheswara Rao (1987b). The resources survey and distribution of marine algae in Kerala coast was conducted by the Central Marine Fisheries Research Institute during 1986 & 1987. Totally 34 species of algae (11 green algae, 8 brown algae and 15 red algae) were recorded (Chennubhotla et al., 1988). Survey of seaweed and seagrasses resources and their distribution in Lakshadweep was carried out by the Central Marine Fisheries Research Institute in 1987. Altogether 114 species of algae (43 green algae, 14 brown algae, 54 red algae and 3 blue-green algae) and 6 species of seagrasses were recorded (Kaliaperumal *et al.*, 1989). The taxonomical study of *Caulerpa* from India was made by Duraiswamy (1989, 1990a & b).

Studies on distribution and standing crop of algae at Muthupet estuary in Tamil Nadu was made by Balakrishnan *et al.* (1992). The distribution and seasonal changes in marine algal flora was investigated for a period of one year during 1983-84 from the intertidal and subtidal region upto 1 m depth at seven localities along Mandapam coast namely Rameswaram, Pamban, Krusadai Island, Thonithurai, Seeniappa Darga, Pudumadam and Kilakkarai. Totally 104 algae (37 green algae, 20 brown algae, 45 red algae and 2 blue-green algae) were recorded from these places with maximum number of 77 species at Krusadai Island and minimum number of 35 species at Rameswaram (Kalimuthu *et al.*, 1992). Resources survey of algae and seagrasses in 63 estuaries and backwaters existing from Madras to Athankarai in Tamil Nadu and Pondichery was made during 1988-89 (Kalimuthu *et al.*, 1995). Among these water bodies only 44 supported vegetation. Totally 36 species of algae (13 green algae, 6 brown alogae, 14 red algae and 3 blue-green algae) and 5 species of seagrasses were recorded from these estuaries.

Taxonomical studies on different genera and orders of marine algae and reports on stray marine algal collections from several localities of Indian coast were made by various workers. So far about 220 genera and 740 species of marine algae have been reported from Indian waters.

## CONCLUSION

Our knowledge on the taxonomy of marine algae occurring in Andaman-Nicobar Islands, deep waters estuaries and back waters is still incomplete. Hence much work in this direction has to be undertaken. The Information on the local marine algal flora of each seaweed growing region is necessary for proper utilisation of various economically important seaweeds by the Indian seaweed industries. Almost all countries have regional algal flora. Such regional floras are not available at present in India. It is necessary to initiate efforts to bring out well illustrated and documented flora for each region in the country.

## REFERENCES

Agadi, V.V. 1983. Intertidal ecology of marine algae along Anjuna coast, Goa. *Seaweed Res. Utiln.*, **6**(1): 27-30.

Agadi, V.V. and Untawale, A.G. 1978. Marine algal flora of Goa coast, *Seaweed Res. Utiln.*, **3**(1 & 2): 56-70.

Anon, 1978. A report on survey of marine algal resources of Tamil Nadu. 1971-76. CSMCRI, Bhavnagar, pp. 1-137.

Anon, 1984. A report on the survey of marine algal resources of Andhra Pradesh 1979-82. CSMCRI, Bhavnagar, pp. 1-8.

Anon, 1988. A report on seaweed resources of Tamil Nadu coast, India. II. Sector—Alantalai to Manapad and Vembar to Nallatanni Tivu, CSMCRI & CMFRI, pp. 1-30.

Anon, 1989. A report on the seaweed resource of the Tuticorin-Tiruchendur coast, Tamil Nadu, India. CMFRI & CSMCRI, pp. 1-19.

Anon, 1993a. A report on the seaweed resources of Tamil Nadu coast, India. III Sector. Valinokkam-Kilakkarai and Manapad-Kanyakumari. 1989-90. CMFRI & CSMCRI. pp. 1-23.

Anon, 1993b. A report on the seaweed resources of Tamil Nadu coast, India. IV Sector. Kilakkarai-Rameswaram Island (Dhanushkodi) 1990-91. CSMCRI & CMFRI, pp. 1-20.

Balakrishnan, M.S. and Kinker, V.N. 1981. A taxonomic account of Indian Ectocarpales and Ralfsiales. *Seaweed Res. Utiln.*, **4**(2): 1-57.

Balakrishnan, S., Ravichandran, M. and Kaliaperumal, N. 1992. Studies on distribution and standing crop of algae at Muthupet estuary, Tamil Nadu. *Seaweed Res. Utiln.*, **15**(1 & 2): 63-68.

Balakrishnan Nair, N., Sobha, V. and Arunachalam,M. 1982. Algae from southern Kerala coast. *Indian J. Mar. Sci.*, **11**(3): 266-269.

Balakrishnan Nair, N., Sobha, V., Chandran, R., Rathiammal, M., Miranda, P.I., Maya, S. and Suryanarayana, H. 1986. Algal resources of Kerala coast II. An up-to-date list of Indian marine algae. *Aguatic Biology*, **6**: 25-52.

Biswas, K. 1945. A general review of the marine algae of the Wester coast of India. *J. Bombay Nat. Hist. Soc.* **45**(4): 515-530.

Boergesen, F. 1930. Some Indian green and brown algae especially from the shores of Presidency of Bombay-I. *J. Indian Bot. Soc.*, **9**: 151-174.

Boergesen, F. 1931. Some Indian Rhodophyceae especially from the Presidency of Bombay-I. *Kew Bull*, No. **1**: 1-24.

Boergesen, F. 1932a. Some Indian Rhodophyceae especially from the Presidency of Bombay-II. *Kew Bull.* **3**: 113-134.

Boergesen, F. 1932b. Some Indian green and brown algae especially from the Presidency of Bombay. *J. Indian Bot. Soc.*, **11**: 51-70.

Boergesen, F. 1933a. Some Indian Rhodophyceae especially from the Presidency of Bombay-III. *Kew Bull.* **3**: 113-142.

Boergesen, F. 1933b. Some Indian green and brown algae from the Presidency of Bombay. *J. Indian Bot. Soc.*, **12**: 1-16.

Boergesen, F. 1934a. Some Indian Rhodophyceae especially from the Presidency of Bombay-IV. *Kew Bull.*, **4**: 1-30.

Boergesen, F. 1934b. Some marine algae from the Northern part of the Arabian Sea with remarks on their geographical distribution. *Kgl. Dansk. Vidensk. Selskab. Biol. Meddel.*, **11**(6): 1-72.

Boergesen, F. 1935. A list of marine algae from Bombay. *Kgl. Dansk. Vidensk. Selskab. Biol. Meddel.*, **12**(2): 1-64.

Boergesen, F. 1937a. Contribution to a South Indian marine algal flora-I. *J. Indian Bot. Soc.*, **16**: 1-56.

Boergesen, F. 1937b. Contribution to a South Indian marine algal flora-II. *J. Indian Bot. Soc.*, **16**: 311-357.

Boergesen, F. 1938a. Contribution to a South Indian marine algal flora-III. *J. Indian Bot. Soc.*, **17**: 205-242.

Boergesen, F. 1938b. Two species of *Scinaia* from South India, *Botaniska Notiser. Lund.*, pp. 183-189.

Boergesen, F. 1946. Remarks on some *Codium* from the Arabian Sea. (M.O.P. Iyengar Commn. Vol.) *J. Indian Bot. Soc. Suppl.* **47**: 1-8.

Chacko, P.I., Mahadevan, S. and Ganesan, R. 1955. A guide to the field study of the fauna and flora of Krusadai Island, Gulf of Mannar. *Contrib. Mar. Biol. Sta., Krusadai Island*, **3**: 1-16.

Chauhan, V.D. 1978. Report on the survey of marine algae resources of Maharashtra coast. *Salt Res. Ind.*, **14**(1): 1-10.

Chennubhotla, V.S.K. and Kaliaperumal, N. 1987. Marine algal flora from selected centres along the Madras coast. *Mar. Fish. Infor. Sorv. T & E Ser.*, **72**: 19.

Chennubhotla, V.S.K., Kaliaperumal, N. and Kalimuthu, S. 1987. Common seaweed products. *CMFRI Bulletin*, **41**: 26-30.

Chennubhotla, V.S.K., Ramachandrudu, B.S., Kaladharan, P. and Dharmaraja, S.K. 1988. Seaweed resources of Kerala coast. *Aguatic Biology*, **7**: 69-74.

Dixit, S.C. 1964. Species list of Indian marine algae determined by Boergesen. *J. Univ. Bombay*, **32**(2-5): 1-23.

Dixit, S.C. 1968. Species list of Indian marine algae-II. J. *Univ. Bombay*, **36**(3-5): 9-24.

Dixit, S.C. 1970. Species list of Indian marine algae-III. *J. Univ. Bombay*, **39**(6): 99-130.

Duraiswamy, R. 1989. Studies on *Caulerpa* Lamouroux from India-I. *Seaweed Res. Utiln.*, **17**(2): 83-106.

Duraiswamy, R. 1990a. Studies on *Caulerpa* from India-II. *Seaweed Res. Utiln.*, **12**(1 & 2): 55-86.

Duraiswamy, R. 1990b. Studies on *Caulerpa* Lamouroux from India-III. *Seaweed Res. Utiln.*, **13**(1): 29-76.

Gopalakrishnan, P. 1969. Some marine algae from the Gulf of Kutch. *Phykos.*, **8**(1 & 2): 61-67.

Gopalakrishnan, P. 1970. Some observations on the shore ecology of the Okha coast. *J. Mar. Biol. Ass. India*, **12**: (1 & 2): 15-34.

Gopinathan, C.P. and Panigrahy, R. 1983. Seaweed resources. In Mariculture potential of Andaman-Nicobar Islands—An indicative survey. *CMFRI Bulletin*, **34**: 47-51.

Iyengar, M.O.P. 1927. Flora of Krusadai Island. *Bull. Madras Govt. Mus. N.S. Nat. Hist. Soc.*, **1**(1): 185-188.

Joshi, H.V. and Krishnamurthy, V. 1971. The species of *Enteromorpha* from India. *Bot. J. Linn. Soc.* **65**(1): 119-128.

Kale, S.R. 1966. On some green algae from the channels and reservoirs of Saurashtra Salt works, Porbandar. *Curr. Sci.* **35**(7): 180-181.

Kale, S.R. 1967. *Enteromorpha gujaratensis*, a new species from Gujarat, India. *Phykos*, **6**(1-2): 29-31.

Kale, S.R. 1968. A new record of *Enteromorpha linza* (L.) J. Ag. for the Indian Ocean. *Proc. 55th Indian Sci. Congress. Banaras* (Abstract).

Kaliaperumal, N. 1991. Seaweed resources in deep waters from Dhanushkodi to Kanyakumari in Tamil Nadu. *CMFRI News letter*, **52**: 5-6.

Kaliaperumal, N., Chennubhotla, V.S.K., Kalimuthus, S., Ramalingam, J.R., Krishna Pillai, S., Subbaramaiah, K., Rama Rao, K. and Subba Rao, P.V. 1995. Distribution of seaweeds of Kattapadu-Tiruchendur coast, Tamil Nadu. *Seaweed Res. Utiln.*, **17**: 183-193.

Kaliaperumal, N., Kaladharan, P. and Kalimuthu, S. 1989. Seaweed and seagrass resources. *CMFRI Bulletin*, **43**: 162-175.

Kaliaperumal, N., Kalimuthu, S., Muniyandi, K., Ramalingam, J.R., Krishna Pillai, S., Chennubhotla, V.S.K., Rajagopalan, M.S., Subba Rao, P.V., Rama Rao, K., Thomas, P.C., Zaidi, S.H. and Subbaramaiah, K. 1996. Distribution of marine algae and seagrasses of Valinokkam-Kilakkarai, Tamil Nadu coast, *Seaweed Res. Utiln.*, **18**(1 & 2): 73-82.

Kaliaperumal, N. and Pandian, G. 1984. Marine algal flora from some localities of southern Tamil Nadu coast. *J. Mar. Biol. Ass. India*, **26**(1 & 2): 159-164.

Kalimuthu, S., Kaliaperumal, N. and Ramalingam, J.R. 1992. Distribution and seasonal changes of marine algal flora from some localities around Mandapam. *Seaweed Res. Utiln.*, **15**(1 & 2): 119-126.

Kalimuthu, S., Kaliaperumal, N. and Ramalingam, J.R. 1995. Distribution of algae and seagrasses in the estuaries and backwaters of Tamil Nadu and Pondichery. *Seaweed Res. Utiln.*, **17**(1 & 2): 79-86.

Kannan, L. and Krishnamurthy, K. 1978. A survey of the algae of the Porto-Novo region (Coromandel Coast, Bay of Bengal). *Seaweed Res. Utiln.*, **3**(1 & 2): 1-4.

Krishnamurthy, V. 1980. The marine algae of Tiruchendur, South India. *Seaweed Res. Utiln.*, **4**(1): 49-58.

Krishnamurthy, V. and Baluswami, M. 1984. The species of *Porphyra* from the Indian region. *Seaweed Res. Utiln.*, **7**(1): 31-38.

Krishnamurthy, V. and Joshi, H.V. 1969. The species of *Ulva* from Indian waters. *Bot. J. Linn. Soc.*, **62**: 123-130.

Krishnamurthy, V. and Joshi, H.V. 1970. A check list of Indian marine algae. *CSMCRI Bhavnagar*, pp. 1-36.

Mahadevan, S. and Nagappan Nayar, K. 1967. Underwater ecological observations in the Gulf of Mannar of Tuticorin. *J. Mar. Biol. Ass. India*, **9**(1): 147-163.

Misra, J.N. 1966. Phaeophyceae in India. ICAR, New Delhi, pp. 1-203.

Misra, J.N. 1967. The Phaeophyceae of the west coast of India. Proc. Sem. Sea Salt and Plants, Bhavnagar, pp. 227-233.

Parija, P. and Parija, B. 1946. Algal succession on a rocky Island named Charai Guha in the Chilka Lake. *J. Indian Bot. Soc.* (M.O.P. Iyengar Comm. Vol.) pp. 375-379.

Qasim, S.Z. and Wafar, M.V.M. 1979. Occurrence of living corals at several places along the west coast of India. *Mahasagar*, **12**(1): 53-58.

Ragothaman, G. 1979. Littoral algal survey of south Gujarat coast (Devka, Golvad and Daman). *Proc. Int. Symp. Marine Algae of the Indian Ocean Region*, CSMCRI, Bhavnagar, India, (Abstract). p. 6.

Ragothaman, G. and Haranath Babu, M.V. 1992. Studies on the marine algae of Devka coast. *Seaweed Res. Utiln.*, **15**(1 & 2): 91-97.

Sarma, Y.S.R.K. and Khan, M. 1980. Algal taxonomy in India. Botanical records and monographs - 2. *Today and Tomorrow Printers and Publishers*, New Delhi. pp. 1-153.

Sobha, V. and Nair, N.B. 1983. Marine algae of South-west coast of India V. In *Marine Plants, Seaweed Research and Utilisation Association, Madras*, pp. 17-24.

Sreenivasa Rao, P. 1970. Systematics of Indian Gelidiales. *Phykos*. **9**: 63-78.

Sreenivasa Rao, P. and Kale, S.R. 1963. Marine algae from little known places of Gujarat coast-I. Algae from Gopnath. *Phykos*. **8**(1 & 2): 71-82.

Sreenivasa Rao, P. and Kale, S.R. 1969. Marine algae from the little known places of Gujarat coast-II. Gopnath. *Proc. 56th Ind. Sci. Congr. Association*, Bombay (Abstract).

Srinivasan, K.S. 1946. Ecology and seasonal succession of the marine algae of Mahabalipuram (Seven Pagodas) near Madras (M.O.P. Iyengar Commn. Vol). *J. Indian Bot. Soc.* (Suppl.) **47**: 267-278.

Srinivasan, K.S. 1960a. Distribution pattern of marine algae in Indian seas. *Proc. Symp. Algology*. ICAR, New Delhi. pp. 219-241.

Srinivasan, K.S. 1960b. On *Gastroclonium iyengarii* a new species from India. *J. Indian Bot. Soc.*, **39**: 156-161.

Srinivasan, K.S. 1960c. Observation on some drift algae at Mahabalipuram coast. *J. Bombay Nat. Hist. Soc.*, **57**(2): 458-461.

Srinivasan, K.S. 1965. Algarum Species ex Indian Oriundae. *Bull. Bot. Surv. India*, **67**(1-4): 188-266.

Srinivasan, K.S. 1966. Conspectus of *Sargassum* from Indian territorial waters. *Phykos*, **5**(1-2): 127-159.

Srinivasan, K.S. 1969. Phycologia indica (Icones of Indian marine algae). *B.S.I. Publ.*, Calcutta, pp. 1-52.

Srinivasan, K.S. 1973. Phycologia indica (Icones of Indian marine algae) *B.S.I. Publ.*, Calcutta. pp. 1-60.

Subbaramaiah, K. 1971. Distribution of marine organisms on the Jalleshwar shore, Veraval. *Seaweed Res. Utiln.*, **1**: 19-29.

Subbaramaiah, K., Nair, M.R.P. and Krishnamurthy, V. 1977. Distribution pattern of marine algae on the shores of Pamban. *Seaweed Res. Utiln.*, **2**(2): 74-77.

Subba Rao, P.V., Rama Rao, K., Nair, M.R.P. and Subbaramaiah, K. 1983. A systematic list of marine algae from Andhra Pradesh and their distribution. In *Marine Plants. Seaweed Research and Utilisation Association, Madras.* pp. 29-34.

Taylor, W.R. 1964. The genus *Turbinaria* in eastern seas. *Jour. Linn. Soc.* London (Botany) **58**: 475-490.

Umamaheswara Rao, M. 1969. Catalogue of marine algae in the reference collection of the Central Marine Fisheries Research Institute. *Bull. Cent. Mar. Fish. Res. Inst.*, **9**: 37-48.

Umamaheswara Rao, M. 1970. The economic seaweeds of India. *Bull. Cent. Mar. Fish. Res. Inst.*, **20**: 1-68.

Umamaheswara Rao, M. 1972a. Coral reef flora of Gulf of Mannar and Palk Bay. *Proc. Symp. Corals and Coral Reefs.* pp. 217-230.

Umamaheswara Rao, M. 1972b. On the Gracilariaceae of the seas around India. *J. Mar. Biol. Ass. India*, **14**(2): 671-696.

Umamaheswara Rao, M. 1978. Seaweed resources of Andhra Pradesh. *Seaweed Res. Utiln.*, **3**(1 & 2): 51-55.

Umamaheswara Rao, M. 1987a. Key for identification of economically important seaweeds. *CMFRI Bulletin*, **41**: 19-25.

Umamaheswara Rao, M. 1987b. Algae of Indian estuaries. *J. Mar. Biol. Ass. India*, **29**(1 & 2): 1-9.

Umamaheswara Rao, M. and Sreeramulu, T. 1970. An annoted list of the marine algae of Visakhapatnam (India). *Bot. Mar.*, **11**(1-4): 122-126.

Untawale, A.G. and Dhargalkar, V.K. 1975. Seaweed resources of Goa coast. *Report. Natl. Inst. Oceanogr. Goa.* pp. 1-10.

Untawale, A.G., Dhargalkar, V.K. and Agadi, V.V. 1983. List of marine algae from India, NIO, Dona Paula, Goa. pp. 1-42.

Untawale, A.G., Dhargalkar, V.K., Agadi, V.V. and Jagtap, T.G. 1979. Marine algal resources of Maharashtra coast. *Technical Report, Natl. Inst. Oceanogr., Goa.* pp. 1-48.

Varma, R.P. 1960. Flora of the pearl beds of Tuticorin. *J. Mar. Biol. Ass. India.* **2**: 221-225.

Varma, R.P. and Krishna Rao, K. 1962. Algal resources of Pamban area. *Indian J. Fish.*, **9**: 205-211.

Recent Trends in Algal Taxonomy
Vol. I. Taxonomical Issues
Editors : Vidyavati and A.K. Mahato
APC Publications Pvt. Ltd., New Delhi, 1999, pp. 287-293

# Cytotaxonomical Approach to Red Algal Taxonomy

K. Rama Rao
CSMCRI Marine Algal Research Station
Mandapam Camp 623 519, Tamil Nadu

## INTRODUCTION

The red algae are a vast group of morphologically and anatomically very complex species. Many prominent Phyco-logists feel that the knowledge of this group of algae is often minimal and based on the limited information in the general botany text books. Among the red algal species belonging to Gelidiales and Gigartinales assume greater significance in view of their economic importance, in protecting marine eco-environment (Adey, 1971; Luning, 1985; Round, 1981 and Steneck, 1986) that the coral reefs could not survive without the cementing action of certain red algal species (Littler and Littler, 1984). Many red algal species provide valuable sources for base products useful in industry worth hundreds of millions of dollars (US) annually (Doty, 1979 and Tseng, 1981). Gelidiales, Gigartinales Ceramiales and Corallinales in that order have played a significant role for both scientific research involving basic and applied, as well as industrial use.

There are about 10,000 described species of red algae (Cole and Sheath, 1990; Dring, 1982; Peterfi and Ionescu, 1977; Bold and Wynne, 1987; Dixon, 1973; Kraft and Woelkerling, 1990 and Thappan, 1980). The classification and taxonomic distribution of species within orders and families of Rhodophyta are unresolved, more so particularly, Gigartinales and Gelidiales. The main problem underlying the red algal classification is that some characteristics that have been used to delimit orders and families, are based on rather very obscure and peripheral structures in the sexual cycle (e.g. the time of formation, location and nature of generative auxiliary cell). Inspite of several recent reviews (Bold and Wynne, 1987; Brawley and Wetherbey, 1981;Gabreilson and Garbey,1986; Kraft and Woelkerling, 1990; West and Hammersend,1981) which provides basic information on morphology, ultrastructure, development and life history they could not provide conclusive evidence on the speciation. Even taking into consideration of the excellent reviews on classification of algae by G.H. Papenfuss and the classical works on classification of algae (Lamouroux, 1813; Agardh, 1820, 1822& 1824; Harvey, 1836 and Kuetzing,1843) and especially of Agardh's *Species algarum* 1824 which are very noteworthy for their use of vegetative, anatomy and reproductive features etc; they would not find solution to taxonomy. The present review is aimed at to focus the importance of cytotaxonomical investigations in solving the many red algal classifications, not withstanding the fact of proximity to high-tech instrumentation and sophisticated infrastructure facilities to go into the deeper aspects of cytological investigations.

## HISTORICAL

After the classical works on taxonomy cited, Schmitz (1892) recongnized four orders in red algae namely. (1) Nemalionales (2) Gigartinales (3) Rhodymeniales and (4) Cryptonemiales, mostly based on the gonimoblast development and post fertilizarion morphogenesis. Later, Kylin (1956) proposed a classification which was subsequently modified by Dixon (1973) and then revised by Pueschel and Cole (1982). Ultimately Cole and Sheath (1990) proposed another classification comprising 15 orders. Krishnamurthy (1990) adopted a classification comprising 17 orders, 4 in Bangiophysidean and 13 in Florideophycidean. Most of these classifications including that of Cole and Sheath (1990) mostly were based on vegetative and reproductive development and to some extent chromosomal, genetical and ultrastructural studies (Gabreilson and Garbey, 1986). Some efforts were also made using the quantitative analysis of the characters while others based on the life history studies and chemotaxonomy as well. There are excellent monographs dealing taxonomical classifications with more traditional morphological studies that will continue to provide the foundation for red algal systematics particularly, generic and familian ranks (Akatsuka, 1986) on Gelidiales (Lindstorm, 1987, 1988 and Lindstorm and Scagel, 1987) on Dumantiaceae (Gigartinales); (Woelkerling, 1987, 1988 and Wynne, 1983) on Delsseriaceae and at the generic ranks (*eg.* Bird *et al.*, 1982) on *Gracilaria* (Kapraun *et al.*, 1983) on Ceramiales (Kraft, 1981) on Gigartinales (Huisman, 1987) and Nemalionales.

## RECENT TRENDS

It is evident from the aforesaid that many of the taxonomical classifications of red algae in general and particularly, those of Gelidiales and Gigartinales were unfortunately, not based on cytological observations of genome characteristics to the desired level not withstanding the fact the biogeography, ecology, cyto-logy are well interlinked in the correct identification of the taxa upto the species level. In some cases these monographic studies are supplemented (Lindstorm, 1988, Adey and Steneck, 1985 and Whetley, 1981) by cytological research (Bird *et al.*, 1982) or by field and culture studies (Kapraun, 1978 and Maggs and Guiry, 1987).

There are some classical investigations to interpret algal taxa of filamentous species of red algae based on cytological observations. These researchers have confined only to investigating into karyological aspects, effects of some chemicals on karyology etc. The noteworthy among them are the cytotaxonomical studies by Patel (1971a, b, c & 1972) on *Chaetomorpha* and in addition there were interesting studies on karyological aspects on different algae (Sarma and Tripathi, 1973; Vedajanani and Sarma, 1978a, b, c & d; Chowdhary, 1973; Chowdhary & Mathew, 1982; Vidyavati & Nagaraju, 1983; Vidyavati *et al.*, 1983; Vidyavati & Sathaiah, 1985; Mogili & Vidyavati, 1985).

## MODERN TRENDS

Though some of the aforesaid investigations led to some taxonomical conclusions leading to exact speciation, some lacunae still exist in arriving at a definite classification. The cytological basis may provide a description of highly stable character for correct identification based on genome character (Patel, 1971a, b, c & 1972) as it is in case of *Chaetomorpha* species. Similarly, the cytological observations in many algae lead to correct classification of algae based on genome characters (Bhatnagar, 1989a & b; Vidyavati and Sathaiah, 1985; Digamber Rao and Vidyavati, 1983; Chowdhary and Mathew, 1982; Mathew and Chowdhary, 1982 a & b; Vedajanani and Sarma, 1978a, b, c & d; Abbas, 1965 and Abbas and Godward, 1965). These studies provide reliable tools for proper perspectives into the basic characters that can be attributed to classification of algae. Without the knowledge of basic genome characteristics the identification of the taxa will be void.

With the latest techniques like cytophotometric estimations of nuclear DNA variation will suggest genome repatterning process which may have accompanied speciation in the taxa (Kapraun and Detcher, 1991). DNA analysis of eukaryotic algal species may also help to distinguish whether individual A, as genetically identical to individual B. It is even possible to hope that differences among organelle genomes of algal classes may provide major clues to the origin and puzzling relationships of the different groups of protists. Further, based on DNA—DNA hybridization, evolutionary relationship between species can assess the level of divergence among geographical samples of benthic Seaweeds (Bot, 1991).

Both the new information and new techniques have enabled to question about the homologues of morphological and developmental features (Cole and Sheath, 1990). With the availability of sufficient model systems, the taxonomic and evolutionary problems can be examined using variety of ultrastructure, chaemotaxonomic and developmental approaches (Pueschel and Cole, 1982; Scott, 1986 and Giff and Coleman, 1984).

## ACKNOWLEDGEMENT

The author gratefully acknowledges Prof. P. Natarajan, Director, Central Salt and Marine Chemicals Researh Institute, Bhavnagar for his help and encouragement in the write-up of this review article. Thanks are also due to my colleagues in rendering assistance in various ways.

## REFERENCES

Abbas, A. 1965. Observations on cytology of *Stigeoclonium fligellifierum* Kuetz, and *Microthamnion Kuetzingianum* Nerg. *Phykos.* **4**: 14-16.

Abbas, A. and Godward, M.B.E. 1965. Cytology of two species of *Chlamydomonas* Ehrb. *Phykos.* **4**: 1-2.

Adey, M.H. 1971. The sublittoral distribution of crustose corallines on the Norwegian coast. *Sarvia.* **46**: 41-58.

Adey, A.H. and Steneck, R.S. 1985. Highly productive eastern caribbean reefs; Synergistic effects of biological, chemical. Physical and geological factors. *NOAA. Underseas Research Program.* **3**: 163-187.

Agardh, C.A. 1820. *Species Algarum.* Vol. I - Part 1. Lund Berling.

Agardh, C.A. 1822. *Species Algarum.* Vol. I - Part 2. Lund Berling.

Agardh, C.A. 1824. *Species Algarum.* Lund Berling.

Akatsuka, I. 1986. Surface cell morphology and its relarionship to other generic characters in one parasitic Gelidiales (Rhodohyta) *Bot. Mar.* **29**: 59-68.

Bhatnagar, S.K. 1989a. Cytological perspectives of Charophyta-I. *Chara globularis* Complex. *Phykos.* **28**: 156-165.

Bhatnagar, S.K. 1989b. Cytological perspectives of Charophyta-II. *Nitella furcata* Complex. *Phykos.* **28**: 166-177.

Bird, C.J. Van der Meer, J.P. and Mclachlan, J. 1982. A comment on *Gracilaria verrucosa* (Huds.) Papenf. (Rhodophyta, Gigartinales). *J. Mar. Biol. Ass.* U.K. **62**: 453-459.

Bold, M.C. and Wynne, M.J. 1987. Introduction to the Algae. Englewood Cliff. N.J. Prentice Hall.

Bot, P.V.N. 1991. Evolutionary relationships between four species of *Cladophora* (Cladophorales—Chlorophyta) based on DNA hybridization, *J. Phycol.* **27**: 617-623.

Brawley, S.H. and Witherbey, 1981. Cytology and ultrastructure in the Biology of Seaweeds, eds. C.S. Lobban and M.J. Wynne, Berkeley, University of California Press. pp. 248-299.

Chowdhary, Y.B.K. 1973. Further observations on cytological races in *Schizomeris leibleinii* Kuetz. *Phykos.* **12**: 5-7.

Chowdhary, Y.B.K. and Mathew, T. 1982. Cytological observation on some members of Oocystaceae. *Phykos.* **21**: 1-9.

Cole, K.M. and Sheath, R.G. 1990. Biology of red algae. Pub. Cambridge University Press, p. 517.

Digamber Rao, B. and Vidyavati, 1983. Nuclear cytology of *Euastrum verrucosum* Ehnerb. *Phykos.* **22**: 125-127.

Dixon, P.S. 1973. Biology of Rhodophyta, New York.

Doty, M.S. 1979. Status of marine agronomy with special reference to the tropics. *Proc. Intl. Seaweed Symp.* **9**: 35-58.

Dring, M.J. 1982. The biology of marine plants. London, ed. Edward Arnold.

Gabreilson, P.W. and Garbey. 1986. Systematics of red algae (Rhodophyta). A reappaisal. *Phycologia.* **26**: 167-174.

Giff, L.J. and Coleman, K.M. 1984. Elucidation of fertilization and development in red alga by quantitative microspectrofluorometry. *Develop. Biol.* **102**: 173-194.

Harvey, W.H. 1836. Algae: In Flora Hibernika. 2nd Vol. ed. J.T. Mackay, Dublin, N. Curry & Sons. pp. 157-256.

Huisman, J.H. 1987. The Scinaia assemblage (Galaxauraceae: Rhodophyta). A reappraisal. *Phycologia.* **26**: 167-174.

Kapraun, D.F. 1978. Field and cultural studies on selected North Carolina *Plyosiphonia* species, *Bot. Mar.* **21**: 143-153.

Kapraun, D.F. and Detcher, J.A. 1991. Cytophotometric estimation of Inter and Intraspecific nuclear DNA content variation in *Gracilaria* and *Gracilariopsis* (Gracilariales, Rhodophyta) *Bot. Mar.* **34**: 139-144.

Kapraun, D.F., Lemus, A.J. and Bula Meyer. 1983. The genus *Plyosiphonia* (Rhodophyta, Ceramiales) in the tropical western Atlantic-I Columbia and Venuzuela. *Bull. Mar. Sci.* **33**: 881-898.

Kraft, G.T. 1981. Rhodophyta: Morphology and classification: In the Biology of Seaweeds, eds. C.S. Lobban, and M.J. Wynne, Berkley, University of Calfornia Press. pp. 6-51.

Kraft, G.T. and Woelkerling, W.J. 1990. Rhodophyta—Systematics and Biology of marine plants. eds. M.N. Cleyton and R.J. King. In press. Melboune, Longman, Cheshire.

Krishnamurthy, V. 1990. Taxonomy of Rhodophyceae: In Rhodophyta eds. T.V. Desikachary, V. Krishnamurthy and M.S. Balakrishnan, Pub. University of Madras, Madras. pp. 188-196.

Kuetzing, F.T. 1843. Phycologia Generalise. Leipzig. Broc. Kraus.

Kylin, H. 1956. Die Gattungen der Rhodophyceen. Ling, Glearup.

Lamouroux, J.V.F. 1813. Essa sur les geenres do la familie de Thalassiophytes, non-articulates—*Ann. Mus. Natl. Hist. Natu.* Prais, **20**: 115-139, 267-293.

Lindstorm, S.C. 1987. Possible sister groups and phylogentic relationships among North Pacific and North Atlantic red algae *Helgol. Meerssunterrs.* **41**: 245-260.

Lindstorm, S.C. 1988. The *Dumantia* J. Agardh: A resurrected tribe of red alga (Dumantiaceae: Rhodophyta) *Phycologia.* **27**: 89-102.

Lindstorm, S.C. and Scagel, R.F. 1987. The marine algae of British Columbia, Northern Washington and Southeast Alaska, division on Rhodophyceae, order Gigartinales, family Dumantiaceae, with an introduction to the order Gigartinales *Can. J. Bot.* **65**: 2202-2232.

Littler, M.M. and Littler, D.S. 1984. Models of tropical reef biogenesis, the contribution of algae: *Prog. Phycol. Res.* **3**: 323-364.

Luning, K. 1985. Meeresbotanik, Stuttgart: G. Theises.

Maggs, C.A. and Guiry, M.D. 1987. *Gelidiella calcicola* sp. nov. (*Rhodophyta*) from the British Isles and Northern France. *Br. Phy. J.* **24**: 88-94.

Mathew, T. and Chowdhary, Y.B.K. 1982a. On the morphology of some chlorococcoid green algae. *Phykos.* **21**: 10-18.

Mathew, T. and Chowdhary, Y.B.K. 1982b. Comparative nuclear cytology of *Scenedesmus Phykos.* **21**: 19-27.

Mogili, T. and Vidyavati. 1985. Mutagenic effect of some analgesic and antipyretic substances on *Cladophora crispata* (Roth) Kuetz. *Phykos.* **24**: 88-94.

Patel, R.J. 1971a. Morphological and cytological studies in *Zygnemopsis godwardense* sp. nov. *Phykos.* **10**: 12-17.

Patel, R.J. 1971b. Cytotaxonomical studies of *Pithophora kewensis* Wittr. *Phykos.* **10**: 18-23.

Patel, R.J. 1971c. Cytotaxonomical studies of British marine species of *Chaetomorpha*-I, *Chaetomorpha linum* Kuetz. and *Chaetomorpha aurea* Kuetz. *Phykos.* **10**: 127-136.

Patel, R.J. 1972. Cytotaxonomical studies of some British marine species of *Chaetomorpha*-II, *Chaetomorpha melagonium* Kuetz. *Phykos.* **11**: 17-22.

Peterfi, and Ionescu, A. 1977. Trotat de Algologie-II. Rhodophyta-Phaeophyta. *Bucuresti Editure Acadamiei Republicai Socialist Romania.*

Pueschel, C.M. and Cole, K.M. 1982. Rhodophycean pit plugs: An ultrastructural survey with taxonomic implications. *Amer. J. Bot.* **69**: 703-720.

Round, F.E. 1981. The ecology of algae, Cambridge, Cambridge University Press.

Sarma, Y.S.R.K. and Tripathi, S.N. 1973. Some observations on the effect of colchicine and maleic hydrazide on the karyology of a green alga *Oedogonium acmandrium* Elfing. *Phykos.* **12**: 28-35.

Schmitz, F. 1892. Florideae: In Syllabus der Planzenfamilein, ed. A. Engler. Berlin, G. Boruntrager. pp. 16-23.

Scott, J. 1986. Ultrastructures of cell divisions in the unicellular red alga *Flintella sanjuinansia. Can. J. Bot.*

Steneck, R.S. 1986. The ecology of Carolline red algal crusts: convergent patterns and adaptive strategies, *Ann. Rev. Ecol. Syst.* **17**: 273-303.

Thappan, H. 1980. The paleobiology of plant protists. San Fransisco, W.H. Freeman.

Tseng, C.K. 1981. Commercial cultivation: In the Biology of Seaweeds eds. C.S. Lobban, and M.J. Wynne, Berkely University of California Press. pp. 680-695.

Vedajanani, K. and Sarma, Y.S.R.K. 1978a. Karyological studies on Indian Conjugales-I. *Spirogyra* link *Phykos.* **17**: 17-26.

Vedajanani, K. and Sarma, Y.S.R.K. 1978b. Karyological studies on Indian Conjugales-II. *Zygnema* Agardh *Phykos.* **17**: 17-26.

Vedajanani, K. Sarma, Y.S.R.K. 1978c. Karyological studies on Indian Conjugales-III. *Mougeotia* Agardh. *Phykos.* **17**: 27-32.

Vedajanani, K. and Sarma, Y.S.R.K. 1978d. Karyological studies on Indian Conjugales-IV Desmids, *Phykos.* **17**: 63-72.

Vidyavati and Nagaraju, B. 1983. Effect of gamma rays on *Closterium acerosum* Ehrenb. *Phykos.* **22**: 171-175.

Vidyavati, Reddy, Y.R. and Nagaraju, B. 1983. Effect of gamma rays on *Euastrum* sp. *Phykos.* **22**: 144-146.

Vidyavati and Sathaiah, G. 1985. Physiomorphological variations in *Cosmarium. Phykos.* **24**: 69-75.

West, J.A. and Hammersend, M.H. 1981. Rhodophyta life histories: In the Biology of Seaweeds, eds. C.S. Lobban and M.J. Wynne, Berkley, University of California Press. pp. 113-193.

Whetley, J.M. 1981. Chloroplast evolution, *New Phytol.* **87**: 33-247.

Woelkerling, W.I. 1987. The genus *Choreonema* in Southern Australia and subfamilial classification within the Corallinaceae (Rhodophyta) *Phycologia.* **26**: 111-127.

Woelkerling, W.I. 1988. The coralline Red Algae: An analysis of the Genera and subfamilies of non-geniculate Corallinaceae (Rhodophyta, Corallinceae) London. British Museum (Natural History) & Oxford. Oxford University Press.

Wynne, M.J. 1983. The current status of genera in the Melessaiaceae (Rhodophyta) *Bot. Mar.* **26**: 437-450.

Recent Trends in Algal Taxonomy
Vol. I. Taxonomical Issues
Editors : Vidyavati and A.K. Mahato
APC Publications Pvt. Ltd., New Delhi, 1999, pp. 295-301

# *Modern Trends in the Taxonomy of Gelidiales*

M.V.N. Panikkar and P. Ampili
S.N. College, Kollam–691001, Kerala

## ABSTRACT

The order Gelidiales is related to Nemalionales and other members of Florideae. The relationship of Gelidiales with other orders of Red algae is only because of their common ancestry. Gelidiales have certain unique characters to maintain the status of a natural order. Detailed morphological and anatomical characters of mature sexual and asexual plants are required for the identification upto the generic or specific level.

## INTRODUCTION

The order Gelidiales contains the most economically important seaweeds used as the raw material for agar production. In most of the developing countries the identity of the seaweed is not known or is uncertain, the economic potential for using this untapped seaweed resource is unrealised (Abbot, 1985) Species of the economically important genera such as *Gelidium*, *Pterocladia* and *Gelidiella* are highly variable. The recent trend is to take as many morphological and anatomical characters as possible for species identification (Abbot, 1985; Akatsuka, 1986; Santelices and Stewart, 1985). Clarification of the taxonomic status of all the species is necessary for efficient management and effective utilization.

## GELIDIALES—GENERAL CHARACTERS

The order Gelidiales includes two families, (Gelidiaceae and Gelidiellaceae), 10 genera and over 140 species. Gelidiales are characterised by a *Polysiphonia*–types of life history with male, female and tetrasporangial plants which are morphologically identical. The thallus of the plants is uniaxial and composed of pericentral cells. The cells are connected by pit connections which have pit plugs with a single cap layer. The cell walls and the intercellular matrix contain agar and other polymers, characteristic of agarophytes.

Male fertile areas are irregular and colourless that occur on the surface of fertile branchlets. Spermatangial initials are produced by the cortical cells and spermatangia are formed by the transverse divisions of the initials. The female reproductive structures develop near the apex of the pinnules and consists of carpogonia. The carpogonia are intercalary and occur along the sides of the central furrow. Nutritive filaments also develop before fertilization.

A fertilized Carpogonium enlarges and becomes multinucleate and it cuts off uninucleate gonimoblast initials. Carposporangia are cut off from the uninucleate gonimoblast cells or the multinucleate fusion cell.

Cystocarps are raised oval to circular in outline and either unilocular or bilocular; carposporangia are borne either in chains on a basal placenta or singly.

Asexual reproduction is by tetrasporangia or bisporangia. Sporangial initials are cut off laterally from the apical cells of cortical filaments by anticlinal division.

## ORDINAL STATUS

Kylin (1923) erected the order Gelidiales to accomodate the family Gelidiaceae from the order Nemalioniales. Such a separation was made on the basis of the difference in their life cycle patterns. The diplobiontic life history of Gelidiales is contrasted with that of the other members of Nemalionales which is exclusively haplobiontic. The validity of such a segregation was challenged by Dixon (1961), Parke and Dixon (1964); Abbot and Hollenberg (1976); Kraft (1981) and Bold and Wynne (1982).

The proposal for inclusion of Gelidiales in Nemalionales is based on the following common features: 1. Absence of auxiliary cells, 2. Absence of a Carpogonial branch, 3. Development of gonimoblast filaments directly from the fertilised carpogonium and 4. Nutritive filaments of Gelidiales and the family Naccariaceae of Nemalionales.

However, Papenfuss (1966) and others strongly opposed the retention of Gelidiaceae under Nemalionales. He supported for giving the status of a natural order for Gelidiales because of the following uniqe characteristics of Gelidiaceae: 1. Diplontic life history. 2. Segmentation of single apical cell. 3. Structure of pit plugs, 4. The presence of nutritive cells, 5. *Gelidium* type of spore germination, 6. Characteristic division of the spermatangia and 7. Compact arrangement of the Cystocarp.

Kylin (1923) commented that the Gelidiales were not closely related to any other orders of Florideae. Fan (1961) suggested a relationship of the Gelidiales with Nemalionales through the Naccariaceae and with Cryptonemales through *Acrosymphyton*. The Family Naccariaceae as suggested by Fan (1961) is not ontogenetically and morphologically related to the nutritive filaments of Gelidiales (Abbot, 1985). Moreover in Naccariaceae a carpogonial branch is present.

As suggested by Hommersand and Fredericq (1988) that set them apart from all other orders presently recognised among the red algae. Close scrutiny of the developmental basis for each of the vegetative and reproductive characters that are recognised in Gelidiales suggested that each had an independent origin and evolutionary history. All the present algologists now consider Gelidiales as a natural order that is related to other orders of Florideae only through the ancestor common to all of them.

The present trend is to consider the biochemical, Cytological and anatomical characters to study the phylogenetic relationships of the various orders in red algae. In this connection, some similarities exist between Gelidiaceae and Gracilariaceae in the structure of polysaccharides. Both are related in the following features: 1. Structure of agar, 2. Behaviour of spermatangial cells. 3. Multinucleate fusion cell, 4. Absence of a typical auxiliary cell.

## FAMILY CONCEPTS

Kylin (1923, 1956) Feldmann and Hamel (1934), Dixon (1961) maintained a single family Gelidiaceae under Gelidiales Fan (1961) erected the family Gelidiellaceae to accomodate Gelidiella which differ from the rest of the genera in the absence of hyphal cells. Maggs and Guiry (1987) suggested that Fan's distinction between families could not be sustained and that they should be merged. According to Santelices (1990) the presence or absence of hyphae is an artificial segregation character and it is necessary to review the classification concepts used by Fan (1961).

## GENERIC CONCEPTS

Fan (1961) revised the order Gelidiales and accepted 10 genera. The taxonomic criteria taken into consideration are: 1. lack of hyphal cells, 2. Branching pattern (monopodial or sympodial), 3. Structure of inner cortex, 4. Nature of the basal portion, 5. Morphology of the pinnule and 6. Structure of the Cystocarp.

A taxonomic key is given below to accomodate all the ten genera.

1. Thallus lacking internal hyphae – *Gelidiella*
1. Thallus with internal hyphae 2
   2. Thallus with sympodial growth – *Acanthopeltis*
   2. Tallus with monopodial growth 3
3. Pinnules multified echinate, imbricated – *Yatabella*
3. Pinnules smooth, not imbricated 4
   4. Thallus with distinct inner cortex 5
   4. Thellus without distinct inner cortex 6
5. Pinnules arise from surface of thallus – *Ptilophora*
5. Pinnules arise from margins of thallus – *Beckerella*
   6. Base prostrate creeping 7
   6. Base discoid – *Suhria*
7. Pinnules arise along the median region – *Prophyroglossum*
7. Pinnules arise from margins 8
   8. Pinnules with peltate head – *Acropeltis.*
   8. Pinnules without prltate head 9
9. Cystocarp bilocular with two ostioles – *Gelidium.*
9. Cystocarp monoculat with one ostiole – *Pterocladia*

After studying the morphological variations, particularly the arrangement of the epidermal cells, two more new genera have been suggested, *Pterocladiastrum* and *Onikusa* (Akatsuka 1981, 1986). Cell divisions and cell elongation are highly variable according to the environment and age of the plant. Therefore, the features that separate *Onikusa* and *Pterocladiastrum* from *Gelidium* and *Pterocladia* are of questionable taxonomic value (Santelices, 1990). *Acropeltis* is very similar to *Gelidium* and the difference is only a growth modification of the frond (Santeices and Montalva, 1983). No taxonomic significance was given to this feature and it should be placed in Synonomy with *Gelidium* (Akatsuka, 1986).

## SPECIES TRENDS

Five of the general, *Acanthopeltis* Okamura, *Porphyroglossum* Kurtzing, *Ptilophora* Kuetzing, *Suhria* J. Agardh ex Endlicher and *Yatabella* Okamura are either monotypic or have fewer than three species (Kylin, 1956; Fan, 1961; Santelices and Stewart, 1985). Thirteen species are listed in *Ptilophora* (Akatsuka, 1987; Norris, 1987). 23 in *Gelidiella* (Maggs and Guiry, 1987), 15 in *Pterocladia* and 90 in *Gelidium* (Kraft, 1981; Santelices, 1990). External and internal morphologies have been the basis of species recognition.

External morphology is highly variable among the species and identification based on morphology created a lot of confusion. It is often difficult to outline sharp specific limits for these species unless patterns of external morphological variation can be established, and these must be evaluated as to their stability if they are to be used as taxonomic characters (Santelices, 1988). The following are the most important morphological and anatomical characters of vegetative and reproductive parts still in use for species identification in Gelidiales.

| | | | |
|---|---|---|---|
| 1. | Size of the plant | – | Height |
| 2. | Colour | – | Dark red/reddish/black. |
| 3. | Growth pattern | – | Tufted, enlangled/sparse. |
| 4. | Branching | – | Upto 4-6 orders, irregular/dichotomous/pin nate/opposite/sub opposite/alternate/pinnae simple/compound, degree of branches to the main axis. |
| 5. | Axes | – | Lanceolate/cylindrical/filifo rm flat/cylindrical at base flat above/compressed. |
| 6. | Outline of axes in cross section | – | Compressed/terete/sub terete. |
| 7. | Width and thickness of the axes in section | – | |
| 8. | Branches | – | Thickness and width of 1-6 order branches, length. |
| 9. | Apices of branches | – | Truncate/blunt/pointed/ex panded. |
| 10. | Holdfast | – | Creeping, Stoloniferous/rhizomatous/f ibrous, erect axes arise at regular/irregular intervals. |
| 11. | No. of layers of Cortical cells. | – | |
| 12. | Size of external cortical cells | – | Length and width. |
| 13. | Size of medullary cells | – | Diameter. |

| | | | |
|---|---|---|---|
| 14. | Distribution of rhizoidal filaments | – | Present/absent, in inner Cortex/outer medulla/inner medulla, abundant/sparse. |
| 15. | Apical part of Cystocarpic pinnule | – | Pointed/blunt/rounded. |
| 16. | Cystocarpic structure | – | Bilocular/unilocular. |
| 17. | Shape of Cystocarp | – | Elongate/ovoid/spherical. |
| 18. | Size of Cystocarp | – | Length, thickness. |
| 19. | Outline of tetrasporangial pinnule | – | Oblong/dialated/obovate. |
| 20. | Arrangement of the tetrasporangial sorus | – | Regular/irregular. |
| 21. | Width of the marginal sterile region of tetrasporangial sorus | – | |
| 22. | Shape of the tetraspares | – | Ovoid/spherical. |
| 23. | Arrangement of Spermatangia | – | Regular/irregular. |

Convertional morphological and anatomical studies are still in practice for studying the taxonomy of Gelidiales. The difficulties in getting mature sexual plants may create problems for correct identification. Simultaneous biochemical studies at the molecular level will also be helpful for proper generic and specific distinctions.

## ACKNOWLEDGEMENTS

The authors are highly thankful to Prof. L. Suresh Babu, Principal, S.N. College, Kollam for encouragements and providing facilities.

## REFERENCES

Abbot, I.A. 1985. Taxonomy of Economic Seaweeds California Sea Grant College Programme. pp. 166.

Abbot, I.A. and G.J. Hollenberg. 1976. Marine algae of California. Stanford University Press, California. pp. 827.

Akatsuka, I. 1981. Comparitive morphology of the Outermost cortical cells in the Gelidiaceae (Rhodophyta) of Japan Nove Hedwigia. **35**: 453-463.

Akatsuka, I. 1986. Surface cell morphology and its relationship to other generic characters in non-parasitic Gelidiaceae (Rhodophyta). *Bot. Mar.* **29**: 59-68.

Akatsuka, I. 1987. Taxonomy and distribution of Beckerella (Gelidiales, Rhodophyta) *Aquat. Bot.* **28**: 257-274.

Bold, H.C. and M.J. Wynne. 1982. *Introduction to algae*. Printice-Hall Inc., USA. pp. 706.

Dixon, P.S.1961. On the classification of the Florideae with particular reference to the position of the Gelidiaceae. *Bot. Mar.* **3**: 1-6.

Fan, K.C. 1961. Morphological Studies of the Gelidiales. Univ. *Calif. Publ. Bot.* **32**: 315-368.

Feldmann, J. and G. Hamel. 1934. Observations sur quelques Gelidiacess. *Rev. Gen. Bot.* **46**: 528-549.

Hommersand, M.H and S. Fredericq. 1988. An investivation of Cystocarp development in *Gelidium pteridifolium* with a revised description of the Gelidiales (Rhodophyta) *Phycologia.* **27**: 272.

Kraft, G.T. 1981. Rhodophyta: Morphology and classification. In C.S. Lobban & M.J. Wynne (eds.), The Biology of the Seaweeds. University of California Press, Berkeley. pp. 6-51.

Kylin, H. 1923. Studien uber die Entwicklungsgeschishte der Florideen. K. Sven. Vetenskapsakad. *Handl.* **63**: 1-139.

Kylin, H. 1956. Die Gattungen der Rhodophyceen Gleerups, Lund. pp. 673.

Maggs, C.A. and M.D. Guiry, 1987. *Gelidiella calcicola* sp. nov. (Rhodophyta) from the British Isles and Northern France. *Br. Phycol. J.* **22**: 417-434.

Norris, R. 1987. *Pterocladia* (Gelidiaceae, Rhodophyceae) a new genus previously known in South Africa, as it occurs in *Natal. S. Afr. J. Bot.* **53**: 39-43.

Papenfuss, G.F. 1966. A review of the present system of classification of the Florideophycideae *Phycologia.* **5**: 247-255.

Parke, M and Dixon, 1964. A revised check-list of British marine algae. *J. Mar. Biol. Ass.* UK. **44**: 491-542.

Santelices, B. 1988. A synopsis of biological data on the seaweed *Gelidium* and *Pterocladia* (Rhodophyta) FAO fisheries *Synopsis.* **145**: 1-55.

Santelices, B. 1990. New and old problems in the taxonomy of the Gelidiales (Rhodophyta) *Hydrobiologia.* **204-205**: 125-135.

Santelices, B. and S. Montalva. 1983. Taxonomic Studies on Gelidiaceae (Rhodophyta) from Central Chile. *Phycologia.* **22**: 185-196.

Santelices, B. and J.G. Stewart. 1985. Pacific species of Gelidium and other *Gelidiales* (Rhodophyta) with keys and descriptions to the common or economically important species. In I.A. Abbot and J.N. Norris (eds). Taxonomy of economic seaweeds with reference to some Pacifc and Caribbean species. California sea grant Progrum, La Jolla, California. pp. 17-71.